Radio Wave Propagation in Vehicular Environments

Related titles on electromagnetic waves

Dielectric Resonators, 2nd Edition Kajfez and Guillon
Electronic Applications of the Smith Chart Smith
Fiber Optic Technology Jha
Filtering in the Time and Frequency Domains Blinchikoff and Zverev
HF Filter Design and Computer Simulation Rhea
HF Radio Systems and Circuits Sabin
Microwave Field-Effect Transistors: Theory, design and application, 3rd Edition Pengelly
Microwave Semiconductor Engineering White
Microwave Transmission Line Impedance Data Gunston
Optical Fibers and RF: A natural combination Romeiser
Oscillator Design and Computer Simulation Rhea
Radio-Electronic Transmission Fundamentals, 2nd Edition Griffith, Jr
RF and Microwave Modeling and Measurement Techniques for Field Effect Transistors Jianjun Gao
RF Power Amplifiers Albulet
Small Signal Microwave Amplifier Design Grosch
Small Signal Microwave Amplifier Design: Solutions Grosch
2008+ Solved Problems in Electromagnetics Nasar
Antennas: Fundamentals, design, measurement, 3rd Edition Blake and Long
Designing Electronic Systems for EMC Duff
Electromagnetic Measurements in the Near Field, 2nd Edition Bienkowski and Trzaska
Fundamentals of Electromagnetics with MATLAB®, 2nd Edition Lonngren *et al.*
Fundamentals of Wave Phenomena, 2nd Edition Hirose and Lonngren
Integral Equation Methods for Electromagnetics Volakis and Sertel
Introduction to Adaptive Arrays, 2nd Edition Monzingo *et al.*
Microstrip and Printed Antenna Design, 2nd Edition Bancroft
Numerical Methods for Engineering: An introduction using MATLAB® and computational electromagnetics Warnick
Return of the Ether Deutsch
The Finite Difference Time Domain Method for Electromagnetics: With MATLAB® simulations Elsherbeni and Demir
Theory of Edge Diffraction in Electromagnetics Ufimtsev
Scattering of Wedges and Cones with Impedance Boundary Conditions Lyalinov and Zhu
Circuit Modeling for Electromagnetic Compatibility Darney
The Wiener–Hopf Method in Electromagnetics Daniele and Zich
Microwave and RF Design: A systems approach, 2nd Edition Steer
Spectrum and Network Measurements, 2nd Edition Witte
EMI Troubleshooting Cookbook for Product Designers Andre and Wyatt
Transmission Line Transformers Raymond Mack and Jerry Sevick
Electromagnetic Field Standards and Exposure Systems Grudzinski and Trzaska
Practical Communication Theory, 2nd Edition Adamy
Complex Space Source Theory of Spatially Localized Electromagnetic Waves Seshadri
Electromagnetic Compatibility Pocket Guide: Key EMC facts, equations and data Wyatt and Jost
Antenna Analysis and Design Using FEKO Electromagnetic Simulation Software Elsherbeni, Nayeri and Reddy
Scattering of Electromagnetic Waves by Obstacles Kristensson
Adjoint Sensitivity Analysis of High Frequency Structures with MATLAB® Bakr, Elsherbeni and Demir
Developments in Antenna Analysis and Synthesis Vol. 1 and Vol. 2 Mittra
Advances in Planar Filters Design Hong
Post-processing Techniques in Antenna Measurement Castañer and Foged
Nano-Electromagnetic Communication at Terahertz and Optical Frequencies, Principles and Applications Alomainy, Yang, Imran, Yao and Abbasi
Nanoantennas and Plasmonics: Modelling, design and fabrication Werner, Campbell and Kang
Electromagnetic Reverberation Chambers: Recent advances and innovative applications Guillaume Andrieu

Radio Wave Propagation in Vehicular Environments

Leyre Azpilicueta, Cesar Vargas-Rosales,
Francisco Falcone and Ana Alejos

The Institution of Engineering and Technology

Published by SciTech Publishing, an imprint of The Institution of Engineering and Technology, London, United Kingdom

The Institution of Engineering and Technology is registered as a Charity in England & Wales (no. 211014) and Scotland (no. SC038698).

First published 2020

The Institution of Engineering and Technology
Michael Faraday House
Six Hills Way, Stevenage
Herts, SG1 2AY, United Kingdom

www.theiet.org

British Library Cataloguing in Publication Data
A catalogue record for this product is available from the British Library

ISBN 978-1-78561-823-9 (hardback)
ISBN 978-1-78561-824-6 (PDF)

Typeset in India by MPS Limited

Contents

About the authors

Leyre Azpilicueta is an associate professor at the School of Engineering and Sciences of Tecnologico de Monterrey, Campus Monterrey, Mexico. She is involved in several professional and editorial activities, acting as a reviewer, advisory board member, associate/guest editor of top-rank journals and TPC member of international conferences. Her research interests include radio propagation, mobile radio systems, ray tracing, channel modelling, wireless sensor networks, electromagnetic dosimetry, IoT networks and devices, 5G communication systems and vehicular communications.

Cesar Vargas-Rosales is a professor and leader of the research group in telecommunications at the Tecnologico de Monterrey, Mexico. He is a member of the Mexican Research National System, the Academy of Engineering of Mexico and the Mexican Academy of Sciences. He is a senior member of the IEEE and the IEEE Communications Society Monterrey Chapter Chair. His research interests span personal communications, 5G, cognitive radio, MIMO systems, reconfigurable networks and more.

Francisco Falcone is an associate professor in the Department of Electrical, Electronic and Communication Engineering at Universidad Pública de Navarra and also with the Institute for Smart Cities, Spain. From January 2018 to May 2018 he was a visiting professor at Kuwait College of Science and Technology. His research interests include artificial electromagnetic media, complex electromagnetic scenarios, wireless system analysis and context-aware IoT environments.

Ana Vazquez Alejos is an associate professor at the Signal and Communications Theory Department at the University of Vigo in Spain. In 2009 she was granted with the Marie Curie International Outgoing Fellowship developed in the New Mexico State University (NM, USA). Her research work includes radio propagation, radio channel modelling, wireless multimedia systems, waveform design, noise codes and radar. She is associate editor and reviewer of several magazines.

Raquel Fernández González, PhD in Economics since 2015 and researcher in Department of Applied Economics in University of Vigo since 2010. He has been a visiting scholar at the University of Porto for 2 years. She has participated in several international research projects while she worked at CETMAR Research Foundation. Her research interests cover a range of topics related to institutional economy, aquaculture economics and natural resources management. She has been a co-author in Chapter 8 of the presented book.

Acknowledgments

The authors would like to acknowledge all people who have helped inspire us to write this book. In particular, Leyre Azpilicueta and Cesar Vargas-Rosales, wish to express their thanks to the School of Engineering and Sciences as well as the Telecommunications Research Group, both at Tecnologico de Monterrey, for their support by providing an environment for the development of this book. Ana Alejos wants to thank the University of Vigo and finally, Francisco Falcone would like to thank the Public University of Navarre and the Institute of Smart Cities.

We all want to thank our teammates, colleagues and those who have assisted us for their important feedback, technical discussions, suggestions and involvement, which have immeasurably improved the research content of the book. Particularly, the collaboration of Mikel Celaya Echarri and Fidel Alejandro Rodríguez-Corbo, from Tecnologico de Monterrey; Peio López Iturri, Imanol Picallo, José Javier Astrain, Jesús Daniel Trigo and Luis Serrano, from the Public University of Navarre; and Hicham Klaina, from the University of Vigo, for their valuable insights, contributions and illustration assistance in order to improve the quality of the book. The authors would also like to extend a special thanks to Raquel Fernández González, from the University of Vigo, for her relevant contribution co-authoring Chapter 8 of the book.

Finally, the authors wish to acknowledge the support provided by project RTI2018-095499-B-C31, funded by Ministerio de Ciencia, Innovación y Universidades, Gobierno de España (MCIU/AEI/FEDER,UE), the research grant ED431C-2019/26, funded by Xunta de Galicia, and the research project TEC2017-85529-C03-3R, funded by Ministerio de Ciencia, Innovación y Universidades, Gobierno de España (MCIU/AEI/FEDER,UE), as well as from AtlantTIC Research Center of University of Vigo.

Acronyms

3GPP	Third-generation partnership programme
3D RL	Three-dimensional ray launching
ACEA	European Automobile Manufacturers Association
ACF	Autocorrelation function
ADAS	Advanced driver-assistance systems
AMPS	Advanced Mobile Phone Service
AoA	Angle of arrival
AoD	Angle of departure
AP	Access point
AS	Angular spread
BS	Base station
CDMA	Code division multiple access
CIR	Channel impulse response
COST	Committee on Science and Technology
CV2X	Cellular V2X
D2D	Device-to-device
DB	Double-bounce
DBR	Double-bounced rays
DoA	Direction of arrival
DoD	Direction of departure
DSRC	Designated short-range communications
EKF	Extended Kalman filter
EM	Expectation maximization
FCC	Federal Communications Commission
FCF	Frequency correlation function
FHWA	Federal Highway Administration
GBDM	Geometry-based deterministic models
GBSCM	Geometry-based stochastic channel model
GBSM	Geometry-based stochastic model

GEMV^2	Geometry-based efficient propagation model for V2V communication
GWSSUS	Gaussian wide-sense stationary uncorrelated scattering
HF	High frequency band
HHI	Fraunhofer Heinrich Hertz Institute
IFT	Inverse Fourier transform
IS-GBSM	Irregular-shaped GBSM
ITS	Intelligent transportation systems
KF	Kalman filter
LFM	Linearly frequency modulated
LoS	Line of sight
LSP	Large-scale parameters
LSS	Large spatial scale
LTE	Long Term Evolution
MAC	Medium access control
MIMO	multiple-input-multiple-output
ML	Maximum likelihood
MMSE	Minimum mean square error
mmWave	Millimetre wave
MPC	Multipath components
MS	Mobile station
MSS	Moderate spatial scale
MUSIC	Multiple signal classification
NAMPS	Narrowband AMPS
NGSM	Non-geometry-based stochastic models
NLoS	Non-LoS
PAR	Peak-to-average ratio
PAS	Power angular spread
PDF	Probability density function
PDP	Power delay profile
PLE	Path-loss exponent
PN	Pseudo-noise
PRBS	Pseudorandom binary sequences
PSD	Power spectral density
RL	Ray launching
RMS	Root mean squared

RS-GBSM	Regular-shaped GBSM
RSSI	Received signal strength indicator
RSU	Roadside units
RT	Ray tracing
RX	Receiver
SB	Single-bounce
SBR	Single-bounced rays
SCM	Spatial channel model
SDR	Software-defined radio
SHF	Super high frequency
SIMO	Single-input-multiple-output
SISO	Single-input single-output
SLS	System-level simulations
SSS	Small spatial scale
TBP	Time-bandwidth product
TDL	Tapped delay line
TX	Transmitter
UAV	Unmanned aerial vehicle
UHF	Ultra-high frequency
V2C	Vehicle-to-cloud communications
V2H	Vehicle-to-home communication
V2I	Vehicle-to-infrastructure communications
V2N	Vehicle-to-network communications
V2P	Vehicle-to-pedestrian communications
V2R	Vehicle-to-road communications
V2V	Vehicle-to-vehicle communications
V2X	Vehicle-to-everything communications
VHF	Very high frequency
VLC	Visible light communications
VNA	Vector Network Analyser
VTD	Vehicular traffic density
WiFi	Standard IEEE 802.11X
WSN	Wireless sensor network
WSS	Wide-sense stationarity
WSSUS	Wide-sense stationary uncorrelated scattering
XPD	Cross-polarization discrimination

Notation

α_k	Attenuation of multipath component k
ψ	Angle of the relative direction of movement between transmitter and receiver
B_C	Coherence bandwidth
d_0	Close-in distance from transmitter, usually a short distance in the far field
d	Separation distance of transmitter and receiver
d_{km}	Separation distance of transmitter and receiver in kilometre
$\delta(t)$	Dirac delta function
f	Frequency
f_d	Doppler shift
$f_{\max}$	Maximum Doppler shift or Doppler spread
G_T	Antenna gain at the transmitter, not in dB
G_R	Antenna gain at the receiver, not in dB
h	Height difference of transmitter and receiver in the shadowing losses
h_T	Transmitter antenna height
h_R	Receiver antenna height
λ	Wavelength
L	Number of multipaths in the channel
M_R	Number of receiving antennas in a MIMO system
M_T	Number of transmitting antennas in a MIMO system
n	Path-loss exponent (PLE)
$PL(d)$	Path loss at a separation distance d
$PL_{dB}(d)$	Path loss in dB at a separation distance d
$P_R(d_0)$	Power received at the close-in distance, usually a measurement.
σ_τ	RMS delay spread
τ_k	Delay of kth multipath component
T_C	Coherence time
θ_k	Signal-phase change for multipath component k
v	Speed of receiver
ν	Parameter used for the calculation of shadowing losses

Chapter 1
Introduction to vehicular communications

Transportation systems are evolving in order to become safer, more efficient and increasing user quality of experience. Transportation infrastructure, vehicles and users are called to interact within vehicular context-aware environments termed as intelligent transportation systems (ITS). In this chapter, we will describe the fundamentals of ITS, and more specifically, the characteristics of vehicular communications in terms of the different types of communication links, systems and architectures employed. An overview of propagation standards as well as applications within multiple ITS domains is also given, as examples of a wide range of future developments to be seen.

1.1 Intelligent transportation systems

One of the main challenges faced by mankind is to achieve sustainable living environments and to increase citizen's wellbeing and active participation in all levels of governance. In this sense, specific issues apply for urban areas, in which forecasts indicate that over 70% of the population will be living by 2050, and for rural areas, in order to reduce drastic migration. In order to face these foreseen challenges, the concept of Smart Cities and Smart Regions is developed, in which the use of information and communication technologies (ICT) and focused policy-making allow multiple system interaction in order to enhance overall efficiency [1]. It is usually considered that the term Smart City was initially coined in 2005 [2], with the aim of increasing efficiency in the operation of multiple complex systems within dense urban areas, hence achieving higher sustainability levels as well as increasing the quality of life of its citizens. The impact and the benefit in the application of Smart City/Smart Region adoption can be observed in multidimensional value benchmarking, with continued efforts in promotion and adoption around the world [3–6]. Within this context, multiple systems are considered, such as

- Energy systems: generation, transportation and management of energy are key points when considering energy-related aspects within Smart Cities and Smart Regions. The concept of Smart Grids in terms of energy distribution and dynamic load and demand profiles, intelligent public lighting systems, microgrids in order to achieve highly sustainable environments or new energy consumption models, such as vehicle-to-grid (V2G) communications in order to manage massive electric vehicle demands, are some of the exponents within energy-related systems in context-aware environments.

- Waste management: handling waste considers multiple aspects, such as waste pickup considering optimal waste collection routes, waste separation, classification of the waste material in adequate recycling processes or the inclusion of waste in related bio-processing schemes, such as obtaining biofuel. In order to enhance waste management, several elements are employed, such as intelligent waste bins and waste containers, with different types of embedded sensors in order to measure occupancy levels of the containers, determining the most adequate routes for the waste recollection trucks or manual waste pickup to be planned [7,8].
- Smart health: in order to provide sustainable health services as well as to increase the quality of life of users, ICTs have been extensively investigated and integrated within the health system. In this way, the adoption of ICT has paved the path initially to e-health systems (introduction of computer-based systems and databases for the handling of items such as electronic health records, electronic prescriptions or remote connection to medical consultation); m-health systems (introduction of mobile connectivity, initially as an alert service of 2G systems and evolved to different levels of biomedical signal handling with the introduction of smartphones); and the latest ICT adoption phase, given by smart health (integration of e-health/m-health within smart city infrastructure). In this way, the city is foreseen as a context-aware environment, in which information form city-based sensors (such as surveillance cameras or particle detectors for pollution analysis) are combined with user information (location, user preferences, and movement) in order to activate elements such as traffic light control and lane management (for agile ambulance displacement) and emergency room resource allocation within a hospital, for example, and accident is detected [9]
- Industry 4.0: the evolution in production, distribution, maintenance and logistics has led towards an industrial transformation process. In this sense, by the combination of elements such as Internet of Things (IoT), cyber physical systems and advanced additive manufacturing systems, the paradigm of Industry 4.0 has been established, evolving from embedded electronics proper of the third industrial revolution, to fully connected industrial environments, with elements such as cooperative robots or digital twins, among others [10].

Transportation systems are one of the most relevant elements in relation to the development of human activity, strongly impacting regional development, as well as the way in which cities have been planned and deployed integrating different transportation infrastructures. Transportation systems enable commercial transactions as well social interactions, providing clear benefits in overall regional and national growth. However, transportation systems account for a relevant percentage of energy consumption demands, contribute to the CO_2 footprint and disseminate particle-based pollutants to the atmosphere (such as NO_X components). The operation of transportation systems leads to traffic congestion, with a direct impact

Photo by Abigail Keenan on Unsplash

Figure 1.1 Smart Cities and Smart Regions combine the use of multiple systems in order to optimize overall operations and transactions within the scenario under analysis

on economic transactions (costs derived in logistic, distribution and operation) and on quality of life (travel time investment, acoustic pollution, etc.). Moreover, transportation systems are also implying huge tolls in terms of fatalities and injuries, resulting in huge personal and economic losses [11,12].

Intelligent transportation systems (ITS) have been proposed in order to enhance their operation in terms of travel time reduction, reducing injuries and casualties and reducing energy consumption as well as environmental impact [11,13]. More specific goals of ITS are outlined as follows [14] and are schematically depicted in figure 1.2:

- Provide drivers notifications of hazards on the road ahead in advance.
- Increase driver safety by providing adequate vehicle separation indications, based on multiple variables, such as traffic conditions, environmental conditions, etc.
- Takes advantage of multiple communication schemes in order to operate in a context-aware transportation environment. Information exchange is provided by dedicated communication systems as well as by general-purpose communication systems, in which users, vehicles and infrastructure can establish communication links.
- Provides information to drivers in relation to location-specific traffic regulations (e.g., applicable speed limit, allowed lane use, etc.).
- Provides auxiliary information, such as the availability of parking spaces, the proximity of service stations or the presence of electric vehicle charging points.
- End-to-end journey planning can be performed, by providing public transportation scheduling information as well as multi-platform ticketing options.

Figure 1.2 Services provided by intelligent transportation systems

- Traffic regulation policies related to specific traffic conditions and operations can be put in place and controlled. This is applicable, for example, in the operation of public buses in excluded zones to general traffic or in the use of reserved lanes for public transportation or emergency vehicles.
- Optimizes transportation and handling of merchandises and goods, providing real-time tracking and other logistical-based services. This can be used both by freight operators and by international authorities for operations related to custom inspections and transactions, as well as for insurance purpose.
- Energy and fuel consumption, as well as pollution levels, can be decreased by the application of route planning techniques, as well as by providing optimized driving parameters, in combination with systems such as cooperative adaptive cruise control.

ITS rely on communication systems in order to provide context-aware transportation scenarios. Operation is provided under specific conditions, such as highly dynamic channels, stringent delay requirements and a large amount of communication links. These conditions impose a specific requirement for the overall vehicular communication systems, from the physical layer, network architecture, system definition and data processing/data analytics perspective. Specific characteristics of vehicular communication channels will be described in the remainder of this chapter, related to vehicular context-aware scenarios and application.

1.2 Vehicular communication environments

In order to provide interactive capabilities within context-aware environments, communication systems are employed, in order to establish multiple types of connections, between users, end-user devices, servers and communication nodes.

Vehicular communications can be employed in order to provide the following use cases [15]:

- Advanced driver assistance systems, increase in safety measures and enabling automated driving.
- Information related to road/traffic conditions.
- Services related to mobility, such as inter-modal travel handling, toll management, parking suggestions, etc.
- Services related to other auxiliary aspects, such as location-based marketing, infotainment or route planning.

Applications in which a high degree of mobility is required makes intensive use of wireless communication systems, which provide inherently ubiquity as well as massive deployment capabilities with a constrained cost. The use of wireless channels provides speedy deployment as well as flexible physical connectivity for potentially a massive number of transceivers. However, several elements affect wireless channel performance which must be taken into consideration, which are schematically depicted in figure 1.3:

- Losses in the received power level as a function of transmitter–receiver distance, as well as the dependence of signal attenuation with the operating frequency, determined by available frequency band allocations.
- Losses owing to attenuation of electromagnetic waves, given by the multiple gaseous components in the atmosphere. Attenuation values are dispersive with frequency, greatly increasing as operating frequency bands span into the millimetre wave range, with values in excess of 10 dB/km up to 100 GHz and in excess of 100 dB/km for operating frequencies beyond the 400 GHz range [16,17].

Channel Attenuation	EM Interaction	Coverage/Capacity
• Distance effect • Frequency of operation • Atmospheric Gases • Hydrometeors	• Reflection, Refraction, Diffraction • Diffuse Scattering • Multipath propagation • Slow Fading/Fast Fading • Doppler shift	• Interference impact (intra-system, inter-system, external sources • User density • Receiver sensitivity thresholds as a function of required transmission rate

Figure 1.3 Characteristics and effects on wireless channel behaviour and coverage/capacity relations in the context of vehicular communications

- Losses owing to electromagnetic wave interaction with surrounding media, mainly given by reflection, refraction, diffraction and diffuse scattering. As a function of the surrounding environment, propagation mechanisms will be given mainly by line of sight propagation components (and hence, by basic propagation losses given by slow fading characterization) or by multipath propagation (described by fast fading characterization). Loss estimation is going to be strongly dependent on an adequate description of the environment, given by the shape, size and electric properties of the constituent materials, which are frequency dispersive [18,19].
- Channel variability owing to traffic density variations and vehicular speed conditions, which strongly modify signal attenuation in path loss calculations and hence coverage/capacity estimations.
- Doppler shift owing to relative differences between transmitter and receiver locations. This effect is particularly relevant in the case of communication channels established between moving elements such as vehicles and static stations, such as purpose-specific infrastructure communication nodes or general-purpose base stations in mobile networks.
- Interference impact on radio channel performance, given by multiple interference sources, such as intra-system, inter-system or external unintentional sources [20]. The surrounding environment also plays a relevant role, especially in the case of industrial applications [21–25].

In this section, the different types of communication schemes will be presented in relation to the type of environment, allocated frequency band, link type and application in which they are employed.

1.2.1 Characteristics of the environment and frequency bands

Wireless communications are essential drivers for the implementation of fully context-aware, interactive vehicular environments. Moreover, different applications within vehicular scenarios (e.g., road situation awareness, location-based marketing information, site-specific traffic regulations and rulings, etc.) exhibit different coverage/capacity requirements. It is worth noting that traffic density also plays a key role in communication parameters, influenced by traffic density as well as by relative transmitter–receiver speed [26,27]. Another particular aspect in relation to vehicular communications is the differences in transceiver antenna location, with low heights in the case of inter-vehicular communications and in general, smaller height of static equipment in the case of vehicle-to-infrastructure communications. These environments exhibit specific characteristics, given mainly by multiple communication agents within the vehicular scenario, inherently high mobility and tight specifications in relation to safety applications. More specifically:

- Communication links can be established between different agents, such as vehicles, road infrastructure, communication networks or pedestrians. In this way, within the framework of vehicle to-everything (V2X) connectivity,

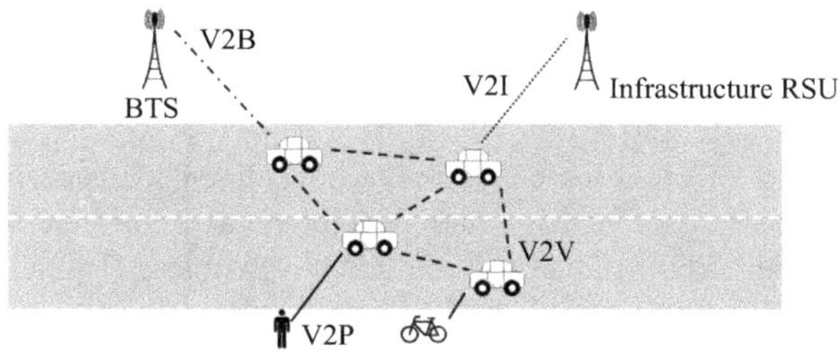

Figure 1.4 Vehicular communication links, which include communication among vehicles (V2V), communication with infrastructure communication nodes (V2I), communication with public land mobile networks (V2B) and communication with pedestrian and vulnerable roadside users (V2P)

different communication links can be defined, which are schematically depicted in figure 1.4:

– Vehicle to vehicle (V2V): direct communication between vehicles is established, enabling functionalities such as fast information exchange among different vehicle subsets, for applications such as danger warnings, adaptive speed control, platoon control or lane management. Communication is enabled by means of an embedded transceiver within the vehicle termed onboard unit (OBU). Antennas are located at relatively low heights as compared to conventional communication channels established in radio broadcasting systems or public land mobile systems. Multiple types of vehicles are also present, which also modify wireless channel characteristics.
– Vehicle to infrastructure (V2I): communication links are established among vehicles and static communication infrastructure, usually within the roadside. The infrastructure can be dedicated, establishing communication between OBU and static dedicated infrastructure telecom equipment called road side units (RSU), with the possibility to provide a connection to Public Land Mobile Networks. In this case, the communication links can be termed as vehicle-to-broadband or vehicle-to-network (V2N) links.
– Vehicle to pedestrian (V2P): communication links can be established between vehicles and pedestrians (or equivalent users, such as bicycles, skateboards, scooters, etc.). This communication scheme enables better information handling and management of vulnerable road users in order to avoid accidents or disturbances of the travel experience of these vulnerable users.
– Intra-vehicular communications: communication links are established within the users/devices with the sole purpose of communicating them.

An example can be the use or wireless personal area network protocols such as Bluetooth in order to connect a mobile phone terminal to the vehicle loudspeaker system.

- Different types of environments can be considered, considering urban/sub-urban/rural areas, user density and developed activities (residential/office/commercial/industrial). In this way, in terms of wireless channel operation the following classification can be established [28]:
  - Dense urban areas, with variable building heights, moderate speed traffic, with large interaction with vulnerable road users, option for inter-modal traffic and fluctuations in vehicle speeds. In terms of radio propagation phenomena, there will be urban street canyons, moderate Doppler shift and indoor/outdoor propagation conditions, among others. Multipath propagation phenomena will be relevant, as well as scattering effects due to non-uniform building facades, among others. Specific conditions will also be present, owing to the existence of urban tunnels, which can require the use of indoor coverage solutions, such as dedicated cells or distributed antenna systems.
  - Suburban areas, with lower commercial activity and average lower building heights. Line of sight (LOS) as well as non-line of sight (NLOS) propagation conditions will be present, with lower levels of multipath propagation compared with dense urban areas. A specific case is given by industrial areas, which exhibit specific issues in relation to vehicle/freight movement, data traffic seasonal behaviour and increased shadowing and interference effects in radio channel performance.
  - Rural areas, with low building/clutter density and lower user densities. Communication links will be mainly operating under slow fading conditions, with LOS propagation, especially with small terrain height variation areas. An additional aspect in relation to rural areas is population and infrastructure sparsity, leading to potentially out of coverage zones, which can demand specific solutions or cooperative/heterogeneous network operation in order to guarantee the quality of service metrics.

As it can be seen, vehicular communication span in coverage areas that can vary from very short-range communications (inside the vehicle, in the range of cm to 1–2 m) to several hundreds of metres between vehicles to several km in the case of communications to external communication nodes, such as road traffic control centres. In order to take advantage of vehicular context awareness, information exchange driven by different events detected by sensors (e.g., street hazard detection, traffic signals, approaching vehicles, etc.) is provided, with multiple messages of different duration, periodicity and link range requirements. Typical values in relation to message exchange times provide message repetition times in the range of approximately 0.5–15 s, depending on the situation and message type [26].

Vehicular communications employ purpose-specific communication systems as well as general-purpose communication systems. The allocated frequency bands span from low-frequency systems employed in closed group tracking systems in

VHF bands to UHF and microwave bands generally used in mobile communication systems and wireless sensor network-based systems.

Considering specific spectrum usage, the channel allocation in the US for direct short-range communications is in 5.850–5.925 GHz, in channels of 10 MHz bandwidth, which can be employed as service channels (SCH), control channels (CCH) or safety channels (SfCH). In the case of Europe, ITS-G5 spans from 5.855 to 5.925 GHz (20 MHz for ITS-G5B, non-safety applications, 30 MHz for ITS-G5A and 20 MHz for ITS-G5D for safety and traffic efficiency). Japan employs frequency allocation for ITS in the 700 MHz range (755.5–764.5 MHz) and from 5.770 to 5.850 MHz [15]. Future 5G systems also rely on the use of mm-wave bands, initially in the 28–70 GHz spectral range. The 63–64 GHz frequency band has been designated to CEPT administrations for ITS applications, with special focus on its use in cooperative adaptative cruise control and platooning applications [29].

1.2.2 Intra-vehicular communications

Initial communication requirements are given within different elements inside the vehicle, thus conforming an intra-vehicle communication scenario. A vehicle can be described as a case of the system of systems, in which each one of the corresponding systems requires information gathering from sensors, command signals for actuators or information transmission. Information processing and communication requirements vary as a function of vehicle type and complexity, with a variable number of onboard embedded control units of up to 70 within one vehicle and up to 2,500 onboard sensors.

Intra-vehicular communication has been traditionally handled by means of wired communications, based on the use of well-established field bus communication technologies, such as controller area network (CAN), local interconnect network (LIN), FlexRay or media-oriented system transport (MOST). Some of their characteristics are detailed below [30]:

- Controller area network (CAN): proposed in the mid-1980s by Bosch GmbH. It employs CSMA/CR access over twisted pair, allowing transmission rates of up to 1 Mbps. It is widely adopted, mainly for transmitting control traffic.
- Local interconnect network (LIN): proposed in the 1990s by the LIN consortium. It employs serial access over a single wire, achieving low transmission rates, up to 19.2 kbps. It was conceived as a lower cost alternative to CAN bus for connection of less critical elements within the vehicle.
- FlexRay: the FlexRay automotive consortium developed the standard from 2000–2009. It proposes the use of twisted pair/optical fibre links with TDMA access, with transmission speeds up to 20 Mbps. It was introduced in 2007 in the BMW X5 model.
- Media-oriented systems transport (MOST): was launched by the MOST Cooperation initiative in 1998, with the aim of providing multimedia and infotainment networking in intra-vehicular scenarios. It is an optical fibre bus, with TDMA access, providing transmission speeds up to 150 Mbps.

It is worth noting that the current evolution is towards the adoption of Ethernet communications within intra-vehicular communication environments, owing to native interconnectivity and increase transmission rate capabilities [31].

1.2.3 Inter-vehicular communications

Vehicles have the capability of generating a great deal of information based on embedded intra-vehicle sensors, which provide automation and control capabilities for multiple embedded electro-mechanical systems. However, inference capabilities are limited in the case of individual vehicle information analysis. This is given by physical factors, such as limitations in the detection range of sensors (e.g., anti-collision radar range, ultra-sound detector range and optical visibility). Moreover, the fusion of data of multiple vehicles provides further insight into multiple aspects, such as road conditions, sudden weather changes or unpredictable events, such as uncommon elements in the road or accidents. In this sense, inter-vehicular communications, also known as vehicle-to-vehicle (V2V) communications, are key enabler in order to increase information gathering in relation to the transportation context of vehicles. In this way, multiple applications can be envisaged, such as [32]

- Early warning of obstacle detection, including potential collision with pedestrians [33], enabling the activation of emergency braking or speed reduction to an individual vehicle or to a vehicle platoon
- Driving assistance in order to increase safety, such as left-turn assistance, does not pass warnings, intersection movement assist or curve speed warning.

One of the main characteristics of V2V communications is the inherently high variability present in the links established, owing to the vehicular movement. Moreover, onboard transceiver operation is also conditioned by antenna placement, which is strongly influenced by the vehicle structure, leading to strong impedance mismatch, modification of radiation diagrams and, in some instances, shadowing and blocking due to the vehicle. Antennas are usually located on the vehicle rooftop, although different locations have been considered as a function of the communication system, such as rear-view mirrors, lateral doors or licence plates, among others. This strongly influences coverage/capacity relations, which has led to the specific study of V2V propagation conditions, which is outlined in section 1.2.6.

The requirements in relation to safety-critical applications demands high reliability and low response time (usually below 100 ms). These conditions cannot be guaranteed with conventional low to mid-range communication systems which can be readily embarked to communicated vehicles, such as conventional IEEE 802.11 systems, such as IEEE 80211a/b/g/n. This has been the main driver for the implementation of the purpose-specific communication framework in vehicular communications, mainly given by direct short-range communications (DSRC) in the US and ITS-G5 in the case of Europe. The underlying physical layer is IEEE 802.11 p (also termed as 802.11-2012), operating within the 5.9 GHz frequency band, which introduces enhancements in terms of maximum transmission rate, guard time, FFT period, preamble duration or subcarrier spacing [34] in order to effectively reduce delay and increase message integrity. Coverage/capacity relations determine that practical V2V links operate within a maximum range of 300 m. Requirements derived from connected and autonomous vehicles indicate that high transceiver node density ($\approx$1,000 nodes per linear km), very low delay ($\leq$ 10 ms) and high BW demands ($>$1 Mbps/node) require new enhanced communication links, which can be feasible with the use of 5G communication networks, combined with DSRC/ITS-G5 communication systems [29].

1.2.4 Vehicle-to-infrastructure (V2I) communications

In order to provide full context awareness, vehicles require communication capabilities with other communication networks, such as dedicated road communication systems or general-purpose communication systems. Vehicle-to-infrastructure (V2I) communication links enable vehicle connectivity, providing an exchange mechanism between vehicles and road infrastructure. In this way, information obtained from road infrastructure sensors or surveillance systems can be provided to vehicles in order to advance weather conditions, traffic density issues or unexpected events such as diverted traffic owing to maintenance procedures, among others [28,35]. Communications will be performed by connecting vehicle OBU with infrastructure RSU and, in the near future, with the possibility of connecting with mobile network communications via LTE V2X or 5G V2X. Applications can be defined as safety or non-safety in relation to information retrieval and actions undertaken by vehicles and users. Examples of safety applications in the V2I domain as defined by the US Department of Transportation are [36]

- Curve speed warning
- Red light violation warning
- Reduced speed zone warning with lane closure
- Spot weather information warning – reduced speed
- Spot weather information warning – diversion
- Stop sign gap assist

In order to perform V2I information exchange, multiple elements from the vehicle as well as from the infrastructure must be engaged:

- Vehicle: onboard data systems and sensors, driver warning system, vehicle wireless data systems and vehicle V2I application component.
- Infrastructure: infrastructure data systems and sensors, traffic signals/traffic signal controller, roadside signage system, infrastructure wireless communication system and infrastructure V2I application component.

Communication exchange will be performed physically via V2I and V2V wireless communication links and logically between the V2I vehicle application component and the infrastructure application component.

1.2.5 Highway, railway, parking and urban applications

The development of ITS has propitiated the implementation of different applications in relation to multiple aspects of transportation in order to increase efficiency, enhance safety, reduce cost, better user quality of experience or reduction of impact in citizens daily life. To achieve these goals, system implementation is focused on aspects such as

- Optimizing transportation routes in order to reduce fuel consumption, reduce transit time and reduce transportation impact in terms of noise and pedestrian/vehicle coexistence
- Implementation of multi-modal transportation schemes in order to combine different transportation systems, focusing towards common information handling platforms
- Enhance long haul freight transportation in terms of logistics and law enforcement/border control operations
- Flexible use of highways with automated toll systems and interactive user information systems
- Preserving the safety of vulnerable users, such as pedestrians, bicycle or motorbike users.

Several examples of such applications are outlined:

- Safety related: within the application scope of ITS in highways and roadways, transportation flow control is one of the most relevant, in direct relation to road safety. Initiative such as the US Department of Transportation Intelligent Network Flow Optimization (INFLO) provide solutions in order to provide queue warning, speed harmonization and to enable the use of cooperative adaptive cruise control (CACC) systems. Specific applications within INFLO are queue warning (Q-WARN), dynamic speed harmonization (SPD-HARM) and CACC [37].
- Multimodal transportation: One of the main applications within ITS is to adequately manage multiple transportation sources in order to provide smooth travel planning, with optimal resource allocation in terms of route time, distance travelled, fuel consumption and travel/toll cost. Different solutions have been implemented, such as bike-sharing handling system for last-mile user connection to transportation systems, in which dimensions of the bike pools as well as their distribution within train stops has been analysed and optimized [38]. Optimal route planning solutions have been implemented based on the dynamic programming of lexicographic analysis in order to consider flexible user route planning with reduced time consumption [39]. Optimization of Bus Rapid Transit systems, based on an IoT approach has been proposed, by using opportunistic signals of users within urban buses by means of Bluetooth communication of their mobile terminals in order to obtain origin/destination matrixes [40]. Integrated dynamic transit operations has been developed as a set of different applications with the aim of improving several aspects of multimodal transportation [37], such as T-CONNECT (transfer time optimization in

transit/non-transit modes), T-DISP (travel options leveraging, with the proposal of different multi-modal routes) or D-RIDE (carpooling application). Another example is given by the multimodal intelligent traffic signal systems (), with the aim of providing normalized next generation traffic signalling, extendable to multiple transportation modes. It provides the following specific traffic signal applications: intelligent traffic signal system, transit signal priority, mobile accessible pedestrian signal system, emergency vehicle pre-emption and freight signal priority.

- Freight transportation: Optimization of freight transportation within a multi-modal context has also been analysed, providing a control platform called COSMO, based on a centralized load balancing method in order to optimize individual transportation routes, effectively tested in the Los Angeles area [41]. Optimized multimodal route planning based on user preferences, implemented by means of the Bayesian preference learning strategy has given as a result the FAVOUR algorithm, improving overall recommender system performance [42]. The USDOT has developed the freight advanced traveller information system, which consists of multiple solutions in order to handle aspects such as dynamic freight handling or optimization of drayage operations [43]. Within the context of the Smart Roadside Initiative [44], law enforcement in relation to freight handling and transportation is considered, with developments such as universal commercial motor vehicle identification, electronic screening/virtual weigh station, wireless roadside inspections and truck parking programs
- Waste management: multiple solutions have been envisaged in order to optimize waste recollection, classification, recycling and management, mainly in the context of dense urban environments. One example has been tested in Fujisawa, Japan, by employing sensorized garbage trucks, which combine multiple types of sensors (contaminant gases, pollen, environmental parameters) and GPS-based location information [45].

1.2.6 Propagation standards

In order to understand the feasibility in the use of communication links in general and in the specific case of vehicular communications, quality of service (QoS) metrics must be satisfied. One of the main indicators of QoS is provided by the estimation of coverage/capacity relations, in which received power levels are compared against receiver sensitivity thresholds, which are dependent on parameters such as bit rate, modulation and coding schemes, or specifications of the employed electronics within transceivers. In this way, a simplified analytical approach can be followed, given by the following expressions:

$$P_{RX} \geq \mathrm{SENS}_{RX} \tag{1.1}$$

$$P_{RX} = P_{TX} - L_{cable-feed\ TX} + G_{Ant\ TX} - L_{prop} + G_{Ant\ RX} - L_{cable-feed\ RX} \tag{1.2}$$

where

- P_{RX}: Power level at the receiver end of the communication link in the log scale
- $SENS_{RX}$: corresponds to the receiver sensitivity threshold, which depends on multiple parameters such as binary transmission rate, modulation and coding schemes and transceiver specifications (such as amplifier noise factors or device phase noise)
- P_{TX}: Transmission power level; radio link level functionalities such as power control can dynamically modify transmit power levels within the corresponding ranges of the communication system employed and the terminal class used.
- $L_{cable\text{-}feed\ TX}$: losses corresponding to the transmission lines and cables employed for the feeding network, antenna matching circuits, power divider, power coupling and diplexer filters, at the transmitter side. Usual cable losses for coaxial cables are in the range of 0.05–1.5 dB/m in the RF range up to 6 GHz. In the case of optical fibres (if electro-optical conversion is employed with remote radio units), attenuation is in the range of 0.0005 to 0.5 dB/m, as a function of the optical fibre type.
- $G_{Ant\ TX}$: Gain of transmitter antenna, which depends on the radiation diagram of the corresponding antenna element. Antenna radiation performance is strongly dependent in the case of onboard vehicular transceivers to the location of the antenna element, as well as on the constituent materials employed.
- $G_{Ant\ RX}$: Gain of receiver antenna, which depends on the radiation diagram of the corresponding antenna element. As in the case of transmit antennas, performance is strongly affected in the case of onboard receiving antennas to the vehicle embodiment.
- $L_{cable\text{-}feed\ TX}$: losses corresponding to the transmission lines and cables employed for the feeding network, antenna matching circuits, power divider, power coupling and diplexer filters, at the receiver side.
- L_{prop}: propagation losses, accounting the different physical phenomena related to the interaction of electromagnetic waves with the surrounding environment, as well as to inherent propagation mechanisms.

An example of propagation losses as a function of distance for a wireless sensor network operating at 868 MHz is presented in figure 1.5. The figure plots received power level as well as path loss as a function of transmitter to receiver linear distance. Results in initial distances exhibit higher variability and values of fading, given by the fact that higher obstruction was occurring given the relative heights between transceivers, with dominating NLOS conditions. At approximately 150 m, propagation losses decay in a smoother manner, in line with LOS conditions. Two different sensitivity values have been considered, corresponding to XBee ZigBee motes operating at a transmission rate of 10 kbps and 80 kbps, respectively. As expected, as the transmission rate increases, sensitivity levels are more stringent, leading to reduced coverage range as compared with lower transmission speeds.

From the link balance equation, it can be seen that propagation losses are the most complex element in terms of estimation, given the high dependence on the

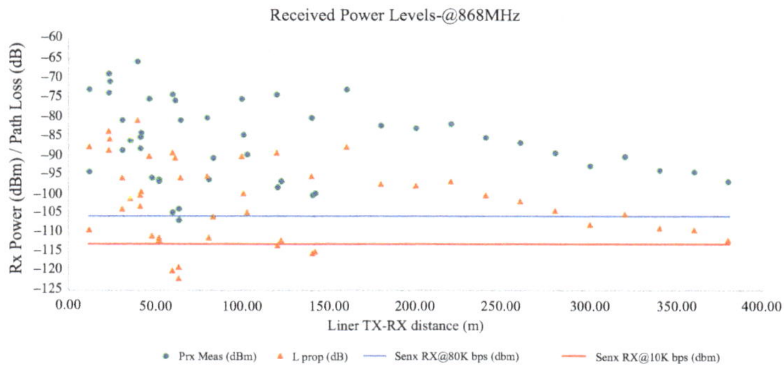

Figure 1.5 Variation of received power levels and propagation losses as a function of linear transmitter–receiver distance, for wireless sensor motes operating at a frequency of 868 MHz

morphology of the scenario, channel dynamics and the inclusion of vehicle structure (impacting on antenna location and metallic vehicle structure). Propagation phenomena include large-scale fading (mainly related to general surrounding environment properties), small-scale fading (given by multipath propagation), diffuse scattering (owing to non-uniform surfaces such as building facades or vegetation canopies) or Doppler shift (given by relative velocity differences between emitter and receiver). Multiple approaches have been discussed in order to model wireless channel behaviour, spanning from empirical models (usually regression based) to deterministic models (geometry based and full-wave electromagnetic simulation). Precise channel modelling requires the use of full-wave electromagnetic techniques, which are computationally intensive and are not feasible in large scenarios. On the other hand, empirical-based models are partially site/region specific and hence exhibit lower accuracy, with non-zero average error and higher standard deviation (which can be partially compensated by means of measurement-based re-calibration of regression coefficients [46]). Channel modelling based on geometric optics with uniform theory of diffraction (GO-UTD) provides a trade-off in terms of computational cost and accuracy by means of ray-launching and ray-tracing approaches [47].

There is a wide range of propagation models, as a function of the application scenario. In relation to vehicular environments, propagation modelling can be classified as [28]

- Geometry-based models: are based on the application of physical optic principle, which is based on the approximation of the impinging wave-front as rays, which represent a finite discretized version of such wave front, in the same direction as the propagation vector. Depending on their complexity and implementation, they can be classified as
  - Ray-tracing or ray-launching models: full implementation of GO-UTD, with different implementations, such as RADII

  - Simplified geometry-based models, which make use of geometric-based information of the environment and combine with measurement or simulation-based information. The GEMV^2 simulator is an example of a well-established simulation tool that combines information obtained from OpenStreetMap in order to distinguish LOS/nLOS link types and perform analytical based path loss estimations [48]. A previous simulation tool considering the VANET simulation scenario called CORNER was implemented, aimed to detect obstacle presence in the propagation path [49].
- Nongeometry-based models: in this case, feature extraction is based on channel measurements, in order to perform the corresponding parameter adjustment. The most commonly employed approach is based on the tapped-delay line model, in which each tap represents a component, with amplitude, phase and time delay given by the specific characteristics of the scenario under consideration.

Recent advances in relation to propagation modelling in vehicular communication context include the consideration of propagation conditions in over ground and underground parking locations [50,51], wideband measurement-based of V2P channel model [52] or rectangular tunnel model characterization for V2V communications [53], among others. Further research is being performed in order to assess propagation modelling in 5G systems considering frequency ranges extending from 0.5 to 100 GHz [54] and other effects such as effects of link transitions from LOS to nLOS conditions [55].

1.3 Vehicular networking

One of the most relevant goals within the ITS framework is to provide a fully context-aware vehicular environment, enabling information exchange among vehicles and different infrastructure elements. This is of particular interest in applications within connected vehicles (safety information exchange) and autonomous driving. In order to provide the required connectivity levels, specific solutions have been proposed in terms of wireless communication system integration (enabling V2V, V2I, V2P and V2B/V2N links) as well as in the required network architecture. As a function of available technology as well as on the specific communication requirements, communication can be established solely between vehicles, between vehicles and specific vehicular communication infrastructure nodes or between vehicles and general-purpose communication networks. As wireless channel conditions exhibit large variability, owing to vehicular dynamics, specific environmental/structural conditions and traffic density, communications can be handled by different systems in a cooperative manner. In this way, message exchange between vehicles can be handled by the V2V communication system, whereas weather information or maintenance-related information can be provided by the V2N communication network and/or the V21 communication network.

In this section, vehicular network architecture from initial vehicle-constrained communications towards cloud-enabled communications will be presented.

1.3.1 Architectures and cooperation

Different approaches have been followed in order to establish context-aware vehicular communication environments. Initial efforts have been proposed in order to apply conventional IEEE 802.11 WLAN networks directly connected with vehicles, termed as Drive-thru Internet, with enhanced operation obtained via macroscopic traffic modelling and ad-hoc clustering techniques, such as ChainCluster [56]. Even though the implementation is attractive in terms of ease of deployment, 802.11 networks exhibit relevant limitations, mainly given by the operation of CSMA/CA access (which introduces excessive delays and potential BW degradation) and the lack of mobility-based functionalities.

In consequence, cooperative mechanisms have been implemented in order to provide robust V2V and V2I communications, by means of using purpose-specific vehicular communication systems, based primarily on wireless access in vehicular environments (WAVE) protocol in combination with DSRC (in the US)/ITS G5 (in Europe). The network is formed by OBU and RSU, which can be complemented with direct wireless access to public land mobile networks (PLMN).

In parallel, developments within LTE PLMN and 5G systems encompass vehicular communication capabilities starting Release 14 of 3GPP, which enable full deployment of network connectivity from 4G (LTE V2X)/5G (5G V2X) enables OBU to eNodeB stations, as well as cooperative schemes between V2V and V2I implementations within WAVE and LTE V2X/5G V2X [15,29].

1.3.2 Ad-hoc and cognitive vehicular networks

Vehicular networks provide a paradigmatic use case in relation to highly reconfigurable and resilient operation. This is particularly true in the case of V2V communications in which point-to-point links can operate in LOS conditions, but information dissemination among the vehicles within communication range will face highly dynamic conditions, with NLOS propagation owing to inter-vehicle obstruction, as well as losses due to other external elements, such as road infrastructure, buildings or land elevations, among others.

Given the variability, the inherent dynamics within the vehicular environments, inter-vehicular communications are given by vehicular ad hoc networks (VANETs), which can be considered as a special case within the more general framework of Mobile Ad Hoc Networks. Some of the properties of VANETs are [34]

- Peer-to-peer ad-hoc communications, by using embarked vehicle OBU
- Specific communication characteristics: intermittent connectivity with highly variable topologies, predictable mobility patterns and requirement for broadcasting/controlled flooding message dissemination

- Limitations in terms of BW availability and potential security issues, owing to shared medium information exchange
- Specific routing approaches, given by inherent channel dynamics.
  - Path cost metrics can be defined as a function of a number of hops or by different link quality estimators, which can be hardware based (such as RSSI or SNR) or software based (packet reception radio, received number of packets, etc.)
  - Specific routing protocols are implemented, considering the highly dynamic and distributed nature of VANETS. Protocols can be classified as
    - Topology-based: consider topology information to implement the routing path. Can be proactive, reactive or hybrid
    - Position-based: consider vehicle location, obtained by means such as GNSS devices
    - Cluster-based: employs a device cluster to reduce message broadcasting
    - Geocast-based: performs multicasting on a specific geographical location
    - Multicast-based: based on the use of a tree or mesh, sends messages to all vehicles within a region
    - Broadcast-based: floods messages to all the network

1.3.3 Internet of vehicles and information processing

ITS are one of the main elements within the context of Smart Cities and Smart Regions, in which we can consider that all agents involved are connected, at different levels, to the Internet. Therefore, vehicular communications can be considered a specific use case of Internet of Things, taking therefore advantage of its inherent properties. Moreover, increasing communication capabilities as well as computational and sensing capabilities of vehicles increases functionality levels of ITS, which can be combined with the inherent scalability of Cloud computing and storage platforms [57,58]. Internet connectivity provides multiple functionalities, related to elements such as infotainment or location-based services, within V2X communications (mainly through V2N, tethered or via direct communications). Functionalities such as 5G MEC will further exploit the use of cloud-based communications and resources in order to take full advantage of 5G network capabilities [29].

In relation to system-level analysis and optimization, data processing techniques based on Big Data analysis can provide relevant information in relation to inference and prediction, as already demonstrated in consumer trend analysis, social network application or smart grids, among others [59]. The benefits in the use of big data within ITS are the following:

- Handling of large amounts of data, with the use of platforms such as Apache Hadoop and Apache Spark.

- ITS operation efficiency improvement, by enabling real-time traffic flow prediction or public transportation service planning, based on traveller journey patterns.
- Enhancement of the ITS safety level, by the prediction of traffic accident occurrence or identify transportation asset issues, such as pavement degradation.

To enable Big Data analytics within ITS, the architecture is divided into three distinct layers: data collection layer, data analytics layer and application layer, schematically depicted in figure 1.6:

Some examples of information analysis-based application in relation to ITS are outlined:

- Trajectory data processing, based on multiple information sources (GPS traces, Bluetooth detectors, WiFi detectors, call detail records), which can be employed for applications such as demand estimation, human behaviour modelling, public transit system planning, traffic performance analysis/prediction, pollution/emission analysis or safety [60]. Visual information of different types can also be employed for multiple applications, including trajectory/movement analysis or weather impact [61,62]
- Estimation of vehicular traffic flow including factors such as seasonal adjustments [63] and travel time estimation [64]
- Data collection and analysis of weather conditions and application to mitigate their effects by active use in roadway maintenance and traffic management [65]
- Logging, storing and analysing naturalistic driving data in order to optimize ITS systems in terms of user behaviour, sensor implementation and required ICT architecture [66]

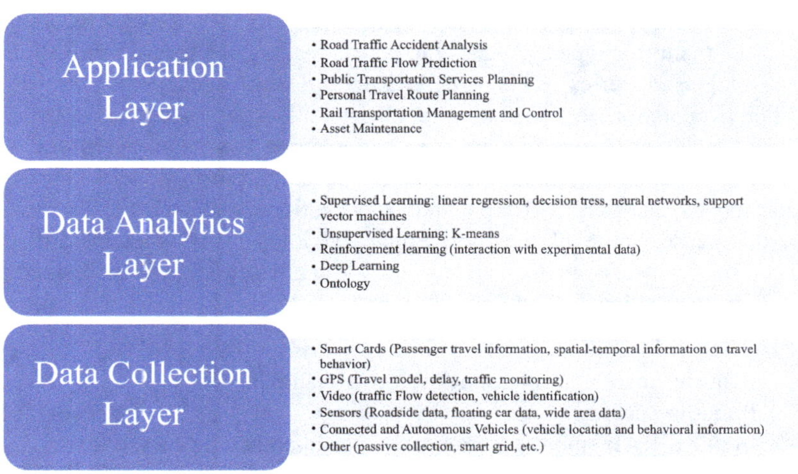

Figure 1.6 Layers in Big Data ITS-driven architecture

- Agent-based computing ITS platforms in applications such as freeway traffic management or urban traffic control [67]
- Identification of passenger itinerary applied to reconfigurable transportation systems to cope with issues such as passenger relocation owing to anomalies given by high congestion, natural disaster, etc. [68]. Bus rapid transit optimization based on ACP (combination of artificial systems, computational experiments and parallel execution) [69]
- Use of multiple sources of social data (i.e., vehicle trajectories, incidence reports, human mobility patterns, social networking or web logs) for the optimization of social transportation systems [70]. Crowd sensing approaches have also been tested in order to optimize public transport systems [71]
- Implementation and optimization of co-drivers (i.e., artificial agents capable of driving with similar behaviour as a human and capable of inferring human intentions) [72]

1.3.4 Performance metrics

The analysis of performance in communication systems is one of the fundamental aspects related to network planning, deployment, operation and maintenance. Different variables are identified in order to perform network planning and network optimization tasks, which have a direct impact on capital expenditure (CAPEX, investment in terms of network deployment and configuration) and in operational expenditure (OPEX, infrastructure running costs, maintenance procedures, etc.). Moreover, overall network investment has a direct impact on QoS and QoE parameters, owing to control in system/network outages and establishment of minimum requirements in traffic handling. In order to analyse performance metrics, key performance indicators (KPIs) are defined. These can be related to different network aspects, such as infrastructure integrity, energy management, transport network availability or access network operation, to name a few [73]

Wireless channel-related performance metrics are dependent on user requirements as well as on the traffic type and density to be handled and the specific network under analysis. In general terms, KPIs will refer to different aspects, such as network accessibility (random access performance, channel allocation success rate, associated signalling functions), mobility (handover performance for intra-system and inter-system cases), retainability (radio access bearer handling, establishment of data transmission contexts) and integrity (throughput, latency, frame loss/packet loss indicators).

In this sense, some of the most employed KPIs in 4G systems and those envisaged for 5G systems, which are currently being tested in different testbeds and with different operational conditions [74], are listed:

- 4G-LTE:
  - radio resource control connection establishment
  - Random access
  - Initial E-RAB (radio access bearer) establishment success rate
  - Handover success rate (intra-frequency, inter-frequency, inter-RAT)
  - Throughput (E-UTRAN, IP)

- 5G (definition of KPI's underway, as per 3GPP TS 28.554):
  - Accessibility:
    - Registered subscribers of network and network slice instance through AMF
    - Registered subscribers of network and network slice instance through UDM
    - Registration success rate of one single network slice instance
    - DRB accessibility for UE services
    - PDU session establishment success rate of one network slice (S-NSSAI)
  - Integrity:
    - Latency of 5G network
    - Upstream throughput for network and network slice instance
    - Downstream throughput for single network slice instance
    - Upstream throughput at the N3 interface
    - Downstream throughput at the N3 interface
    - RAN UE throughput
  - Utilization:
    - Mean number of PDU sessions of network and network slice instance
    - Virtualized resource utilization of network slice instance
    - PDU session establishment time of network slice
  - Retainability:
    - QoS flow retainability
    - DRB retainability
  - Mobility:
    - NG-RAN handover success rate
    - Mean time of inter-gNB handover execution of network slice

1.4 Standards and current technologies

In order to provide wireless connectivity within the ITS ecosystem, multiple wireless communication systems can be employed in order to provide service to the variety of communication links that can be established. Communication systems can be classified in a variety of ways, e.g., as a function of coverage area and radio resource level functionalities (personal area networks, short-range, local-range, wide-area communications, availability of handover functionalities, i.e., PLMN); as a function of their spectral allocation (sub-6 GHz, above 6 GHz) or their dedicated nature (general-purpose communication systems, specific vehicular communication systems) [75]. Operation of these systems can be independent, and they can also be configured in order to provide a collaborative framework, within the paradigm of heterogeneous network operation, in which channel assignment and handling is performed in order to provide optimal resource allocation whilst maximizing quality of experience (QoE) parameters [76].

In this section, an overview will be provided in relation to general-purpose (PLMN, LTE V2X and 5G V2X) as well as purpose-specific communication systems (DSRC and C-ITS) that are currently employed, or their use is envisaged in the near future in relation to the multiple communication links required in vehicular communication scenarios.

1.4.1 Purpose-specific vehicular communication standards

Context-aware vehicular environments require communication capabilities among vehicles as well as with different infrastructure elements, in which V2V and V2I connectivity play a prominent role. Wireless access in this context has been provided in the early 2000s by means of WiFi-based communications, with support from cellular networks for specific applications. Owing to inherent restrictions in terms of allowed delay as well as with the required reliability, especially for safety-related applications, several modifications have been implemented, mainly on the physical and MAC layer of WiFi (looking into mitigation of effects such as inter-carrier interference). Standardization efforts have given rise to several standards, with DSRC-WAVE (US) and C-ITS (Europe) currently being the most representative [77,78].

DSRC employs IEEE 802.11p (currently defined as 802.11-2012, an evolution from 802.11a) at the PHY and MAC layers. The main characteristics in relation to IEEE 802.11p are outlined below [79]:

- Transmission rate up to 27 Mbps
- Spectral allocation from 5.850 to 5.925 GHz
- Channel BW of 10 MHz
- Guard time: 8 μs (double of 802.11a)
- FFT period: 1.6 μs (double of 802.11a)
- Preamble duration: 6.4 μs (double of 802.11a)
- Subcarrier spacing: 0.15625 MHz (half of 802.11a)
- Enhanced distributed channel access (EDCA), which is based on CSMA/CA
- Outside the context of a BSS (OCB) mode for immediate data exchange

Upper layers in this context are defined by the IEEE 1609 protocol, which provides the following functionalities:

- 1609.2: Security services, providing security services that span from the physical layer to the transport layer and providing authentication and optional encryption services. It makes use of a certificate authority and a public key infrastructure (PKI).
- 1609.3: Network services, which defines the wave short message protocol (WSMP).
- 1609.4: Multichannel operations, defined as an MAC layer extension to enable efficient channel switching. Two main types of channels are employed: CCH and SCH.

- 1609.6: Remote management services, providing specific over-the-air management and data message formats to enable WAVE device remote management.
- 1609.11: Over-the-air electronic funds collection, defined as the services and secure message formats necessary to support secure electronic payments.
- 1609.12: Identifier allocation, which provides identifier use description and allocated values.

The combination of DSRC and IEEE 1609 protocol is known as WAVE. Wireless connectivity between OBU and RSU employs WSMP, with optimized header size and capable of message multiplexing to upper layers. WSMP provides services towards the facilities layers in order to provide effective V2X information exchange. Standard J2735, defined by SAE, provides the syntax and semantics of V2X. The most relevant message type is the Basic Safety Message, which is sent out periodically (a maximum period of 10Hz, which can be controlled as a function of wireless channel load), containing information such as position, velocity and vehicle status.

In the case of Europe, the C-ITS standard has been implemented, which follows a similar structure to WAVE-DSRC, with specific differences. The physical and MAC layers employ the ITS-G5 standard similar to 802.11p. Frequency bands are allocated within the 5.9 GHz, subdivided in ITS-G5A (30 MHz primary band for safety applications from 5.875 to 5.905 GHz), ITS-G5B (20 MHz for non-safety applications from 5.855 to 5.875 GHz), ITS-G5C (45 MHz shared with Road Local Area Network from 5.725 to 5.470 GHz) and ITS-G5D (20 MHz for future C-ITS applications from 5.905 to 5.925GHz). Shared medium access is provided by means of CSMA/CA with access categories and EDCA, similar to IEEE 802.11p/DSRC.

The networking and transport layer for safety applications in C-ITS is provided by GeoNetworking, an ad-hoc routing protocol that makes use of geographical coordinate information in the message forwarding process. In between the routing layer and the facilities layer is the basic transport protocol, a transport layer protocol that enables matching ITC component requirements with the geographical information provided by GeoNetworking. Non-safety applications can be provided by conventional IPv6/TCP networking/transport layers, similar to WAVE-DSRC. The facilities layer provides V2X. The most relevant messages employed are

- Cooperative aware message: periodically sent, provides information related to vehicle status (such as location/speed) as well as with communication capabilities (vehicles located within single-hop distance).
- Distributed environmental notification message: event triggered, providing information such as road hazard detection.

Different types of applications in the V2X context can be provided on top of the facilities layer such as [78]

- Intersection information service (intersection status)
- Topology service (static topologic information)

- In-vehicle information service (mandatory/advisory road signage information)
- Signal control service (traffic light control)
- Infrastructure notification service (information on road hazards, road works, etc.) infrastructure awareness service (notification on the existence of infrastructure, such as tolls and estimation of traffic situation and traffic flow).

As in the case of WAVE 1609.2, in C-ITS, security aspects span multiple layers, with standards TS 103,097 employed for security and TS 102,941 for privacy. It is worth noting that in all cases, system implementation takes into consideration that vehicular data processing is compliant with user privacy, not providing general user tracking mechanisms [80]. Future evolutions consider the use of distributed ledger technology such as blockchain, for trust and privacy-related aspects [81]

A schematic of the protocol stack for WAVE-DSRC and C-ITS is depicted in figure 1.7.

1.4.2 Mobile communications infrastructure (2G to 5G)

Public land mobile networks provide a straightforward solution to enable wireless communication links in vehicular environments owing to widely deployed infrastructure, massive user adoption and advanced functionalities related to mobility management and system interoperability, among others. PLMN networks have evolved during the last 40 years from analogue circuit-switched operation to full IP multi-service flexible platforms. Performance has been enhanced in terms of increasing link capacity (reaching Gbps values, aiming toward tens of Gbps in 5G systems and beyond) as well as in reducing latency values (with the aim of reaching 1 ms values). Evolution is schematically depicted in Figure 1.8 from 1G to 5G systems.

General characteristics of mobile communication systems are the following:

- Use of the licensed spectrum, with frequency bands spanning from UHF to the microwave frequency bands, with foreseen use of millimetre wave bands.
- Mobility is organized following a cellular network structure. Cells can be configured to have different cell sizes, spanning from large coverage umbrella cells of several km^2 to small high capacity femtocells, of several m^2 which operate in quasi-static conditions (figure 1.9)

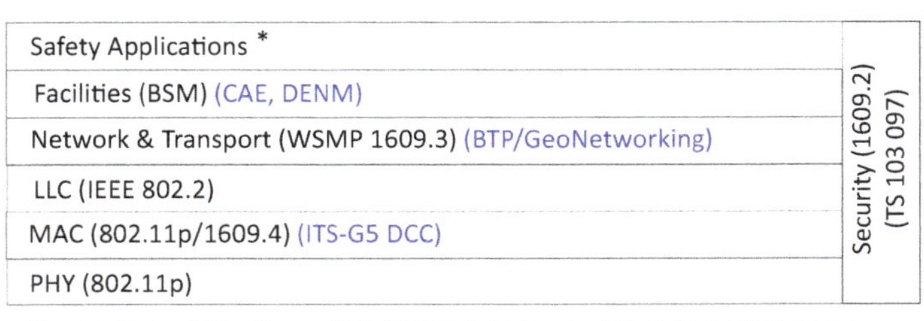

Figure 1.7 Protocol stack (DSRC WAVE in black/C-ITS in blue)

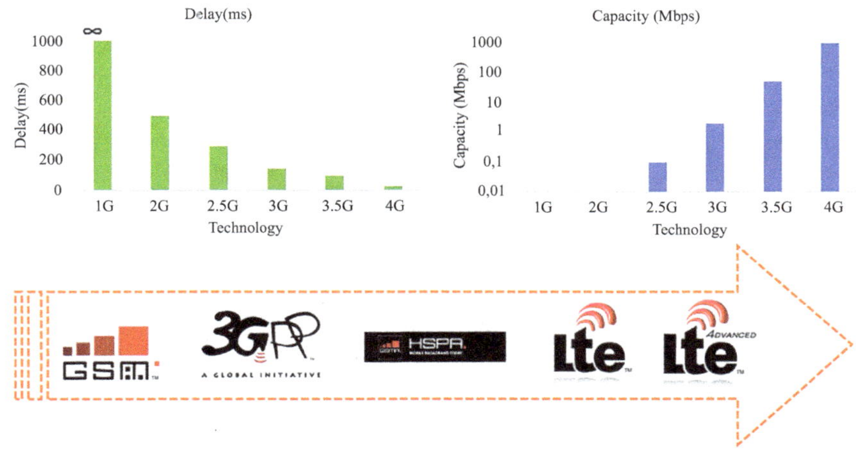

Figure 1.8 Mobile communication systems have evolved in terms of increasing capacity (over 1 Gbps link capacity) and reducing latency (current latency thresholds in the 10–20 ms range)

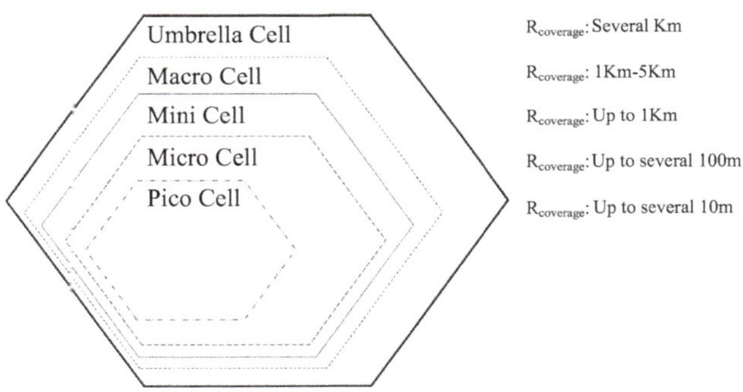

Figure 1.9 Types of coverage cells usually employed in mobile communication systems. Larger cell sizes provide extended coverage range with lower capacity per individual links, whereas smaller cells (e.g., femtocells) provide high capacity links to a small number of users under quasi-static conditions of operation

- Cell mobility is enabled by handover mechanisms, which can be triggered by signal level thresholds, quality measurements (such as BER or FER values) or by traffic load mechanisms.
- Multiservice network-oriented applications can be provided, given by the use of multiple radio access technologies, as well as by the evolution of the core network (CN). In this sense, cooperative networking techniques can optimize overall network operation by allowing optimal network selection, which is triggered by different handover mechanisms

Data transmission capabilities have been provided initially by the introduction of generalized packet radio service, as an embedded packet data transmission solution over the original frequency division multiple access/time division multiple access (FDMA/TDMA) 2G networks, in the case of GSM. In this way, circuit switching (CS) and packet switching (PS) cores were simultaneously available, with mobility control separated from circuit-switched voice calls and packet switch traffic, allowing transmission rates up to 170 kbps in downlink and latency values in the range of several hundreds of ms. 3G networks also operate with dual network cores, with CS and PS connections provided from the radio network controller (RNC). Transmission rates and latency values improve the range of 50 Mbps and below 100 ms, respectively, with the introduction of high-speed packet access (HSPA). One of the relevant elements introduced by HSPA is the inclusion of a new MAC sublayer (hs-MAC), which enables the use of faster retransmission techniques, without the need of establishing connections to the RNC, reducing latency. Capacity increase and significant delay reduction are achieved in the case of 4G systems, namely, long-term evolution (LTE), in which substantial modifications have been implemented, such as end-to-end IP connectivity, the use of time division duplex access schemes or higher-level adaptive modulation and coding schemes.

Specific communication solutions for vehicular communications, considering V2X scenarios, were already foreseen in Release 14 of 3GPP, owing to higher achievable transmission bit rates as well to lower delays. Some of the offered specifications are the following [15]:

- Vehicular speeds up to 160 km/h (absolute speed) and up to 280 km/h (relative speed)
- Driver response time in the order of 4 s
- Enabled message exchange, with periodic messages with payloads in the order of 50B–300B and event-triggered messages up 1,200B in length
- Latency parameters with maximum values of message transfer of 100 ms between two terminals and 1,000 ms for messages sent via a network server

The current evolution in mobile communications is given by 5G systems, with enhanced performance in terms of capacity handling and energy efficiency. In this way, 5G systems provide multi-Gbps capacity (eMBB, enhanced mobile broadband), ms delay (uRLLC, ultra-reliable low latency communications), massive simultaneous link provision, reduced computational complexity of terminals (MEC, multi-access edge computing) and ad-hoc quality of service requirements (network slicing). This is possible owing to some of the following functionalities specified within 5G:

- Adaptability for different frequency bands (sub-6 GHz/above 6 GHz), with different BW and with variable propagation conditions (600/700 MHz–3.3 GHz/ 4.5 GHz-mm wave) (28 GHz/29 GHz/40–60 GHz) (scalable numerology)
- Individual control of the user plane/control plane in order to provide higher adaptability and scalability (in applications with very high node density)
- Software-based implementation of the network edge (radio access network – RAN) as well as in the core network (CN)

- Enabling mechanisms in order to precisely differentiate traffic and quality of service requirements (network slicing)
- Support for communication schemes such as multi-connectivity (e.g., HetNet operation) or device-to-device (D2D) communications, which are controlled by the network
- Support for multiple network architectures (e.g., fronthaul and backhaul)
- Energy-efficient network operation by providing smaller network nodes and mechanisms such as radio over fibre

LTE V2X communication systems provide capabilities that are in the order of V2V DSRC-based systems. 5G-based communication systems, owing to functionalities given by high reliability, reduced delay and high transmission rates, can provide adequate communication solutions for advanced use cases [29], which are schematically depicted in figure 1.10. Moreover, further functionalities such as non-orthogonal multiple access [82] or statistical signal transmission [83] can enhance 5G operation within vehicular scenarios.

The adoption of 5G V2X communications to enable context-aware vehicular scenarios is foreseen to provide the following benefits [29]:

- Transportation safety: worldwide statistics indicate that 1.25 million people die from traffic accidents, whilst between 20 million and 50 million people suffer non-fatal injuries. Estimation from the National Highway Traffic Safety Administration, V2X-enabled application could reduce the severity of up to 80% of non-impaired crashes.

Use Case	Description	Communication Requirements
Intent/Trajectory Sharing Driving	Information from each vehicle is shared in order to provide trajectory coordination, as well as driving intention	▪ High BW (support for burst transmission) ▪ Latency in the order of 10 ms ▪ 99.99% message reliability
Extended Sensors	Capability to obtain information beyond information capabilities of individual on-board vehicle sensors.	▪ High BW (support for burst transmission) ▪ Latency below 10 ms ▪ High message reliability, in the order of 95% ▪ High connection density
Platooning	Capability to form and manage a set of coordinated sequential vehicle configuration.	▪ Latency in the order of 25ms for end to end communication, with values in the order of 10ms for high automation degree ▪ 90% message reliability, 99.99% for high-level automation ▪ Relative longitudinal position accuracy of less than 0.5m ▪ Dynamic range control
Remote driving	Capability to remotely control vehicles by means of V2N communication, for applications	▪ Data rate up to 1 Mbps (DL) and 25 Mbps (UL) ▪ Ultra high reliability, greater

Figure 1.10 Advanced use cases given by 5G V2X communication capabilities

- Traffic efficiency: multiple solutions within ITS frameworks in cities and regions can be implemented, such as traffic flow optimization or hazard protection, among others.
- Infrastructure savings: the use of V2X communications enables traffic information collection and the subsequent application of deterministic traffic congestion algorithms. The outcomes can be used by city/road administrations, as well as insurance providers.
- Smart green environments: information exchange in V2X can be employed in order to reduce traffic jams and optimize overall driving, reducing emissions and overall energy consumption requirements.

1.4.3 Evolution in wireless connectivity

Wireless connectivity within the V2X framework entails the use of multiple communication systems in a cooperative/HetNet configuration spanning from dedicated vehicular communication systems such as WAVE/C-ITS to vehicular communications supported by PLMN (such as LTE V2X and 5G V2X). While 5G is undergoing final development and initial network planning and deployment stages, requirements for future applications, such as holographic user interaction, fully autonomous vehicle operation or optimized converged network operation, have triggered research and outlook towards future beyond 5G networks [84].

In this sense, communications in increased millimetre wave bands (such as IEEE 802.15.3d-2017 standard, with frequency allocation in the 275–325GHz) and in THz-optical domains (such as visible light communications) are being analysed [85]. Report M.2417 from ITU studies the use of the 275–450GHz spectrum for land mobile services [86], with applications in close proximity mobile communications, intra-device communications or wireless links in data centres. Regulation is being developed as well as in laboratory-based experiments and measurement campaigns to account for the impact in V2V communications, such as side-lane effects [87]

There are, however, multiple challenges that have to be addressed in relation to increased spectrum use, such as [88]

- THz transceiver design, with challenges in both THz source generation (based on down conversion, quantum cascade lasers or up conversion, among others) and signal detection. Path loss values are very large owing to factors such as high attenuation from water vapour and other gases.
- Ultra-massive MIMO antenna array implementation. The reduction in wavelength enables the implementation of arrays with a vast amount of radiating elements, which pose challenges in terms of factors such as antenna coupling. Novel approaches are being analysed, such as hybrid beamforming in the THz domain [89].
- Requirement for new approaches in information theory application for the analysis of items such noise impact in the communication channel or the use of new waveforms.
- Requirement for new propagation models and channel models in THz frequency bands.

- Design on appropriate MAC layers or interference handling mechanisms.
- Backhaul network with sufficient capacity for THz link demands.

References

[1] EU Commission, Initiative on Smart Cities, https://ec.europa.eu/info/eu-regional-and-urban-development/topics/cities-and-urban-development/city-initiatives/smart-cities_en

[2] Bernard Cathelat, 'Smart Cities: Shaping the Society of 2030', United Nations Educational, Scientific and Cultural Organization and Netexplo, 2019

[3] European Smart City, Medium Size City Profiling and Benchmarking, www.smart-cities.eu

[4] ASEAN Smart Cities Network, https://asean.org/asean/asean-smart-cities-network/

[5] UNECE United Smart Cities, https://www.unitedsmartcities.org/

[6] Japan Smart Cities, https://www.urenio.org/wp-content/uploads/2018/03/Japans-Smart-Cities-DeWit.pdf

[7] Pardini, K., Rodrigues, J.J.P.C., Kozlov, S.A., Kumar, N., and Furtado, V., 'IoT-based solid waste management solutions: a survey', *Journal of Sensor and Actuator Networks* 2019, 8, 5.

[8] T. Anagnostopoulos, A. Zaslavsky, K. Kolomvatsos, *et al.*, 'Challenges and opportunities of waste management in IoT-enabled smart cities: a survey', *IEEE Transactions on Sustainable Computing*, 2017, 2(3), 275–89, 2017

[9] A. Solanas, C. Patsakis, M. Conti, *et al.*, 'Smart health: a context-aware health paradigm within smart cities', *IEEE Communications Magazine*, 2014, 52(8), 74–81

[10] Eva Geisberger and Manfred B. (Eds.), *Living in a Networked World. Integrated Research Agenda Cyber-Physical Systems (agendaCPS) (acatech STUDY)*, Herbert Utz Verlag, Munich, 2014.

[11] Studies on Intelligent Transportation Systems, EU Commission, https://ec.europa.eu/transport/themes/its/studies/its_en

[12] Key Performance Indicators for Intelligent Transport Systems – Final Report, EU Commission, https://ec.europa.eu/transport/sites/transport/files/themes/its/studies/doc/its-kpi-final_report_v7_4.pdf

[13] C. Suthaputchakun, Z. Sun, and M. Dianati, 'Applications of vehicular communications for reducing fuel consumption and CO_2 emission: the state of the art and research challenges', *IEEE Communications Magazine*, 2012, 50, 108–15

[14] G. Rafiq, B. Talha, M. Pätzold, *et al.*, 'What's new in intelligent transportation systems? ', *IEEE Vehicular Technology Magazine*, 2013, 8, 45–63

[15] V2X Cellular Solutions, 5G Americas, October 2016, https://www.5gamericas.org/v2x-cellular-solutions/

[16] E.E. Altshuler, V.J. Falcone, Jr., and K.N. Wulfsberg, 'Atmospheric effects on propagation at millimeter wavelengths', *IEEE Spectrum*, 1968, 5, 83–90

[17] G.A. Siles, J.M. Riera, and P. García-del-Pino, 'Atmospheric attenuation in wireless communication systems at millimeter and THz frequencies', *IEEE Antennas and Propagation Magazine*, 2015, 57(1), 48–61

[18] T.K. Sarkar, Z. Ji, K. Kim, A. Medour, and M. Salazar-Palma, 'A survey of various propagation models for mobile communication', *IEEE Antennas and Propagation Magazine*, 2003, 45(3), 51–82

[19] S. Salous, V.D. Esposti, F. Fuschini, *et al.*, 'Millimeter-wave propagation characterization and modeling toward fifth-generation systems', *IEEE Antennas and Propagation Magazine*, 2016, 58, 115–27

[20] J. Pahl, *Interference Analysis. Modelling Radio Systems for Spectrum Management*, Wiley, Chichester-Sussex, 2016.

[21] P. Stenumgaard, J. Chilo, J. Ferrer-Coll, and P. Ängskog, 'Challenges and conditions for wireless machine-to-machine communications in industrial environments', *IEEE Communications Magazine*, 2013, 51, 187–92

[22] R. Candell, K.A. Remley, J.T. Quimby, *et al.*, 'Industrial Wireless Systems Radio Propagation Measurements', NIST Technical Note 1951, 2017, https://doi.org/10.6028/NIST.TN.1951

[23] P. Cardieri, 'Modeling interference in wireless ad hoc networks', *IEEE Communications Surveys and Tutorials*, 2010, 12(4), 551–72

[24] S. Savazzi, V. Rampa, and U. Spagnolini, 'Wireless cloud networks for the factory of things: connectivity modeling and layout design', *IEEE Internet of Things Journal*, 2014, 1(2), 180–94

[25] M. Cheffena, 'Industrial wireless sensor networks: channel modeling and performance evaluation', *EURASIP Journal on Wireless Communications and Networking*, 2012, 2012, 297

[26] Andrews, S., 'Vehicular communications requirements and challenges', In: Eskandarian, A. (eds), *Handbook of Intelligent Vehicles*, Springer, London, 2012

[27] L. Liang, H. Peng, G.Y. Li, and X. Shen, 'Vehicular communications: a physical layer perspective', *IEEE Transactions on Vehicular Technology*, 2017, 66(12), 10647–59

[28] W. Viriyasitavat, M. Boban, H.-M. Tsai, and A.V. Vasilakos, 'Vehicular communications. survey and challenges of channel and propagation models', *IEEE Vehicular Technology Magazine*, 2015, 55–66

[29] Cellular V2X Communications towards 5G, 5G Americas, March 2018, https://www.5gamericas.org/cellular-v2x-communications-towards-5g/

[30] S. Tuohy, M. Glavin, E. Jones, M. Trivedi, and L. Kilmartin, 'Next generation wired intra-vehicle networks: a review', *2013 IEEE Intelligent Vehicles Symposium*, June 23–26, 2013, Gold Coast, Australia, 777–82

[31] P. Hank, S. Müller, O. Vermesan, and J. Van Den Keybus, 'Automotive ethernet: in-vehicle networking and smart mobility', *Design, Automation & Test in Europe Conference & Exhibition (DATE)*, Grenoble, March 2013

[32] Vehicle-to-Vehicle Communications: Readiness of V2V Technology for Application, National Highway Traffic Safety Administration, US Department of Transportation, August 2014.

[33] C. Flores, P. Merdrignac, R. de Charette, F. Navas, V. Milanés, and F. Nashashibi, 'A cooperative car-following/emergency braking system with prediction-based pedestrian avoidance capabilities', *IEEE Transactions on Intelligent Transportation Systems*, 2019, 20(5), 1837–46

[34] F. Drouhin, and S. Bindel, 'Routing and data diffusion in vehicular ad hoc networks', In S. Femmam (Ed.), *Building Wireless Sensor Networks, Application to Routing and Data Diffusion*, London: ISTE Press and Elsevier, Chapter 4, 2017

[35] F. Arena, and G. Pau, 'An overview of vehicular communications', *Future Internet* 2019, 11, 27, doi:10.3390/fi11020027

[36] Vehicle-to-Infrastructure (V2I) Safety Applications Performance Requirements, Vol. 1, Introduction and Common Requirements, 2015, https://rosap.ntl.bts.gov/view/dot/3592

[37] Dynamic Mobility Applications, Intelligent Transportation Systems Joint Program Office, US Department of Transportation https://www.its.dot.gov/research_archives/dma/dma_development.htm

[38] G. Tang, S. Keshav, L. Golab, and K. Wu, 'Bikeshare pool sizing for bike-and-ride multimodal transit', *IEEE Transactions on Intelligent Transportation Systems*, 2018, 19(7), 2279–89

[39] K.G. Zografos, and K.N. Androutsopoulos, 'Algorithms for itinerary planning in multimodal transportation networks', *IEEE Transactions on Intelligent Transportation Systems*, 2008, 9(1), 175–84

[40] L.F. Herrera-Quintero, J.C. Vega-Alfonso, K.B.A. Banse, and E.C. Zambrano, 'Smart ITS sensor for the transportation planning based on IoT approaches using serverless and microservices architecture', *IEEE Intelligent Transportation Systems Magazine*, 2018, 10, 17–27

[41] Y. Zhao, P.A. Ioannou, and M.M. Dessouky, 'Dynamic multimodal freight routing using a co-simulation optimization approach', *IEEE Transactions on Intelligent Transportation Systems*, 2019, 20(7), 2657–67

[42] P. Campigotto, C. Rudloff, M. Leodolter, and D. Bauer, 'Personalized and situation-aware multimodal route recommendations: the FAVOUR algorithm', *IEEE Transactions on Intelligent Transportation Systems*, 2017, 18 (1), 92–102

[43] Intelligent Transportation Systems Joint Program Office, US Department of Transportation, https://www.pcb.its.dot.gov/eprimer/module13.aspx#cva

[44] U.S. Department of Transportation, Research and Innovative Technologies Administration, ITS Joint Program Office, 'Smart Roadside Initiative Concept of Operations', SAIC, May 21, 2012

[45] A. El F. Seghrouchni, F. Ishikawa, L. Herault, and H. Tokuda, *Enablers for Smart Cities*, ISTE Ltd and John Wiley & Sons, London, 2016

[46] A.F. Molisch, *Wireless Communications*, Wiley, New York, 2010

[47] Z. Yun, and M.F. Iskander, 'Ray tracing for radio propagation modeling: principles and applications, *IEEE Access*, 2015, 3, 1089–1100

[48] M. Boban, J. Barros, and O.K. Tonguz, 'Geometry-based vehicle-to-vehicle channel modeling for large-scale simulation', *IEEE Transactions on Vehicular Technology*, 2014, 63(9), 4146–64.

[49] E. Giordano, R. Frank, G. Pau, and M. Gerla, 'CORNER: a radio propagation model for VANETs in urban scenario', *Proceedings of the IEEE*, 2011, 99(7), 1280–94

[50] R. Sun, D.W. Matolak, and P. Liu, '5-GHz V2V channel characteristics for parking garages', *IEEE Transactions on Vehicular Technology*, 2017, 66(5), 3538–3547

[51] M. Yang, B. Ai, R. He, *et al.*, 'V2V channel characterization and modeling for underground parking garages', *China Communications*, 2019, 93–105

[52] Glora Makhoul, Raffaele D′Errico and Calude Oestges, 'Wideband Measurement Based Vehicle to Pedestrian Channel Models', IEEE Transactions on Vehicular Technology, Early Access

[53] N. Avazov, S.M. Riazul Islam, D. Park, and K.S. Kwak, 'Statistical characterization of a 3-D propagation model for V2V channels in rectangular tunnels', *IEEE Antennas and Wireless Propagation Letters*, 2017, 16, 2392–95

[54] T.S. Rappaport, Y. Xing, G.R. MacCartney, Jr., A.F. Molisch, E. Mellios, and J. Zhang, 'Overview of millimeter wave communications for fifth-generation (5G) wireless networks—with a focus on propagation models', *IEEE Transactions on Antennas and Propagation*, 2017, 65(12), 6213–30

[55] T. S. Rappaport, G. R. MacCartney, Jr., S. Sun, H. Yan, and S. Deng, 'Small-scale, local area, and transitional millimeter wave propagation for5G communications', *IEEE Transactions on Antennas and Propagation*, 2017, 65, (6474–90)

[56] H. Zhou, L. Gui, Q. Yu, and X. Shen, 'Cooperative Vehicular Communications in the Drive-thru Internet', Springer, 2015

[57] W. He, G. Yan, and L. D. Xu, 'Developing vehicular data cloud services in the IoT environment', *IEEE Transactions on Industrial Informatics*, 2014, 10(2), 1587–95

[58] K. Wang, and Z. Shen, 'Artificial societies and GPU-based cloud computing for intelligent transportation system management', *IEEE Intelligent Systems*, 2011, 22–28

[59] L. Zhu, F. R. Yu, Y. Wang, B. Ning, and T. Tang, 'Big data analytics in intelligent transportation systems: a survey', *IEEE Transactions on Intelligent Transportation Systems*, 2019, 20(1), 383–98, 2019

[60] N. Markovic, P. Sekula, V. Laan, G. Andrienko, and N. Adrienko, 'Applications of trajectory data from the perspective of a road transportation agency: literature review and the Maryland case study', *IEEE Transactions on Intelligent Transportation Systems*, 2019, 20(5), 1858–69

[61] G. Andrienko, N. Andrienko, W. Chen, R. Maciejewski, and Y. Zhao, 'Visual analytics of mobility and transportation: state of the art and further research directions', *IEEE Transactions on Intelligent Transportation Systems*, 2016, 18(8), 2232–49

[62] S.-C. Huang, B.-H. Chen, and Y.-J. Cheng, 'An efficient visibility enhancement algorithm for road scenes captured by intelligent transportation systems', *IEEE Transactions on Intelligent Transportation Systems*,2014, 15(5), 2321–32

[63] W. Huang, W. Jia, J. Guo, *et al.*, 'Real-time prediction of seasonal heteroscedasticity in vehicular traffic flow series', *IEEE Transactions on Intelligent Transportation Systems*, 2018, 19(10), 3170–80

[64] P. Kachroo, and S. Sastry, 'Travel time dynamics for intelligent transportation systems: theory and applications', *IEEE Transactions on Intelligent Transportation Systems*, 2016, 17(2), 385–94

[65] K. C. Dey, A. Mishra, and M. Chowdhury, 'Potential of intelligent transportation systems in mitigating adverse weather impacts on road mobility: a review', *IEEE Transactions on Intelligent Transportation Systems*, 2015, 16(3), 1107–19

[66] A. Bender, J.R. Ward, S. Worrall, *et al.*, 'A flexible system architecture for acquisition and storage of naturalistic driving data', *IEEE Transactions on Intelligent Transportation Systems*, 2016, 17(6), 1748–61

[67] B. Chen, and H. Cheng, 'A review of the applications of agent technology in traffic and transportation systems', *IEEE Transactions on Intelligent Transportation Systems*, 2010, 11(2), 485–97

[68] A. Viswanath, E. E. S. Baca, and A. M. Farid, 'An axiomatic design approach to passenger itinerary enumeration in reconfigurable transportation systems', *IEEE Transactions on Intelligent Transportation Systems*, 2014, 15(3), 915–24

[69] X. Dong, i, 'A parallel transportation management and control system for bus rapid transit using the ACP approach', *IEEE Transactions On Intelligent Transportation Systems*, 2014, 18(9), 2569–74

[70] X. Zheng, W. Chen, P. Wang, *et al.*, 'Big data for social transportation', *IEEE Transactions on Intelligent Transportation Systems*, 2016, 17(3), 620–30

[71] K. Farkas, G. Feher, A. Benczur, and C. Sidlo, 'Crowdsending based public transport information service in smart cities', *IEEE Communications Magazine*, 2015, 158–65

[72] M. D. Lio, F. Biral, E. Bertolazzi, *et al.*, 'Artificial co-drivers as a universal enabling technology for future intelligent vehicles and transportation systems', *IEEE Transactions on Intelligent Transportation Systems*, 2015, 16 (1), 244–63

[73] D. Kakadia, J. Yang, and A. Gilgur, *Network Performance and Fault Analytics for LTE Wireless Service Providers*, Springer, London, 2017

[74] T. Doukoglou, V. Gezerlis, K. Trichias, *et al.*, 'Vertical industries requirements analysis & targeted KPIs for advanced 5G trials', *European Conference on Networks and Communications*, Valencia, Spain, June 2019

[75] K. Dar, M. Bakhouya, J. Gaber, M. Wack, and P. Lorenz, 'Wireless communication technologies for ITS applications', *IEEE Communications Magazine*, 2010, 156–62.

[76] X. Zhao, X. Li, Z. Xu, and T. Chen, 'An optimal game approach for heterogeneous vehicular network selection with varying network performance', *IEEE Intelligent Transportation Systems Magazine*, Fall, 81–92

[77] A. Festag, 'Standards for vehicular communication – from IEEE 802.11p to 5G', *Elektrotechnik & Informationstechnik*, 2015, 132(7), 409–16

[78] A. Festag, 'Cooperative intelligent transport systems standards in Europe', *IEEE Communications Magazine*, 2014, 52, 166–172

[79] J. Gozalvez, M. Sepulcre, and R. Bauza, 'IEEE 802.11p vehicle to infrastructure communications in urban environments', *IEEE Communications Magazine*, 2012, 50, 176–183

[80] D. Eckhoff, R. German, C. Sommer, F. Dressler, and T. Gansen, 'SlotSwap: strong and affordable location privacy in intelligent transportation systems', *IEEE Communications Magazine*, 2011, 49, 126–133

[81] V. Ortega, F. Bouchmal, and J. F. Monserrat, 'Trusted 5G vehicular networks blockchains and content-centric networking', *IEEE Vehicular Technology Magazine*, 2018, 13, 2–8

[82] B. Di, L. Song, Y. Li, and Z. Han, 'V2X meets NOMA: non-orthogonal multiple access for 5G-enabled vehicular networks', *IEEE Wireless Communications*, 2017, 24, 14–21

[83] T. Xu, M. Zhang, and T. Zhou, 'Statistical signal transmission technology: a novel perspective for 5G enabled vehicular networking', *IEEE Wireless Communications*, 2017, 24, 22–29

[84] M. Chiani, E. Paolini, and F. Callegati, 'Open issues and beyond 5G', *The 5G Italy Book in 2019: A Multiperspective view of 5G, CNIT*, ISBN: 9788832170030, 2019

[85] J. Costa, 'Studies on the use of frequency bands above 275 GHz by land-mobile and fixed service applications', *ITU News Magazine*, 2019, (5), 85–89

[86] Technical and Operational Characteristics of Land-Mobile Service Applications in the Frequency Range 275–450 GHz, Rep. ITU-R M.2417-0, https://www.itu.int/pub/R-REP-M.2417

[87] V. Petrov, J. Kokkoniemi, D. Moltchanov, J. Lehtomäki, M. Juntti and Y. Koucheryavy, 'The impact of interference from the side lanes on mmWave/THz band V2V communication systems with directional antennas', *IEEE Transactions on Vehicular Technology*, 2018, 67(6), 5028–5041

[88] S. Mumtaz, J. M. Jornet, J. Aulin, W. H. Gerstacker, X. Dong, and B. Ai, 'Terahertz communication for vehicular networks', *IEEE Transactions on Vehicular Technology*, 2017, 66(7), 5617–5624

[89] S. A. Busari, K. M. S. Huq, S. Mumtaz, *et al.*, 'Generalized hybrid beamforming for vehicular connectivity using THz massive MIMO', *IEEE Transactions on Vehicular Technology*, 2019, 68(9), 8372–8383

Chapter 2
Wireless channel properties for vehicular environments

In any wireless communication environment, one needs to know the main characteristics that produce impairments to the signals being transmitted. In this chapter, we introduce and define those features and parameters that need to be considered to characterize wireless environments. Fundamentals of propagation are presented together with commonly used, traditional and empirical path loss models, as well as small- and large-scale propagation parameters. Now, the vehicular environment is more complex than other wireless communication systems due to the spatial and time-varying nature of the environment, so we present how the properties of wireless channels need to adapt when vehicular to anything vehicle-to-everything (V2X) communications are characterized by propagation and channel models. Mobility is one of those unique features in the vehicular environment that produces non-stationary environments. The chapter ends with an introduction to the use of multi-antenna systems for vehicular communications.

2.1 Wideband vs. narrowband channels

In the past, communications channels could be considered narrowband or wideband depending on the available amount or range of frequencies for transmission, and then we would design our signals so that their bandwidth would be smaller than the range of frequencies to be used. Such a range would be small or large depending on several different features such as regulation, technology, and number of users, among others. When such a range was large compared to what was previously used, it was generally referred to as a wideband channel. This changed when different communications standards defined the use of techniques such as spread spectrum and the broadband services were introduced. Also, the need to include different services with different characteristics demanded higher bandwidths for transmission.

One example of wideband and narrowband considered was when the first-generation (1G) cellular service was operating. Advance mobile phone system (AMPS) was the service provided with 30 kHz channels for each user, and when more capacity was needed (number of simultaneous users connected), one standard generated was the narrowband AMPS (NAMPS) that provided with 10 kHz channels to each user. But now, that 30 kHz channel is not a wideband channel. There have been definitions of a narrowband channel with an upper limit of 25 kHz of bandwidth, and its advantages are that the noise in such a channel contains less energy than in a wideband channel. Some definitions such as for ultra-wide band signals are for the bandwidth to be at least 25% of the value of the carrier frequency. Thus, the context of the system being used, and the signals being transmitted also influence the definition of a narrowband or wideband channel.

In wireless communications, the definition of a channel being narrowband or wideband depends generally on two characteristics. First, the time duration of the symbols being transmitted, and second, the small-scale propagation characteristics of coherence bandwidth and delay spread. The propagation of a signal will produce at the receiver the superposition of several delayed and attenuated versions of the original signal through the process of multipath propagation. The original signal will be seen at the receiver as occupying more time than the one originally used when transmitted, and this is called delay spread. The delay spread when studied in the frequency domain corresponds to the coherence bandwidth, which is the range of frequencies where the channel response is maximally flat. In general, wireless communication channels have many different behaviours that will produce different gains at different frequency ranges, what is called frequency selectivity.

When the bandwidth of the signal or symbol to be transmitted is smaller than the coherence bandwidth of the channel, the whole spectrum of the signal will be affected (amplitude and phase changes) almost identically producing what is called flat fading which is a narrowband channel, i.e., no frequency selectivity and almost no dispersion or delay spread produced to the signal. On the other hand, if the bandwidth of the signal or symbol is such that is more than the coherence bandwidth, then the channel will be wideband, there would be frequency selectivity, different frequency components of the signal spectrum will be affected by different changes in amplitude and phase.

2.2 Fundamentals of propagation in wireless channels

The characterization of wireless communication channels is fundamental for the understanding of how impairments affect the signals being transmitted, especially for future generations of wireless communications. This characterization includes statistical modelling of the channel, real-field measurements and simulations to produce an integral framework for wireless communications. When designing a communication system, one needs to know the communication channel to be used to get acquainted with the channel impairments to which the signal to be transmitted is exposed. This signal may be affected by additive noise and may be changed by phenomena that cause distortion among other problems. Wireless links present problems that in some cases are severe and exclusive. For example, the transmitted signal will find several obstacles or scatterers while traversing the channel on which the signal will bounce off and lose power due to the reflectivity of materials. While the signal bounces off surfaces with different characteristics, it arrives at the destination from different directions creating the multipath effect. This multipath creates fadings to the signal received that might combine constructively or destructively affecting the power received and generating self-interference when delayed and distorted versions of the same signal arrive. Also, mobility in vehicular environments will produce frequency shifts and phase distortion at the receiver site by the Doppler effect. All the effects are considered and studied jointly through the concept of multipath effects. Thus, the mobile radio channel needs to be characterized in order to know those impairments and then design your system to cope with such signal impairments.

Some parameters used to characterize the mobile radio channel are grouped in two main classes, the large-scale and the small-scale propagation effects. The large-scale characterization models introduce mainly the quantification of power loss of the signal as it traverses the wireless communication channel. The second class is for small-scale conditions (time and distance) and is related to the environment, mainly through the multipath effect of the signal. Small-scale propagation effects are evidenced through sudden changes in power reception. The study of these two classes can be seen in [1–4].

Generally, three tools are combined to generate the characterization of a wireless complex environment such as the vehicular. First, a set of measurement campaigns is important to obtain, at least using carrier signals in the frequencies under study. Second, simulations can be included where different techniques may be used such as considering the use of three-dimensional (3D) modelling for a specific environment. Third, the analytical or statistical model comes from the analysis of the measurement data obtained and the simulations conducted. The model may consist of several components, such as the statistical description of the channel through probability density functions (pdf), large- and small-scale parameters, closed-form mathematical expressions based on statistical descriptions of how random variables and processes interact or empirical formulas derived from statistical behaviour of the data that has been compiled from the measurement campaigns.

2.2.1 Path loss models

Commonly, the mobile wireless channel consists of frequencies in the ultra-high-frequency band (UHF) which is from 300 MHz to 3 GHz and the super high-frequency band with frequencies ranging from 3 to 30 GHz. Most of the standards developed for vehicular communications so far consist of the frequencies below 6 GHz. When a system uses these frequencies, the wireless links must have up to a certain percentage a component that is called the line-of-sight (LOS) component.

Every wireless channel affects a signal being transmitted by causing distortion and adding noise and interference. Some of these phenomena effects are linear and nonlinear changes of amplitude, phase and frequency of the signal. Also, the wireless channel can be studied as a system described by a frequency response, which consists of frequency selectivity that distorts the signal producing time spreading of the signal.

A path loss model consists of a formula (statistical or empirical model) that describes how the power of a signal is lost as the signal is transmitted through a channel of certain frequencies. Although the wireless environment is difficult to characterize, there are path loss models that describe how the average power loss behaves. In order to obtain this average, we start with a simple model that describes the received power of a signal, P_R, as a function of the transmitting power, P_T, the distance between transmitter and receiver, d, and a parameter which can be specified or estimated from measurements that is called the path loss exponent (PLE), n. The model that gives the power received at a separation distance d is the following

$$P_R(d) = P_T d^{-n} \tag{2.1}$$

It is common to use decibels (dB) or decibels with respect to one Watt (dBW) to present results when characterizing the power loss of a signal. Recall that dB is obtained from quantities of power P by using the expression dB $= 10\log_{10}P$. Thus, from (2.1), we can obtain the power received in dB, $P_{R,dB}$, at a separation distance d as follows:

$$P_{R,\mathrm{dB}}(d) = P_{T,\mathrm{dB}} - 10n\log_{10}d \tag{2.2}$$

where $P_{T,\mathrm{dB}}$ is the transmission power in dB. See that (2.2) can be seen as the equation of a line with slope $10n$, y-intercept $P_{T,\mathrm{dB}}$, and independent variable $\log_{10}d$. Also, note that for this model, we are not able to use a distance of zero. All the parameters will have units to consider. For (2.1), the transmission power will be in Watts, and the distance in meters, although we can consider the distance to be in km that would be represented (for example, for d km) as $d_{\mathrm{km}} \times 10^3$ m and (2.2) would be modified as follows:

$$P_{R,\mathrm{dB}}(d) = P_{T,\mathrm{dB}} - 3n - 10n\log_{10}d_{\mathrm{km}} \tag{2.3}$$

where we emphasize the use of km by using the sub-index km in (2.3). When the resulting power received in Watts has very small values, it is common to use decibels with respect to a milliwatt, i.e., $10\log_{10}\left(P/10^{-3}\right) = \mathrm{dB} + 30$, to obtain

$$P_{R,\mathrm{dBm}}(d) = P_{T,\mathrm{dB}} - 3n - 10n\log_{10}d_{\mathrm{km}} + 30 \tag{2.4}$$

Note that in (2.4) we can associate the scalar 30 with the transmission power to get

$$P_{R,\mathrm{dBm}}(d) = P_{T,\mathrm{dBm}} - 3n - 10n\log_{10}d_{\mathrm{km}} \tag{2.5}$$

Figure 2.1 shows an example of received power for typical values of PLE.

From Figure 2.1, one can see that received power is lost as distance increases, and that the values of PLE determine the rate at which this loss occurs.

Path loss PL measures the loss in dB of the signal while being transmitted through a wireless channel with separation distance d. The equation for path loss is given by the difference of the power at which the signal was transmitted and the

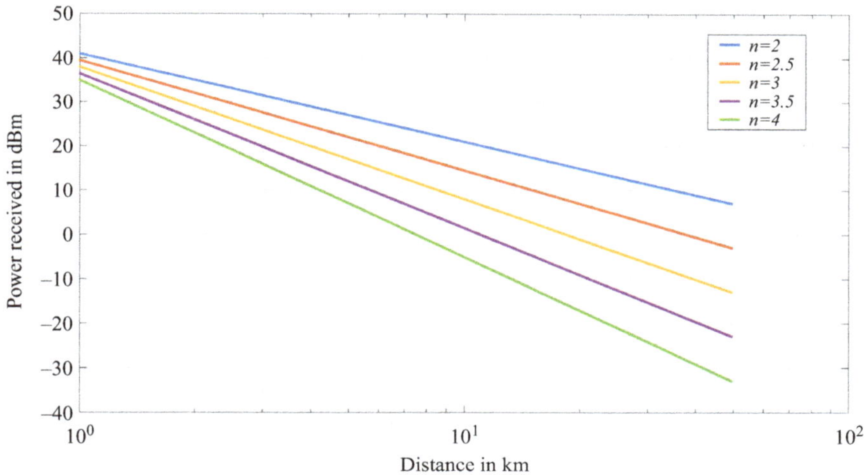

Figure 2.1 Received power in dBm with $P_T = 50$ W and several PLE values

power at which it is received, i.e.

$$PL_{dB}(d) = P_{T,dB} - P_{R,dB}(d) \tag{2.6}$$

See that path loss in (2.6) is in dB, and that the loss in Watts would be given by

$$PL(d) = \left(\frac{P_T}{P_R(d)}\right) \tag{2.7}$$

Path loss can also be obtained in dBm from (2.6) by adding 30 or by including in (2.6) the transmitted power in dBm.

2.2.1.1 The free-space propagation model

There are classical models that are used commonly to characterize a wireless link. Some of the most known are the free-space model and the two-ray model.

The free-space model [4] considers a scenario where transmitter and receiver establish a link and there is nothing in between. If we know the transmission power P_T, the antenna gains for transmitter and receiver G_T and G_R, respectively, the transmission frequency f or wavelength λ, and the separation distance d, then we have that the power received at a separation distance d is

$$P_R(d) = \frac{P_T G_T G_R \lambda^2}{(4\pi)^2 d^2} \tag{2.8}$$

Since wavelength and frequency are inversely related ($\lambda = c/f$, where c is the speed of light at 3×10^8 m/s and the wavelength in units of meters), the received power not only decreases as distance increases, but also decreases as frequency increases. Note that the free-space model follows a second power law with respect to the separation distance, so in Figure 2.1, the line corresponding to the PLE $n = 2$ is the closest to this model depending on the values of the gains and the frequency used for transmission.

The path loss when the free-space model is used with frequency in MHz and distance in km is given by, see [4],

$$PL_{\text{dB}}(d_{\text{km}}) = 32.4417 + 20\log_{10}(f_{\text{MHz}}) + 20\log_{10}(d_{\text{km}}) \tag{2.9}$$

Recall that received power as shown in Figure 2.1 is given by straight lines (semi-logarithmic plot is used). In the case of path loss as shown in (2.9), we also have straight lines with slopes of 20 dB per decade the of distance traversed. Equation (2.9) is obtained considering that the antenna gains are unitary. See that the frequency affects the y-intercept of the lines, as shown in Figure 2.2. Note that in this path loss calculation, the value of the transmission power does not play a role, and the higher the line, the more loss the signal has, which corresponds to a higher frequency.

One needs to take into account that the simple model of path loss in (2.6) and that of the free space in (2.9) predict the loss of power of the signal at a distance d from the transmitter, where the distance has to be in what is called as the far-field distance d_f, or Fraunhofer region, see [4]. This far-field distance is related to the wavelength used and the maximum physical length of the antenna L, so that

$$d_f = \frac{2L^2}{\lambda} \tag{2.10}$$

If propagation models can be simplified when one has measurements, specifically when a measurement has been obtained at a known distance from the

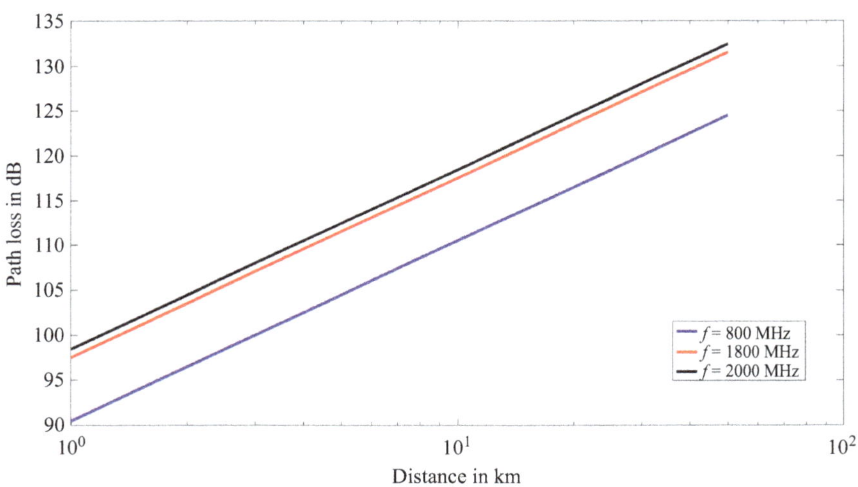

Figure 2.2 Path loss using the free-space propagation model

transmitter d_0, or close-in distance, and such measurement is trusted, then the received power at a distance d calculated by any model and the received power at the close-in distance calculated by the same model can be used in a ratio so that the result (regardless of the model chosen) is given by

$$P_R(d) = P_R(d_0)\left(\frac{d_0}{d}\right)^n \tag{2.11}$$

For the free-space model, (2.11) would have $n=2$, and $P_R(d_0)$ is the trusted measurement of the received power at the close-in distance. The same model that considers the close-in distance, but expressed in decibels is

$$P_{R,\text{dB}}(d) = 10\log_{10}[P_R(d_0)] + 10n\log_{10}(d_0) - 10n\log_{10}(d) \tag{2.12}$$

In Section 2.2.4, these models based on averages will be extended to consider randomness of the channel from problems and situations discussed in this chapter.

2.2.2 Channel impairments

The main impairments that a signal being transmitted through a wireless channel will encounter are

(a) **Noise:** as in any system, noise is present, and in its general form it is considered to be a signal which is Gaussian or Normal distributed due to the superposition of a large amount of unknown and undesired signals being received, and the central limit theorem. Noise is assumed to be additive and white Gaussian noise (AWGN), and its mitigation or cancellation is performed by the modulation–demodulation process of the system. One can study the errors caused by AWGN and determine the performance of the system using optimum techniques design such as maximum a-posteriori probability and maximum likelihood (ML), see [5,6].
(b) **Path Loss:** The power that is lost in decibels of a signal between transmitter and receiver caused mainly by the distance traversed through the channel. It is common to have path loss models from propagation models, and the latter can be of the statistical or the empirical forms as discussed in this chapter. These path loss models measure in principle the average power loss in dB of the signal at the receiver compared to the power in dB at which it was transmitted. It is defined by a parameter known as the path loss exponent (PLE) that depends on the environment. This PLE needs to be estimated from measurements to have a more accurate path loss model.
(c) **Fadings:** Represent losses of a transmitted signal caused mainly by obstacles, scatterers and multipath effects. There are two main classes of fadings:
    (i) *Fast fading*: It represents the losses caused by the reception of multiple components that are superimposed at the receiver. The different versions of the received signals have different phases and amplitudes which combined might be destructive or constructive.

(ii) *Slow fading or shadowing:* It is the power loss caused by obstacles between transmitter and receiver which affects the reception (even partially) of the LOS component.

(d) **Doppler effect:** It causes losses due to mobility (relative speed and direction) between transmitter and receiver. It causes a frequency shift of the signal and is related to coherent time.

(e) **Interference:** This corresponds to undesired signals that are generally known for which the system is aware of the users generating them, some of those signals are of the same kind as the one being transmitted. These signals carry information that may be confused by the receivers causing errors.

(f) **Self-interference:** In wireless communications channels, since signals are subject to multipath propagation, the receiver will have as input the signal of interest and multiple copies of it that are delayed, attenuated and distorted versions of the signal of interest. Even if there are no other signals present, this multipath effect causes interference by the same signal, hence combining with the LOS component, as shown in Figure 2.3.

In the following sections, we discuss the causes of some of these impairments, as well as the parameters that are estimated in order to characterize them.

2.2.3 *Diffraction and reflection*

Propagation models and path loss models help us to predict the power received at a separation distance *d* between transmitter and receiver. Up to this point, we have only talked about the free-space propagation model which is not the best model since it only considers an empty scenario with two antennas, but when we have objects and scatterers in the scenario, for example, a floor or walls, between transmitter and receiver, we need different models to predict the power we receive. These models need to capture the effects of such objects that produce reflections, diffraction and scattering to the signal being transmitted.

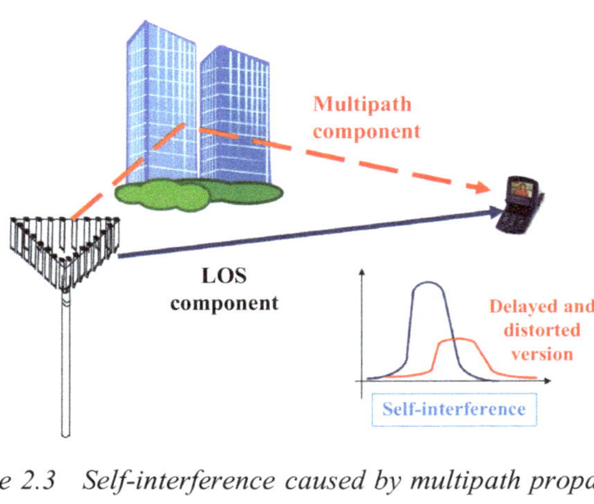

Figure 2.3 Self-interference caused by multipath propagation

Reflection. As shown in Figure 2.3, reflection produces multipath, and the reflectivity characteristic of the surface on which the signal bounces off, will determine the effects that distort such signal. Reflections will produce at the input of the receiver, a set of distorted and delayed versions of the signal being transmitted. The wavelength of the signal must be such that it is small compared to the actual size of the object on which it is bouncing off, see [1,4,7].

The most known and used propagation model for reflection is the two-ray model, where a signal is transmitted, and the receiver gets a LOS component and a multipath component. The latter coming from the signal bouncing off the floor. The transmitter and receiver are separated a distance d, and the signal bounces off the floor at some point that the signal's phase is changed when it arrives at the receiver. The total electrical field is calculated in order to determine the power received, see [4], and depending on the angle at which the signal bounces off from the floor, the change of phase will be. Considering that the transmitting and receiving antennas have heights in meters of h_T and h_R, the two-ray propagation model is approximated by the following equation

$$P_R(d) = \frac{P_T G_T G_R (h_T h_R)^2}{d^4} \tag{2.13}$$

See that the two-ray propagation model depends inversely on the distance to the fourth power, which affects the values of power received that will be smaller than those from the free-space propagation model, hence having plots with lines that have slopes of 40 dB per decade instead of 20 dB per decade. Hence, the consideration of a surface such as a floor, will closely produce losses at a rate of 40 dB per decade of distance. There have been other models that consider more rays, one can see [7] for the ten-ray model. Recall that we can still use the model considering a close-in distance, hence (2.11) with $n=4$ can be used for the two-ray model.

When a propagation model needs to be used, and no PLE is known, one can decide to use the worst case, which it would correspond to use (2.11) with $n=4$. In some scenarios where more obstacles and scatterers exist, the PLE might need to be increased, see [1].

Recall that path loss is the difference between transmission power in decibels and received power in decibels. The path loss of the two-ray model when antenna gains are unitary and $h_T = 50$ m and $h_R = 2$ m is shown in Figure 2.4, see how the two-ray model estimates more losses in power received per decade of distance compared to those shown in Figure 2.2 for the free-space model.

Diffraction. It occurs when the transmitter and receiver have in the middle an obstacle that prevents the LOS from being received, hence the obstacle causing a shadow where the receiver is and thus the name of shadowing effect. Diffraction is the propagation phenomenon through which the signal is received, and it is caused by the effect on the edges of the obstacle. When the signal front finds the obstacle, the edges will produce secondary sources whose positions are determined by the Fresnel zones, see [1,4,8]. The LOS which is partially blocked will not reach the receiver, and the Fresnel zones will determine the first component that will be received that passes over (or aside) the obstacle, as shown in Figure 2.5.

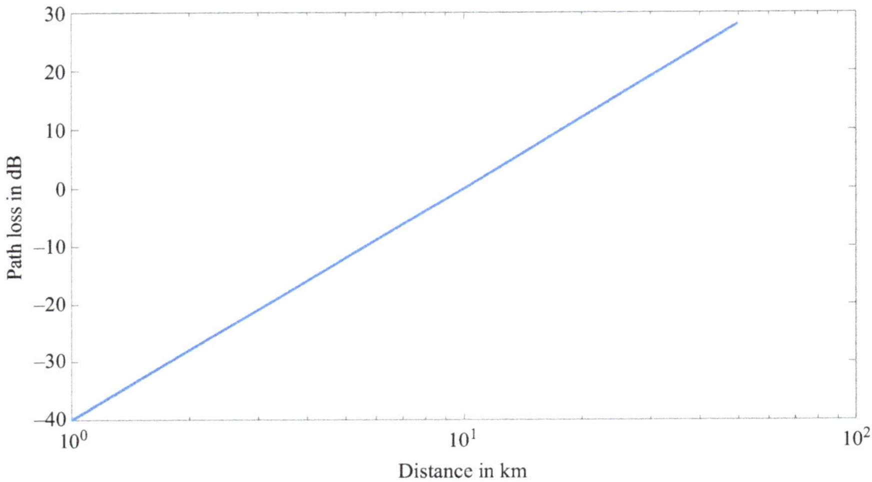

Figure 2.4 Two-ray path loss propagation model

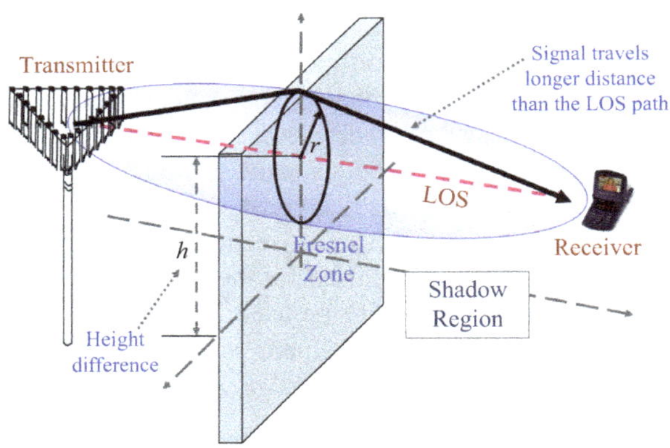

Figure 2.5 Propagation by diffraction and Fresnel zones

To calculate the losses due to diffraction, geometric analysis is conducted, see [4], where the excess path length, i.e., the extra distance traversed by the component reaching the receiver from above the obstacle with respect to the LOS component. Once the excess path length is obtained, the phase difference is calculated. The distance from the transmitter to the obstacle is denoted as d_1, and the distance from the obstacle to the receiver is defined as d_2, then parameter v is calculated first as follows:

$$v = h\sqrt{\frac{2(d_1 + d_2)}{\lambda d_1 d_2}} \tag{2.14}$$

After getting parameter v, the *gain* (which represents the loss in dB) can be calculated as follows:

$$G_{\text{dB}}(d) = \begin{cases} 6 + 9v - 1.27v^2, & 0 \leq v \leq 2.4 \\ 13 + 20\log_{10}(v), & 2.4 < v \end{cases} \tag{2.15}$$

In [4], there is an alternate form to calculate this gain, both forms are valid approximations of the power loss caused by shadowing.

A close model to the two-ray model is the Egli model, see [9]. It is close because it has an inverse relation to the separation distance to the fourth power as that shown in (2.13); however, it does include some other parameters such as frequency in order to try to cope with irregular terrain. The model was initially used for TV signals in LOS transmission in UHF and VHF bands. Using the same definitions and notation defined so far, the model is given by

$$P_R(d) = \frac{P_T G_T G_R (40 h_T h_R)^2}{f_{\text{MHz}}^2 d^4} \tag{2.16}$$

The Egli model does consider that irregular terrain and the assumption that the transmitter is fixed (cellular base station) and the receiver is mobile, although is limited in scenarios where the environment has obstacles like trees or shrubs.

The Longley–Rice model, even though widely used for VHF and UHF frequencies, does help predict power received for frequencies up to 20 GHz. It also considers irregular terrain. The model uses statistical characterizations instead of formulas, and it does consider atmospheric effects, see [9].

2.2.4 Coping with randomness

The propagation models discussed in previous sections estimate the received power or the power loss on average. If one considers the average propagation models, the coverage of such signals being transmitted will be deterministic defining circles around the emitters, but in real scenarios, these coverage areas are not circular, and the edges of such geometries are limited by the environment. The most used model to capture these effects at the coverage boundaries is the log-normal model, which consists in adding a random effect using a Gaussian random variable in dBs. This is explained in the following paragraphs.

Consider the propagation model with a close-in distance and its dB equivalent which is described by (2.11) and (2.12), respectively, and that we rewrite here

$$P_R(d) = P_R(d_0)\left(\frac{d_0}{d}\right)^n$$

$$P_{R,\text{dB}}(d) = 10\log_{10}[P_R(d_0)] + 10n\log_{10}(d_0) - 10n\log_{10}(d) \tag{2.17}$$

Now, define the average path loss by using a bar over the variable or function as follows:

$$\overline{PL_{dB}(d)} = P_{T,dB} - \overline{P_{R,dB}(d)}, \tag{2.18}$$

where $\overline{P_{R,dB}(d)}$ is the same as (2.17). By using (2.17) in (2.18), we get

$$\begin{aligned}\overline{PL_{\text{dB}}(d)} &= P_{T,\text{dB}} - \overline{P_{R,\text{dB}}(d_0)} - 10n\log_{10}(d_0) + 10n\log_{10}(d) \\ &= \overline{PL_{\text{dB}}(d_0)} - 10n\log_{10}(d_0) + 10n\log_{10}(d)\end{aligned} \tag{2.19}$$

We can rewrite (2.19) as follows:

$$\overline{PL_{\text{dB}}(d)} = \overline{PL_{\text{dB}}(d_0)} + 10n\log_{10}\left(\frac{d}{d_0}\right) \tag{2.20}$$

Now, if one wants to have some random effects considered within the propagation model, we can add a random variable to (2.20) to get

$$PL_{\text{dB}}(d) = \overline{PL_{\text{dB}}(d)} + X \tag{2.21}$$

where X is considered to be a Gaussian random variable with zero mean value and standard deviation σ in dB that represents the random variations caused by fadings. The first term in (2.21) can be calculated from any model as those discussed or from real measurements. Among the random variables commonly used besides the Gaussian are the Nakagami, Ricean, and Rayleigh, see [4,8].

Randomness also includes some small-scale propagation effects, which are studied through parameters that describe on average such impairments. For example, it is well known that the Doppler effect causes frequency shifts of the carrier signal as relative movement of transmitter and receiver occurs. For a receiver moving at a speed v in some direction ψ with respect to the straight-line joining transmitter and receiver and receiving a signal with wavelength λ, the Doppler shift (amount of Hz that the carrier frequency is shifted) is given by

$$f_d = \frac{v}{\lambda}\cos(\psi) \tag{2.22}$$

Note that as the frequency increases, the Doppler shift also increases, and also see that the maximum value of Doppler shift (Doppler spread) is obtained when ψ is zero, i.e., the relative movement is such that transmitter and receiver are moving directly towards each other. The speed of the receiver has the same effect when increasing.

Doppler shift has a counterpart in the time domain which is called coherent time, which is the amount of time where the impulse response of the channel does not change significantly. In (2.22), one can see that the maximum Doppler shift will be given when the angle is zero, hence denote and define this as $f_{\max} = \max\{f_d\}$ then the coherent time will be approximated by

$$T_C \simeq \frac{1}{f_{\max}} \tag{2.23}$$

Another suggestion is to obtain the coherent time using the following, see [4],

$$T_C \simeq \frac{9}{16\pi f_{\max}} \tag{2.24}$$

Figure 2.6 shows the Doppler spread for a vehicular scenario that will be introduced in Chapter 5. The vehicles are moving at a speed of 80 km/h and the frequency being used is of 5.9 GHz that corresponds to that used by the standard IEEE802.11p. Yellow colour corresponds to areas where transmitter and receiver are moving towards each other hence a positive shift in the frequency.

Other two parameters of interest are the coherence bandwidth and the delay spread. Both are related to the power delay profile (PDP), which is a measurement of the received power as different multipath components of the same signal arrive at the receiver, i.e., power in dB as a function of time. It is generally obtained through measurements and using channel sounders with network analysers. In general, a carrier signal at the desired frequency of study is transmitted, and the receiver will measure the power of such signal as time evolves. The spatial characteristics of the environment play a role defining the different multipath components that will arrive to the receiver, together with their power, phase change and delay with respect to the principal component. Figure 2.7 shows the PDP obtained through simulation from (a) the interior of a vehicle behind the front passenger seat and (b) from the rooftop of a vehicle in a street within an urban scenario. These examples will be explained in detail in chapters 4 and 5. See that the received power is measured as a function of time, and that the time axis has very small

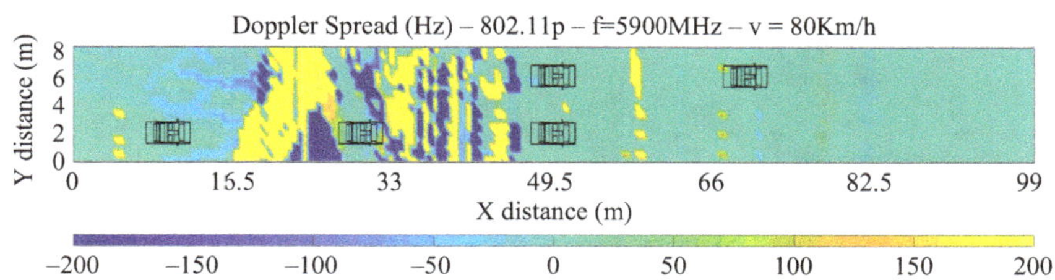

Figure 2.6 Doppler spread in a tunnel

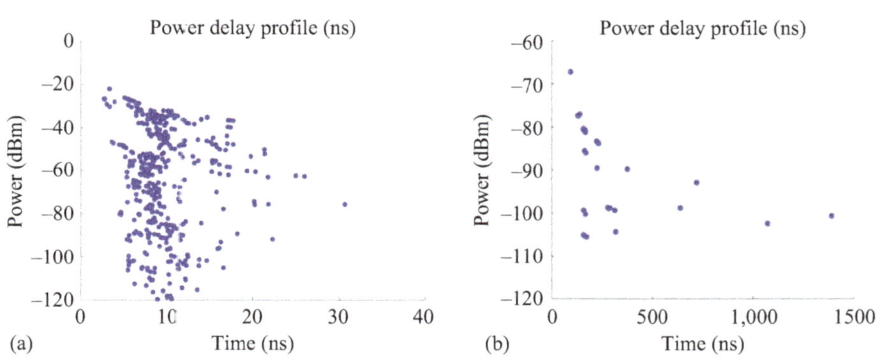

Figure 2.7 Power delay profile at the interior of a vehicle and at the rooftop of a vehicle

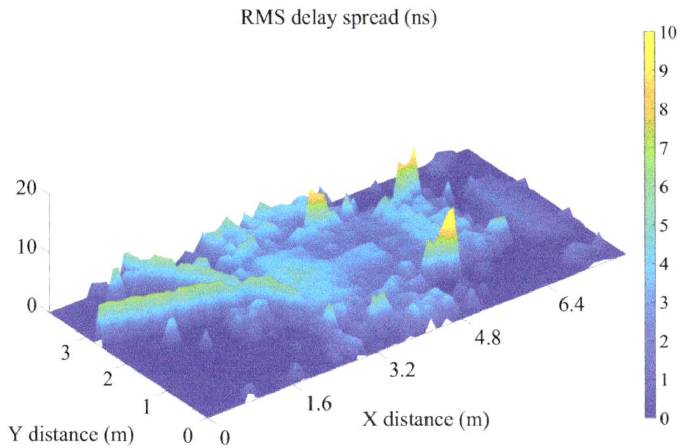

(a)

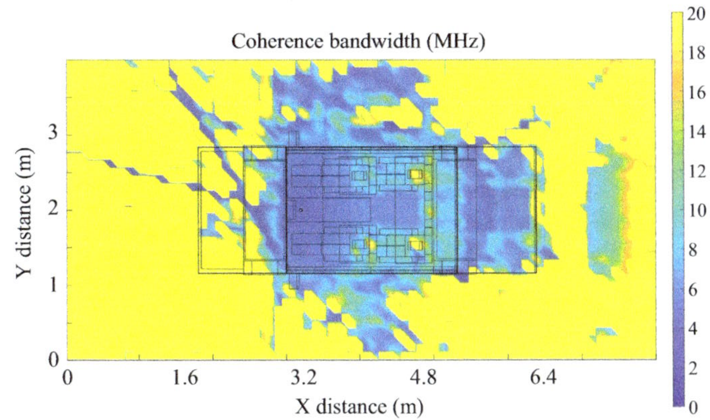

(b)

Figure 2.8 Delay spread and coherence bandwidth inside a vehicle values (nanoseconds) of time. The small values correspond to the time taken by the multiple paths to arrive at the received, and since propagation of the signal is considered to occur at the speed of light, then the values must be very small.

The PDP helps to determine the delay spread that measures the average time when the impulse response of the channel does not change. The received power varies because the signal is being received through several multipath components that generate a superposition of signals at the receiver. This superposition might be constructive or destructive depending on the different values of amplitude, delay and phase that the combining components have. The more components are received with significant power, the more time the receiver will have the signal present, which translates to more time where such signal will be received and that potentially could overlap over the next signal being transmitted. These increasing of time that the receiver perceives due to the multipath effect is what we know as the rms delay spread. The higher the value of the rms delay spread, the smaller the value of the usable bandwidth of the signal. One can see [4,7,8] on how to obtain the rms delay spread, σ_τ, from the PDP. Another parameter is the coherence bandwidth, B_C, which is the amount or range of frequencies through which the frequency response components of the channel are highly correlated. It is considered the maximally flat range of frequencies where our signal needs to be transmitted, i.e., the design of the signal to be transmitted must have a bandwidth that is smaller than the coherence bandwidth in order to get across the channel to the receiver without being critically distorted. The coherence bandwidth is obtained generally from the spectral density which is the Fourier transform of the PDP, and as we mentioned, it is the frequency domain counterpart of the delay spread. Figure 2.7 shows the delay spread (Figure 2.7(a)) and the coherence bandwidth (Figure 2.7(b)) obtained inside a vehicle at 1 m height when ZigBee at 2.4 GHz was used. This example will be explained in detail in chapter 4. Note that they have an inverse relationship, i.e., the dark blue areas in the delay spread corresponding to practically no multipath effects due to no reception of the signal, have in the coherence bandwidth the highest values (yellow colour) that indicates that such frequency is not being received and if a system uses 2.4 GHz in the areas in yellow, it will not be interfered.

The wider B_C is, the smaller σ_τ is and vice versa. If we see the channel as a filter with bandwidth determined by the coherence bandwidth, we have that the smaller the coherence bandwidth, the more selective the channel frequency response is, producing frequency selective fading. Also, the narrower the channel is, the smaller the bandwidth of our signal or symbol needs to have in order to traverse such channel, and as a consequence the larger the duration of our symbol, which translates to having a smaller symbol per second rate at the transmitter.

2.2.5 Estimating parameters for propagation models: an example

In this section, we introduce an example of propagation prediction models when real measurements are available. Even though we have data from measurements with randomness, we establish a prediction model that provides average loss, and then we estimate parameters of a random variable in order to cope with randomness.

Consider a scenario where you place a transmitter fixed in a position and you take measurements around the transmitter of received power and the separation distance of the measurement point. Assume that you generate data given by measured power in dB and distance from the transmitter in meters.

First, you need to obtain the PLE from your data using two methods, one is linear regression on the data as distance varies, and two is the ML method. When you plot the data in dB or dBm as a function of distance in a logarithmic scale, then your data will have a tendency towards a straight line with a negative slope as distance increases. The slope divided by 10 is the PLE that we need.

The second method consists in considering one model as those discussed previously, and the measurements. You need to define a function that will measure the error of the model being considered with respect to the real measurements. The model will make predictions of average values of power received at different distances. Assume that the model is given by

$$\overline{P_R(d)} = \overline{P_R(d_0)}\left(\frac{d}{d_0}\right)^{-n} \tag{2.25}$$

where $\overline{P_R(d_0)}$ is the power received at distance d_0 which is a known distance usually close to the transmitting antenna where you have a measurement you can trust defined as $\overline{P_R(d_0)}$. The PLE is n, which is the unknown in this equation. Then, in dB we have

$$\overline{P_{R,\mathrm{dB}}(d)} = \overline{P_{R,\mathrm{dB}}(d_0)} - 10n\log_{10}\left(\frac{d}{d_0}\right) \tag{2.26}$$

where $\overline{P_{R,\mathrm{dB}}(d)}$ is the power received at distance d in dB. Now, define the average error function based on a minimum mean square error (MMSE) criterion and we will use the measurement data with distances d_i for $i = 1, 2, \ldots, N$, where N is the total number of distances that you have, as follows:

$$J(n) = \frac{1}{N}\sum_{i=1}^{N}\left[P_{R,\mathrm{dB}}(d_i) - \overline{P_{R,\mathrm{dB}}(d_i)}\right]^2 \tag{2.27}$$

where $P_{R,\mathrm{dB}}(d_i)$ is the measurement of the received power in dB at a distance d_i from the transmitter, $\overline{P_{R,\mathrm{dB}}(d_i)}$ is the received power calculated by the model in (2.26) by substituting the corresponding value of d_i. For this second method to obtain the PLE, we take the derivative of (2.27) with respect to PLE n and make that result equal to zero and solve for n, then we get

$$n = \frac{-\sum_{i=1}^{N}\left[P_{R,\mathrm{dB}}(d_i) - \overline{P_{R,\mathrm{dB}}(d_0)}\right]\log_{10}\left(\frac{d_i}{d_0}\right)}{10\sum_{i=1}^{N}\left[\log_{10}\left(\frac{d_i}{d_0}\right)\right]^2} \tag{2.28}$$

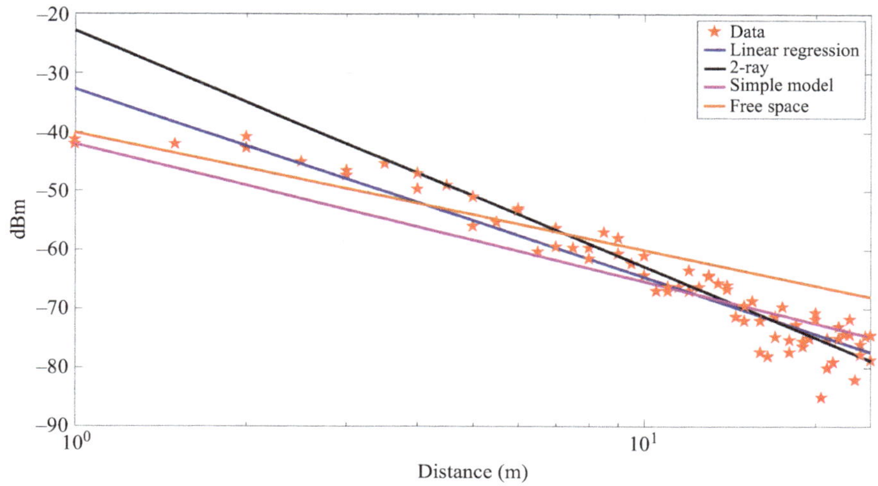

Figure 2.9 Received power measured and predicted by several models

As one can see from (2.28), we need the value of $\overline{P_R(d_0)}$ to solve for the PLE, the other elements in the equation are the distances and power measurements obtained. Once you solve and obtain the value of n, you can consider the simple model in (2.25) to see how close it is to your data. In the same plot, you can also show the straight line obtained by the linear regression model, and you can also include other models such as the free-space model and the 2-ray model.

We obtained a set of measurements in an area and obtain the values of PLE by the two methods just discussed and compare in a plot to other prediction models. This is shown in Figure 2.9.

From Figure 2.9, one can see that the linear regression (blue line) does have a very good fit to the data shown with red stars. The linear regression gave as a result in this example $y = -31.94x - 32.69$, then we take the slope value -31.94 and divide by 10 to get the PLE by linear regression of n=3.194

The second method that uses ML, obtains a value of n=2.36 and this is used in (2.26) and plotted in the same figure (simple model). See also that in the same figure, we show the lines corresponding to the free-space model with PLE of 2, and the 2-ray model with PLE of 4. Note that in this case the line that represents best the data measurements is the linear regression.

2.3 Vehicular channels for V2X communications

The study of channel models has the objective of generating statistical models that explain or characterize the behaviour of a given scenario. These models are helpful in predicting values of channel parameters of interest. Simulators for channel models can have different approaches including deterministic and random approaches and try to mimic the real behaviour of the channel parameters that must be verified with real measurements. Simulators can use the statistical models to produce channel realizations where those parameters of interest can be quantified by affecting statistically the simulated transmission of a known signal as the real channel does. The use of proper models and accurate simulations allows us to design or plan wireless communications systems before fabrication. The objective of producing accurate channel models that predict behaviour and values of parameters is a difficult task which becomes more difficult when mobility and non-stationary effects are involved in the scenarios under study such as those found in vehicular communication scenarios.

One of the earlier works on radio propagation for vehicular communications is [10], where charts of different parameters are obtained. Propagation under the influence of the curvature of the Earth, as well as under some atmospheric phenomena, is also described. The analysis is started considering the free-space propagation model followed by the propagation within a flat scenario to later be extended to consider the curvature of the Earth.

Technology has evolved vertiginously, and now vehicular communications are characterized in more elaborate ways than those in [10]. The dynamic environment found in vehicular scenarios presents specific challenges due to its features, such as antennas with low heights (vehicular and infrastructure antennas), diverse mobility conditions (speed and ways to move), which are related to diverse density conditions (number and size of vehicles), and a need for communication in different ways such as vehicle-to-vehicle (V2V), vehicle-to-infrastructure (V2I), where the I refers to cell towers, roadside units (RSU) or network access points (AP), or vehicle-to-everything (V2X), where X refers to pedestrians, bicycles and many different possibilities.

Vehicular channel models can be separated depending on the propagation model they employ, see [11]. The performance of channel models can be studied through their complexity, scalability and requirements to model specific scenarios in mobile communications, e.g., V2X in open spaces, in highly dense cities, or in tunnels to mention some.

Although vehicular channel models are in principle wireless communications models and are based on propagation models as those discussed, they have different features that make them unique. As introduced above, the different environments within the vehicular communications paradigm are so rich that you will find different models for each one of those scenarios, e.g., tunnels, highways, parking lots, parking garages, bridges, etc. Another particular feature to vehicular communications is the amount and kind of objects that need to be considered as part of the network or as scatterers affecting propagation. All of these features put together in the form of V2X communications make a challenging set of problems for which solutions need to be found in order to have success planning, designing and deployment of vehicular communications network.

Another important issue that makes vehicular communications unique is the need to treat outdoor, indoor and indoor-to-outdoor communications. The propagation models change parameters that need to be estimated among these three scenarios. The intra-vehicle communication will be analysed in Chapter 4 and the inter-vehicular communication in Chapter 5. The definition of channels for vehicular scenarios consists of different combinations due to the different types of communications and issues such as links between static and mobile devices under different complex environments.

Models that have been used throughout the years for cellular communications help us to understand how to elaborate channel models. However, those models are generally not directly applicable to a vehicular scenario. Some of the notable differences are the mobility with relative high speed between transmitters and receivers, the physical characteristics such as distances that are shorter, heights that in several scenarios communication is not necessary with a tower but with another vehicle or a RSU, and the possibility of different scenarios where a mixture of frequencies need to be used to establish communications, where standards such as WiFi, Bluetooth and Designated short-range communications (DSRC) are among some that can be used by the vehicles.

More issues that are specific to vehicular environments are the general term 'devices', where it could mean very different kinds of network elements such as a smart phone, a vehicle, a bicycle, an RSU among others that make propagation modelling difficult since signals will have different characteristic for all these 'devices' depending on the physical environment surrounding them, hence the channel model that could be obtained for a vehicle to bicycle links, does not necessarily describe the behaviour of a V2V link. Recall that what we get as a benefit is a description of the methodology to follow to obtain channel descriptions for different environments, which in general does not change drastically.

2.3.1 Spatial, spectral and time variation of channels

Channel models are obtained after some characterization of the environment and signal propagation. These models are usually presented as the combination of several of the following:

- Impulse or frequency responses of the environment
- Statistical descriptions through probability density functions (pdf) of the signal behaviour when traversing the channel
- Small- and large-scale propagation parameters such as path loss exponent, Doppler spread, coherence bandwidth, among others.

These combinations in all cases depend on three features, the spatial, the spectral and the temporal variations. Even when one deploys a transmitter and receiver with fixed positions, measurements of parameters such as the received power will not be the same because the environment is constantly changing due to movement of objects, weather conditions, and materials of those objects. So, the spatial characterization of scenarios is of critical relevance to be able to formulate a channel model that is accurate matching the real conditions. The objects in the environment will determine multipath components present at the receiver, and also the possible absence of the LOS component. The number of objects (or density), the material they are made of, and their positions in the environment will affect the number of multipath received components, the amplitude and phase of the signal being transmitted, and the delays of the components being received with significant power levels. Figure 2.10 shows three different environments in vehicular communication scenarios. Also, one can see that any transmitter–-receiver communication in each of those scenarios will have impairments associated to the spatial description of the environment that is defined not only by fixed objects, but also by the moving objects such as vehicles and human beings. In some cases, communication will not be achieved due to obstacles impeding the reception of the signal.

Figure 2.11 is an example of roadside scatterers producing multipath effects in the communications of two vehicles moving in opposite directions in the same street. Depending on the materials the scatterers are made off, when the signal component reaches the scatterer and bounces off, it will go through a process of propagation that produces multipath components depending on the

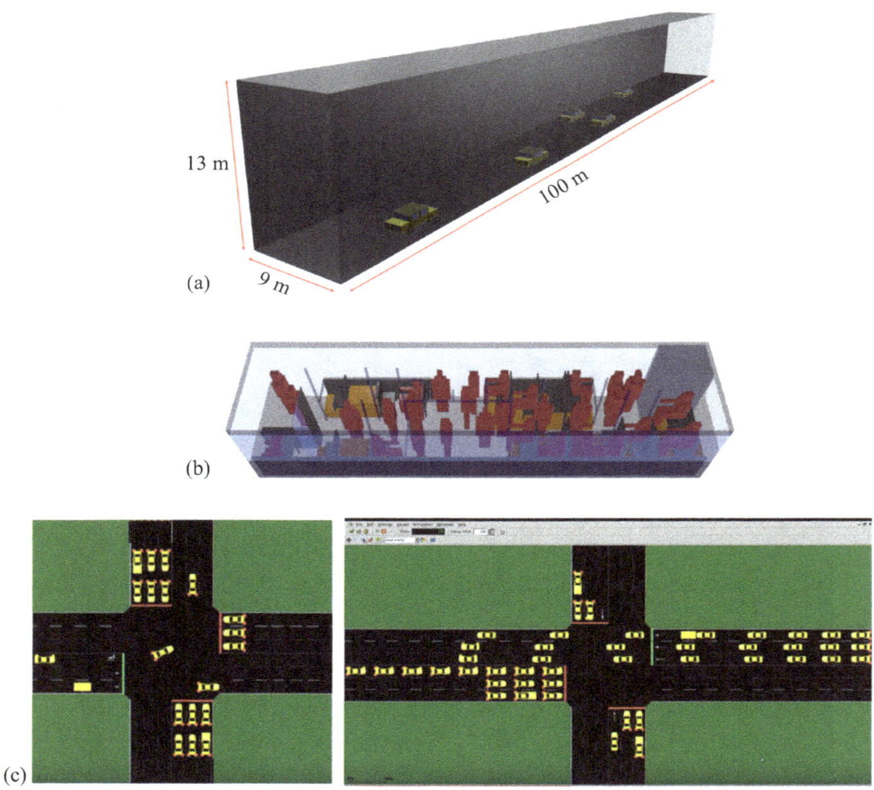

Figure 2.10 Different spatial environments, (a) tunnels, (b) inside a bus, and (c) urban environment

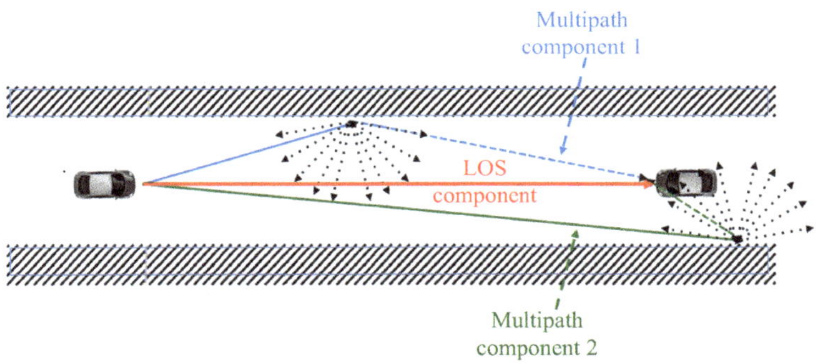

Figure 2.11 Urban environment with roadside scatterers

roughness of the surface. Each of these components will be affected in amplitude and phase.

The temporal variation can also be seen from Figures 2.10 and 2.11, where signals will be bouncing off different surfaces to reach the receiver, and so producing the multipath effect. Since each of those components coming from the surface of a scatterer will traverse longer distances than the LOS component, we will have the superposition of delayed copies of the signal at the receiver. Also, since the distances traversed by all the different multipath components are different, the amplitudes and the phases will be different. In Figure 2.11, the LOS component will arrive first, followed by multipath component 1 and then multipath component 2. If we assume that signal $s(t)$ was transmitted, then a simple channel model for this scenario would be given by the following impulse response

$$h(t) = \sum_{k=0}^{2} \alpha_k \delta(t - \tau_k) e^{-j\theta_k} \tag{2.29}$$

where α_k is the attenuation of multipath component k, θ_k is the signal phase change of the k-th multipath component and τ_k is the delay caused to the signal by the distance traversed by multipath component k, where k=0 is for the LOS component. Note that the model given in (2.23) is a deterministic model where we need to know in advance the delay, phase and attenuation that affect the signal in each component. See that this is also a superposition of multipath components, and that phase changes will produce either, a destructive or a constructive superposition. Note that the signal being received in this case would be given by

$$y(t) = \sum_{k=0}^{2} \alpha_k s(t - \tau_k) e^{-j\theta_k} \tag{2.30}$$

The spectral variation is determined by the frequency response or the magnitude of the frequency response of the channel. These can be obtained by using (2.29) and the Fourier transform and its magnitude. Every scenario considered will have different number of multipath components in the superposition and each component will have different attenuation, delay and phase change that will produce a different spectrum or frequency response of the channel. If we add the mobility in vehicular scenarios, then we will have also the effect of the Doppler spread of the spectrum of our signal depending on the relative movement of transmitter and receiver, hence changing the spectral characteristic of the channel.

2.3.2 Stationary and non-stationary environments

The characterization of a channel is one of the required steps in the design of a system in order to minimize errors. This characterization can be performed with measurements campaigns or statistical modelling of the channel phenomena capable of reproducing with accuracy the channel behaviour. Simulations are also a part of these steps to obtain an efficient and optimized system.

Recall that the wireless communications channel can be considered as a time-varying system with multipath and propagation effects. Variability in time is caused by mobility of the transmitter or receiver, as well as mobility of the objects or scatterers around them where the signal bounces off from, creating the multipath components. The superposition of all these phenomena are difficult to model with deterministic approaches, hence statistical methods are suitable to achieve a description of the behaviour. An environment that does not change in time is considered to be stationary, whereas an environment that changes will be considered non-stationary. These changes in time could be due to the mobility of the device being studied or even if the device is fixed, it could be due to movement of the surrounding objects. This movement could be produced by wind, or by mobile objects and people around.

When one deals with an environment that is changing such as the wireless channel for a vehicular communication, signals that are received are affected as times evolves. Each signal which is considered a random process becomes non-stationary and this behaviour can be captured in different statistical parameters. Scatterers around the V2V communication link determine the multipath components, and as vehicles move, these multipath components will change in quantity and in their phase, delay and attenuation, an example of a V2V communication scenario is shown in Figure 2.12.

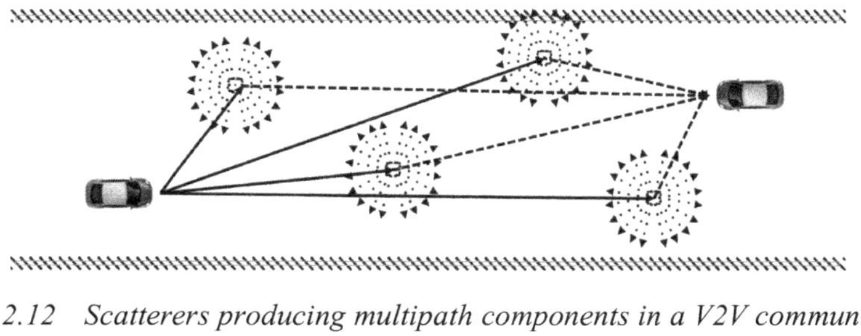

Figure 2.12 Scatterers producing multipath components in a V2V communication

Several models have been developed to capture the non-stationary effects of vehicular channels. Some of those models are based on geometrical descriptions of the environment such as that shown in Figures 2.10–2.12, see, for example, [12–16], among many others.

In [17], a non-stationary model is proposed for V2V communication, where the channel is divided into small areas where wide sense stationarity is determined, then known stationary models are used in each of these areas adjusting the parameters of those models in order to capture the variations from area to area. The PDP is obtained to develop the impulse responses of Gaussian wide sense stationary uncorrelated scatterers. The model is based on statistical modelling of information obtained by measurements in order to characterize the non-stationary behaviour.

Among the techniques for modelling channels in non-stationary vehicular environments based on geometrical statistical modelling, time–frequency characterization of the dispersive effects of multipath channels that capture fading effects is one of the current trends in research [15].

The propagation conditions of the non-stationary environment that shape the channels for vehicular communications, change rapidly in agreement to the scenario under consideration, such as in urban high-density environments through highways with changing flows of vehicles and high-speed transportation systems. In all these environments, stationarity or specifically wide sense stationarity (WSS) of the random processes or signals involved is only confirmed for short periods of time. Models considering the non-stationarity of the channel have been concentrating on the time modelling of the angles of departure and arrival of the signals that are produced as signals bounce off interfering objects. Other models are focused on the plane wave propagation principles that characterize the time-varying behaviour of the signal as it propagates through the environment and that is represented by the different delays that the signal undergoes at the receiver coming from different the scatterers in the environment, see [15].

The impulse response of the channel when considering non-stationary environments in [15] is given by

$$h(t;\tau) = \sum_{k=1}^{L} \alpha_k \delta(\tau - \tau_k(t)) \Pi\left(\frac{t - \frac{2t_0+T_0}{2}}{T_0}\right) e^{j\theta_k} \tag{2.31}$$

where α_k is the attenuation of the signal for multipath component k, $\Pi\left(\frac{(t - \frac{(2t_0+T_0)}{2})}{T_0}\right)$ is a windowing function consisting of a rectangular pulse of unitary amplitude centred at time $(2t_0 + T_0)/2$ and of width T_0 and θ_k is the change of phase and contains the phase shift for the k-th multipath component, the phase rotations of the signal as it traverses the channel and bounces off the k-th scatterer. The phase rotations depend on time t. Time t_0 is when the transmitter started the transmission. The channel impulse response is obtained when transmission took place τ seconds before, and it is being evaluated at time t. The multipath components are determined by using a geometry-based stochastic model (GBSM) for vehicular environments. This impulse response model works for both planar and spherical wave propagation.

We can see the contrast of this impulse response with respect to the previously discussed in (2.29) where we have additional parameters to account for the phase shift and the windowing factor, as well as extra time parameters for several elements in the equation. Most of the proposed channel model that captures non-stationary behaviour in [15] is focused on obtaining an accurate model of the phase term as time varies. For the statistical properties of the non-stationary scenario proposed, the model contains the pdf of the envelope and phase shift, as well as expressions for Doppler profiles using time–frequency diagrams. Results were presented for mobile to mobile (M2M) communication environments using a Rayleigh random variable for fadings in the channel. The carrier frequency and the bandwidth of the signal influence the non-stationary behaviour in the frequency domain. In the time domain mobility has the main influence.

In [18], the focus is on the power spectral density (PSD) and the objective is to reproduce the real channel phenomena based on knowledge of the stationary or non-stationary characteristics of the channel. They start with a classical approach by using Clarke's model that provides the principles of multipath propagation of a signal in a wireless link. The channel is represented by its impulse response as a superposition of complex exponentials with a specific time-delay for each multipath component. The PSD is worked in the time domain through the use of the autocorrelation function and the pdf of the envelope of the signal from the channel. The main objective in this work is to produce channel models that capture the effects of the time-varying characteristics of the channel together with fading effects and that are able to describe stationary and non-stationary scenarios.

The model for the channel is based on the consideration of the in-phase and quadrature components of the signal being transmitted, and how these are affected by the changes in amplitude (attenuation), phase shifts and Doppler shifts. For the mathematical model, an infinite number of multipath components are considered for the channel. With this, the channel impulse response has the following form:

$$h(t;\tau) = \sum_{i=-\infty}^{\infty} \sum_{k=0}^{\infty} \alpha_{i,k}(t)\delta\left(t - \tau_{i,k}\right)e^{j\left(\theta_k + 2\pi f_{i,k} t\right)} \tag{2.32}$$

where $f_{i,k}$ is the Doppler frequency shift. This impulse response represents the general non-stationary channel. With this model and one for the stationary case, authors in [18] proceed to the generation of a simulation by using Monte Carlo methods using windowing based on the square root raised cosine function. They form continuous wide sense stationary processes by the concatenation of uncorrelated random processes. The model provides a unified view of the stationary and non-stationary channel models that are able to produce channel realizations that are defined by PSD and pdf which are time varying.

The channel modelling for V2V communications from the perspective of capturing the fading phenomena in a non-stationary environment is presented in

[19]. The model is based on clustering multipath components in the Doppler shift delay profile. The specifics of the results presented are on a scenario where there is obstruction of a large vehicle so that LOS is not present in the V2V communication. Relationships between delay and Doppler shift are determined to characterize the non-stationary behaviour.

2.3.3 Mobility considerations

Mobility in vehicular environments changes the conditions in such a way that for some instants of time those changes, e.g., power received, are sudden and drastic. Figure 2.13 shows a vehicular scenario for a highway with several lanes in both directions with scatterers on the side road and with more vehicles that are considered scatterers from the point of view of the transmitter and receiver. In the figure, transmitter and receiver are moving in the same direction.

2.3.3.1 Case of study

In this section, we present a case of study characterizing Doppler effect in a vehicular scenario. Doppler has been characterized using statistical descriptions with pdf and spectrum results in [20,21].

The Doppler spectrum describes how the energy is distributed along the frequency domain in a communication where transmitter and receiver move relative to each other. Because the measurements campaigns require so much effort, the amount of data collected is very limited and sometimes there is not enough time or resources to consider a rich variety of propagation conditions. Few methods to obtain the Doppler spectrum in vehicular communication channels have been

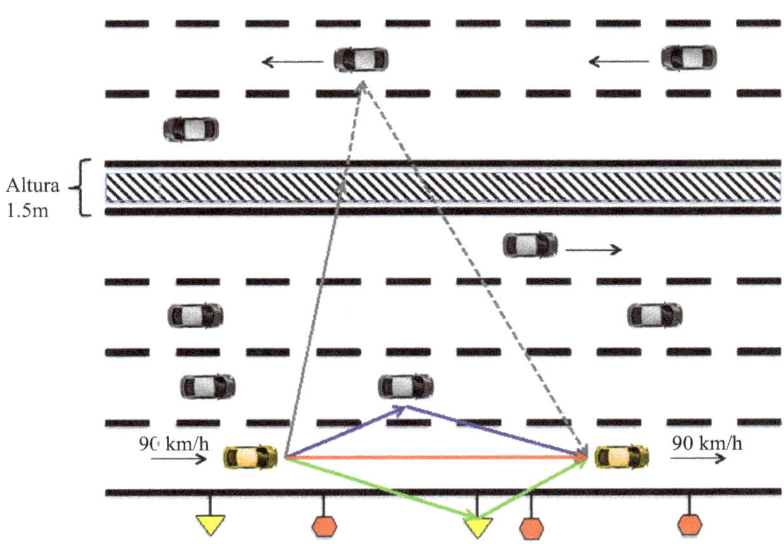

Figure 2.13 Description of an urban environment consisting of a highway

reported, but they are mostly limited to V2V scenarios where there is no relative motion between vehicles. Thus, in this section, an analysis of the Doppler spectrum in V2V communication channels is presented which can be applied to a wide variety of scenarios. We first describe the scenario conditions, then show the method to calculate the Doppler shift and the received power considering scatterers along the road and the line of sight.

Intelligent transportation systems (ITS) is a set of emerging technologies to manage efficiently traffic resources and to increase the safety of transportation systems. One of the key elements of ITS is the communication standard usually referred to as the designated short-range communications (DSRC). The characteristics of a DSRC communication channel are very diverse depending on the relative motion between the transmitter–receiver and the characteristics of the environment or road (lane width, scatterer density, and vehicle speed). Figure 2.14 illustrates a simplified scenario.

In Figure 2.14, vehicle A is transmitting with frequency f_0; however, due to several scatterers along the road, vehicle B receives several replicas of the signal, each with different phases and delays. The scatterers are assumed to be placed in two lines parallel to the vehicle road. Thus, δ_1 defines how far is the line of scatterers from the vehicles and $\boldsymbol{x}$ indicates the distance between the scatterer's position and vehicle A, and it is uniformly distributed. The distance between vehicles is defined by l, v_A (correspondingly v_B) is the constant speed at which vehicle A is moving. Figure 2.15 shows the Doppler shift associated with the LOS component in Figure 2.14 and calculated with (2.22) as the vehicle separation distance l varies, for different vehicular speeds.

According to the federal communications commission (FCC), devices participating in a V2V communication will normally have communication when their distance is between 100 and 400 m. Thus, the maximum separation between vehicles was assumed to be 100 meters and is defined by $l_{\max}$. Table 2.1 shows the

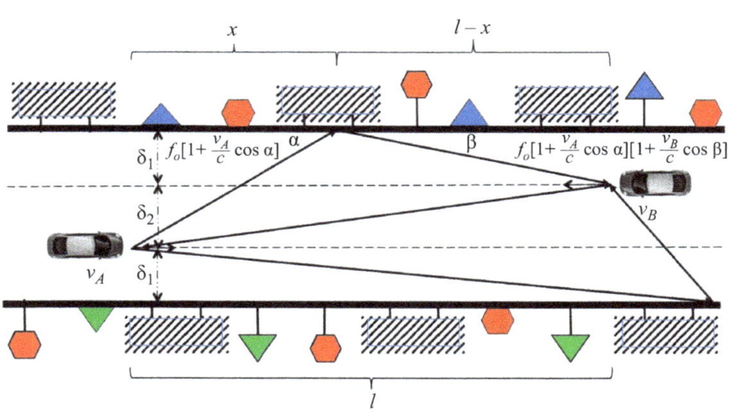

Figure 2.14 Description of urban environment with roadside scatterers

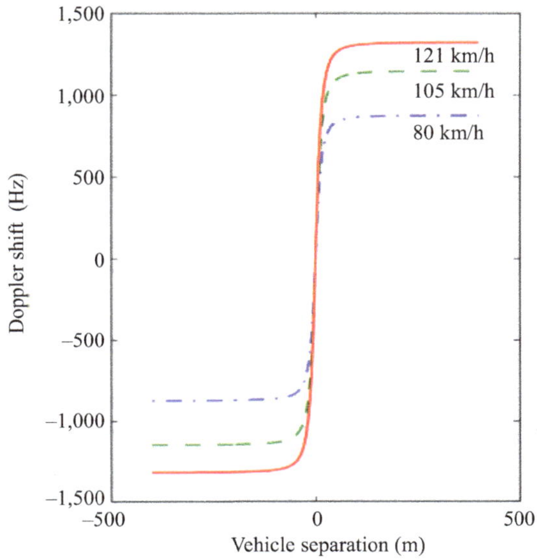

Figure 2.15 Doppler shift as vehicle separation varies

Table 2.1 Device classification by the FCC

Class	Transmission power (dBm)	Maximum distance (m)
A	0	15
B	10	100
C	20	400
D	28.8	1,000

classification of the wireless communications devices according to the FCC, the vehicular communications belongs to class C.

To study the Doppler spectrum, we first need to calculate the Doppler shift f_d related to every scatterer considered in the scenario, which are in different positions. The Doppler shift in (2.22) is used to obtain the following:

$$f_d = f_0 \left[\frac{v_A}{c} \frac{x}{\sqrt{x^2 + (\delta_1 + \delta_2)^2}} + \frac{v_B}{c} \frac{l - x}{\sqrt{(l - x)^2 + \delta_1^2}} \right.$$

$$\left. + \frac{v_A}{c} \frac{x}{\sqrt{x^2 + (\delta_1 + \delta_2)^2}} \cdot \frac{v_B}{c} \frac{l - x}{\sqrt{(l - x)^2 + \delta_1^2}} \right] \tag{2.33}$$

In this scenario, Doppler shift has two contributions, first when the signal from vehicle A reaches the scatterer at position x, and the second when the reflection reaches vehicle B. Now, assume that both vehicles are moving towards each other in a collision course, i.e., $\delta_2 = 0$, also assume that the position of the scatterer at the top is moved such that it is located in three different areas, one to the left of vehicle A's position ($x < 0$), two when $0 < x < l$, and three when $x > l$, i.e., beyond vehicle B's position. Fix the value of l=100 m, and for each value of x, calculate the Doppler shift using (2.33) for the three different areas, and with the separation distance of the reflected component for each value of x, calculate the power received at vehicle B in dBm using the free-space model in (2.8)–(2.12), then you can obtain a graph as that shown in Figure 2.16

In Figure 2.16, and assuming that the speed of the vehicles and the frequency of transmission remain constant, one can see that the position of the scatterer relative to the vehicles' positions does change the Doppler shift values. This is because the position of the scatterer affects the angle at which signals interact and eventually will affect the angle at which the signal is received. Also, it can be seen that power received decreases and Doppler shift increases when the scatterer is farther away in the areas outside the LOS link between the vehicles.

We can conduct the same experiment for different values of the distance between vehicles l, and the behaviour will be similar as it is shown in Figure 2.17 for distances between l=0 and l=100 m.

2.4 Single and multi-antenna communications

All the models that have been discussed previously are of systems that have single antennas for transmitter and receiver. Scenarios where multiple antennas are

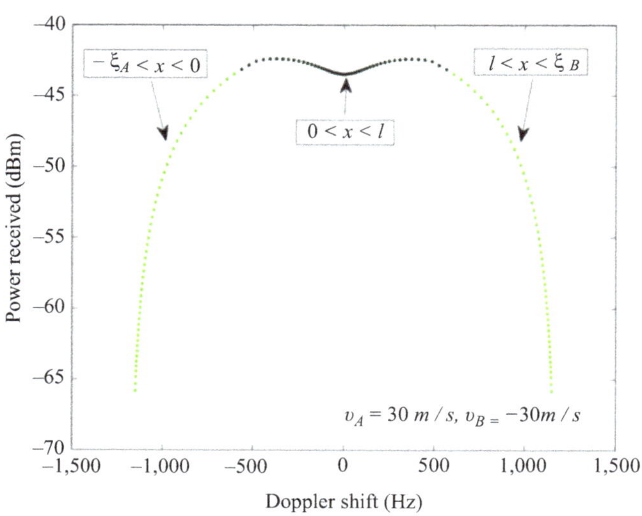

Figure 2.16 Received power in the presence of scatterers for l=100 m as a function of the Doppler shift

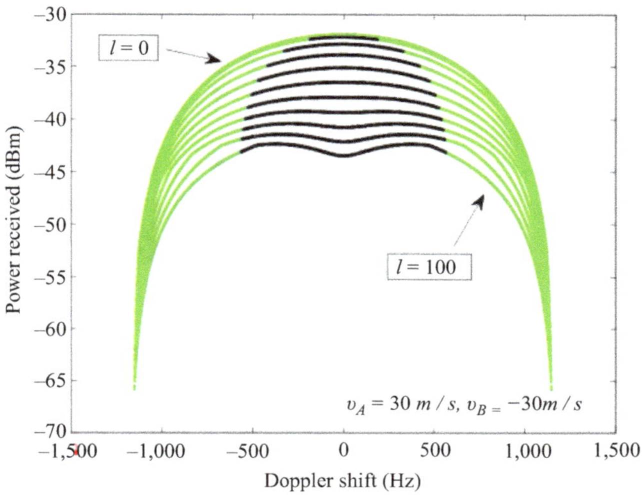

Figure 2.17 Received power in the presence of scatterers for l=0 to 100 m as a function of the Doppler shift

incorporated could be of the single-input-multiple-output (SIMO) kind or the multiple-input-multiple-output (MIMO) kind.

MIMO communication systems are key for the evolution towards the next-generation cellular communications, because they employ diversity techniques to get a reliable communication or spatial multiplexing to get high data rates and capacity enhancement, without the use of additional transmission power or bandwidth. MIMO systems use antenna arrays instead of a single transmit and receive antennas, to provide better spectral efficiency and link reliability than conventional wireless systems and operate by transmitting multiple signals in the same frequency band and at the same time over multiple antennas. Measurements of a MIMO channel through the use of channel sounders are discussed in chapter 6. MIMO systems have the challenge of non-stationary environments as well because the use of antennas closed in proximity increases the correlations of the multipath components from the scatterers degrading performance. In order to avoid such correlations, it is important to see the spatial–temporal correlations and optimize the system with them to improve performance.

A MIMO system is more likely to be in the base stations of the cellular service due to the size of the antennas, although progress has been made towards smaller antennas for portable devices. A vehicle is also a candidate to have several antennas installed, so V2X communication and V2V communication can be carried out by MIMO systems within their limitations. Figure 2.18 shows a representation of a multiuser cellular service where the base station has an antenna array system and each vehicle has one antenna, then the whole system could be considered a MIMO system.

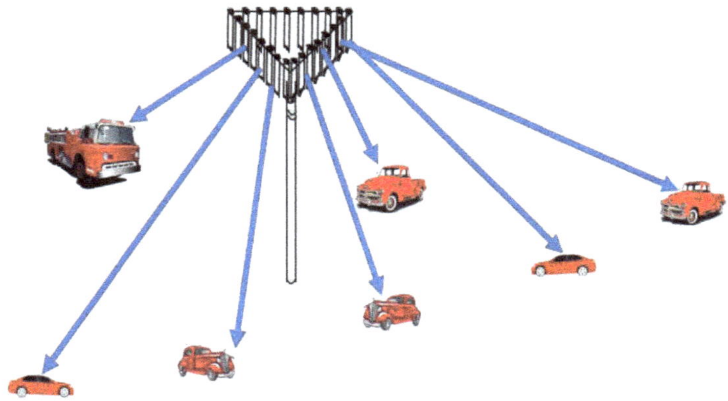

Figure 2.18 Representation of a multiuser MIMO system

As discussed in previous sections, non-stationarity models consider that the assumptions of WSS are present for short periods of time, and the channel models for V2V communications should capture the time-variant nature of the channel through spatial and temporal correlations. In [16], one finds a channel model for V2V communication that is non-stationary and that has good accuracy for Doppler shifts. It is compared to some models based on the GBSM framework for V2V communications. Temporal and spatial correlations are formulated and studied through simulations using GBSM.

The channel model for a MIMO system in a typical V2V communication channel consists of multiple propagation path components for each transmitter–receiver pair of antennas. As in the previous discussion, the multipath components can be clustered, and these clusters will be time-variant. Assuming that the transmitting antennas are M_T and the receiving antennas M_R, the channel response will consist of a matrix of M_RxM_T elements, where the (i, j)-th element of the matrix is given by

$$h_{ij}(t,\tau) = \sum_{k=1}^{L} \alpha_k(t)\delta(t - \tau_k(t)) \sum_{m=1}^{M_L} e^{-j\theta_{k,m}(t)} \quad (2.34)$$

where M_L is the number of groups or clusters of multipath components of the total of L existing multipaths. The exponent in (2.34) contains a random initial phase which is uniformly distributed, and the phase of each antenna and scatterer in the scenario, see [16] for further development of the model.

A MIMO system as the one just described is shown in Figure 2.19, where each of the arrows between pairs of antennas is described by (2.34).

In [22], a channel model for wideband V2V communication using MIMO and a 3D environment is proposed. A statistical analysis is performed considering single, double, triple and higher-bounced rays. Results are validated through a measurement campaign carried out in a setup of a MIMO channel sounding

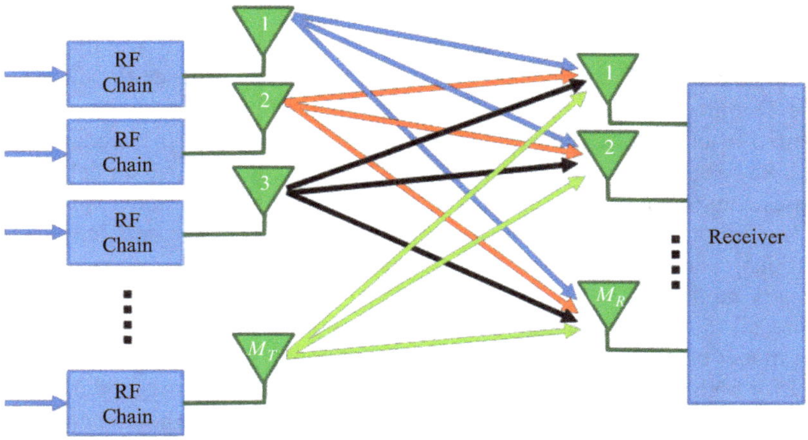

Figure 2.19 Representation of a MIMO system

system where vehicles travel in an urban scenario. Also, a parameter estimation algorithm is included, which is based on the maximum likelihood stochastic estimation algorithm, and it is used to extract the values of parameters from the measured data, which is processed for comparison purposes. Among the results obtained, we have that the statistical properties of triple and higher-bounced rays are close to those of the double-bounced rays, hence they can be modelled using double-bounced rays. Another important result is that the statistics of single-bounced rays and those of double-bounced rays are different because the angles of departure are independent of the angles of arrival for double-bounced rays, whereas for single-bounce rays the angles of arrival depend on the angles of departure.

It is important to have models that are compliant with the standards, that is the case of the 3GPP technical report (TR) for physical layer aspects and channel models for long-term evolution (LTE), see [23]. In [24], a channel model for the 3GPP release 12 3D MIMO system is presented. Small-scale propagation parameters are used to obtain the received power from the reference signals and these are compared to simulations performed.

In [25], a V2V non-stationary channel model for 3D MIMO scenarios is also proposed. It is based on GBSM models and for wideband communications specifically for tunnels. Cylindrical models are used to describe the vehicles within the tunnel, and half ellipses for the tunnel structure. The signals received are considered to come from LOS as well as from single and double-bounced rays. Mobility is considered in the model, producing time-varying phenomena that makes the channel non-stationary. The impulse response model proposed is composed of the superposition of three terms, the impulse response for the LOS component, the impulse response for the single-bounced rays, and the impulse response of the double-bounced rays. For the model, spatial correlation functions dependent on the carrier-to-interference ratio are calculated, also Doppler shift and its power spectral density are derived since communication will be established with multipath components and mobility, which influence the distribution of the Doppler PSD. Correlation functions in the frequency domain are also presented in order to know the selectivity of the channel to the scenario established. Power delay profile is also included in the study from which delay spread statistics can be obtained. For the non-stationary effects, the model is divided in time in such a way that for an interval of time the channel satisfies the WSS conditions, and then different time intervals are joined to create the non-stationary model.

In [26], a MIMO non-stationary channel model is presented. It uses a scattering model based on rings and it is based on GBSM. Mobility is included in the model for both transmitters and receivers, causing to have in the model angles of arrival and departure dependent upon time and making the whole channel time-varying. Spatial and temporal correlation functions are involved in the statistical study of the channel. The channel impulse response is obtained from the plane wave propagation principle, and the impulse response model for a pair of antennas of the MIMO system is given basically by a superposition of complex exponentials where the exponent contains the terms for the phase shift which is uniformly distributed, and the phase change as mobility is included. With respect to previous models, this is simplified not only on the channel impulse response expression where attenuation is not a function of time, but also on the limitation to a two-dimensional (2D) scenario instead of a 3D as the models discussed previously.

References

[1] Bertoni, H.L., *Radio Propagation for Modern Wireless Systems*, Prentice-Hall PTR: Upper Saddle River, NJ, USA, 2000.

[2] Sklar, B., 'Rayleigh fading channels in mobile digital communication systems. I. Characterization', *IEEE Communications Magazine*, 1997, 35(7): 90–100.

[3] Sklar, B., 'Rayleigh fading channels in mobile digital communication systems. II. Mitigation', *IEEE Communications Magazine*, 1997, 35(7): 102–109.

[4] Rappaport, T., *Wireless Communications: Principles and Practice*, Prentice Hall Communications Engineering and Emerging Technologies Series, 2nd edn, Prentice-Hall PTR: Upper Saddle River, NJ, USA, 2002.

[5] Proakis, J.G., *Digital Communications*, 4th edn, McGraw-Hill Education, New York, NY, 2001.

[6] Wozencraft, J.M., and Jacobs, I.M., *Principles of Communication Engineering*, Wiley, New York, NY, USA, 1965.

[7] Goldsmith, A., *Wireless Communications*, 1st edn, Cambridge University Press, New York, NY, 2005.

[8] Stuber, G.L., *Principles of Mobile Communication*, 4th edn, Springer, Cham, Switzerland, 2017.

[9] Phillips, C., Sicker, D., and Grunwald, D., 'A survey of wireless path loss prediction and coverage mapping methods', *IEEE Communication Surveys and Tutorials*, 2013, 15, 255–270.

[10] Bullington, K., 'Radio propagation for vehicular communications', *IEEE Transactions on Vehicular Technology*, 1977, VT26(4): 295–308.

[11] Viriyasitavat, W., Boban, M., Tsai, H.-M., and Vasilakos, A.V., 'Vehicular communications: survey and challenges of channel and propagation models', *IEEE Vehicular Technology Magazine*, 2015, 10(2): 55–66.

[12] Chelli, A., and Pätzold, M., 'A non-stationary MIMO vehicle-to-vehicle channel model derived from the geometrical street model', *IEEE Vehicular Technology Conference,* San Francisco, CA, USA, 2011.

[13] Borhani, A., Stuber, G.L., and Pätzold, M., 'A random trajectory approach for the development of non-stationary channel models capturing different scales of fading', *IEEE Transactions on Vehicular Technology*, 2017, 66(1): 2–14.

[14] Li, Y., He, R., Lin, S., *et al.*, 'Cluster-based nonstationary channel modeling for vehicle-to-vehicle communications', *IEEE Antennas and Wireless Propagation Letters*, 2017, 16, 1419–1422.

[15] Gutierrez, C.A., Gutierrez-Mena, J.T., Luna-Rivera, J.M., Campos-Delgado, D.U., Velazquez, R., and Pätzold, M., 'Geometry-based statistical modeling of non-WSSUS mobile-to-mobile Rayleigh fading channels', *IEEE Transactions on Vehicular Technology*, 2018, 67(1): 362–377.

[16] Zhu, Q., Yang, Y., Chen, Z., *et al.*, 'A novel 3D non-stationary vehicle-to-vehicle channel model and its spatial–temporal correlation properties', *IEEE Access*, 2018, 6, 43633–43643.

[17] Senadji, B., and Levy, A.J., 'A statistical model for the simulation of time-varying multipath mobile radio propagation channel', *Proceedings of the ICASSP'94 IEEE International Conference on Acoustics, Speech and Signal Processing,* Adelaide, SA, Australia, 1994.

[18] Parra-Michel, R., Vazquez Castillo, J., Vela-Garcia, L.R., Kontorovich, V., and Peña-Campos, F., 'A channel model and simulation technique for reproducing channel realizations with predefined stationary or non-stationary PSD', *IEEE Transactions on Wireless Communications*, 2018, 17(8): 5409–5424.

[19] Xu, Z., Gan, M., and Zemen, T., 'Cluster-based non-stationary vehicular channel model', *2016 10th European Conference on Antennas and Propagation (EuCAP)*, Davos, Switzerland, April 2016.

[20] Amaro-Ramos, A., Munoz-Rodriguez, D., and Vargas-Rosales, C., 'Geometrical channel model for vehicle-to-vehicle systems', *AUE-International Journal of Electronics and Communications*, 2014, 68(8): 779–782.

[21] Amaro-Ramos, A., 'Caracterización del Efecto Doppler en Canales de Comunicación Vehiculares en Ambientes ITS', *M.Sc. Thesis*, Tecnologico de Monterrey, May 2013.

[22] Zajic, A.G., Stuber, G.L., Pratt, T.G., and Nguyen, S.T., 'Wideband MIMO mobile-to-mobile channels: geometry-based statistical modeling with experimental verification', *IEEE Transactions on Vehicular Technology*, 2009, 58(2): 517–534.
[23] 3GPP TR 36.873 V2.0.0 (2014-03), 'Study on 3D channel model for LTE (Release 12)'.
[24] Jassal, A., Khanfir, H., and Martinez Lopez, S., 'Preliminary system-level simulation results for the 3GPP 3D MIMO channel model', *2014 IEEE 80th Vehicular Technology Conference* (*VTC2014 Fall*), Vancouver, BC, Canada, September 2014.
[25] Jiang, H., Zhang, Z., Wu, L., Dang, J., and Gui, G., 'A 3D non-stationary wideband geometry-based channel model for MIMO vehicle-to-vehicle communications in tunnel environments', *IEEE Transactions on Vehicular Technology*, 2019, 68(7): 6257–6271.
[26] Ma, Y., Yang, L., and Zheng, X., 'A geometry-based non-stationary MIMO channel model for vehicular communications', *China Communications*, 2018, 15(7): 30–38.

Chapter 3
Modelling and simulation of vehicular communications

In this chapter, we present the transition from the characterization and modelling of traditional wireless channels, mainly static environments, to the highly mobile, non-stationary and prone to a blockage vehicular communication environment. We introduced the main modelling approaches that have been used for simulation and evaluation of vehicular channels and present a discussion on the advantages and disadvantages of the approaches concluding with the basics of ray-tracing techniques that are used currently.

3.1 Channel-specific properties for vehicular communications

It is known that a great percentage of the propagation research historically has been focused on wireless communications channels where objects are static, or the system is based on the classical cellular scenario and stationarity properties are dominant. With respect to vehicular communications, there are several differences that need to be accounted for. For example, a diverse and different set of scenarios need to be characterized and studied such as those of urban, suburban and open spaces, and also those in specific situations such as tunnels, high and low vehicular density, and high and low mobility. Also, different kinds of vehicles (cars, planes, trains, buses, etc.) need to be included when affecting propagation. Scenarios with buildings, vegetation, and smaller objects such as road signs are also important to integrate in the modelling methods.

With respect to different channels, one needs to include different scatterers such as those given by the floor (pavement) or roadside walls or traffic signs, as well as other vehicles and large objects. An important characteristic of these channels is the time-varying nature of their parameters given by non-stationarity random processes that need to be described using the data obtained either through simulations or measurement campaigns. In many scenarios in vehicular communications, small-scale propagation parameters take an important role in the established links of the vehicles, since delay spread, Doppler shift and coherence bandwidth affect the time duration of those links and the quality at which the signal is delivered.

Another important issue that needs to be considered in vehicular communications is the study of three-dimensional (3D) scenarios, since transmitters and receivers are positioned at different heights, and the effects of propagation on the signal are caused at many different heights depending on the scenario under study. The tools such as simulators must consider this to be able to represent real scenarios with 3D objects affecting the signals.

In this section, we introduce some of the most important differences that need to be considered in vehicular communication channels with respect to classical propagation studies.

3.1.1 High-speed environments

As discussed in chapter 2, Doppler shift is one of the small-scale propagation parameters that are essential to characterize in vehicular scenarios since it depends directly upon the relative speed of the transmitter and receiver, which is critical for link availability on vehicle-to-vehicle (V2V) scenarios.

Figure 3.1 shows the amount of Doppler shift in Hz of a signal carrier in a V2V line-of-sight (LOS) link when both vehicles are moving towards each other at different constant speeds in a two-lane road. One can see that the shift of the frequency could be close to 1,500 Hz at vehicle speeds above 121 km/h. Also, as the vehicles have more separation between them, the more Doppler shift, with an abrupt, although monotonically, change between 0 and 100 m of separation [1,2].

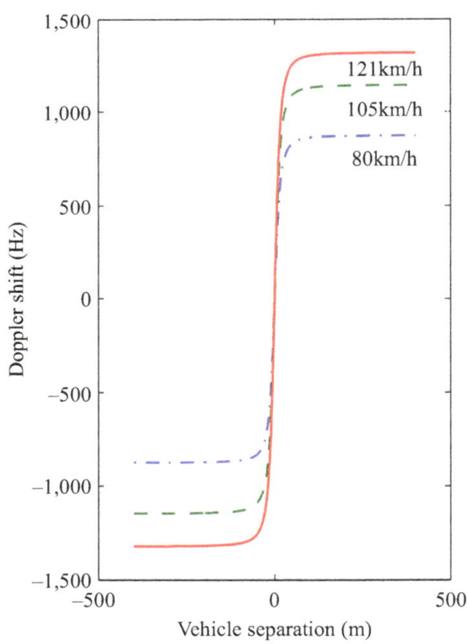

Figure 3.1 Doppler shift as inter-vehicle separation and vehicle speed vary

The phenomenon of Doppler shift is increased if scatterers and reflection is included in the propagation of the signal, rendering in time spreading and Doppler spectrum that depend on the environment. The frequency shift phenomenon is part of any of the communication links such as V2V and vehicle-to-infrastructure (V2I) communications, which need to be characterized through simulation and statistical modelling techniques. Fadings in these scenarios have reflected paths that are shorter since the scatterers such as other vehicles or road signs will be close to the receiver.

Although Figure 3.1 is obtained considering a planar scenario, i.e., transmitter and receiver antennas are at the same height, there have been methods where 3D models have been considered to obtain a characterization of the Doppler shift. Also, these ideas consider not only V2V communications, but also V2I. In [3], a single- and double-bounce ray signal model is developed to study the Doppler spectrum with a geometry-based (GB) scattering model where transmitter and receiver are moving relatively to each other. Autocorrelation functions are obtained of the single and the double-bounce signals. The model includes the angles of departure and arrival in the azimuth and elevation angles of the scenario. The bounce considered is a reflection of the signal on a scatterer located higher than transmitter and receiver, for the double-bounce signal, both scatters are located higher than transmitter–receiver plane. The Doppler power spectra obtained through measurements are compared for the 3D and 2D models.

Two-dimensional models have been used for other scenarios where the Doppler shift is of interest, see [1]. Figure 3.2 shows the same scenario set-up for the Doppler shift of Figure 3.1, except that besides including the LOS component, there are scatterers on the floor (four scatterers) and two on the roadside.

Considering the scenario as that shown in Figure 3.2, the received power by all the multipath components received at the vehicles due to LOS, roadside scatterers and the scatterers on the floor when the separation distance of the vehicles is 100 m is shown in Figure 3.3.

Figure 3.3 shows the received power as a function of the Doppler shift for the scenario in Figure 3.2, where vehicles are moving in the opposite direction. The

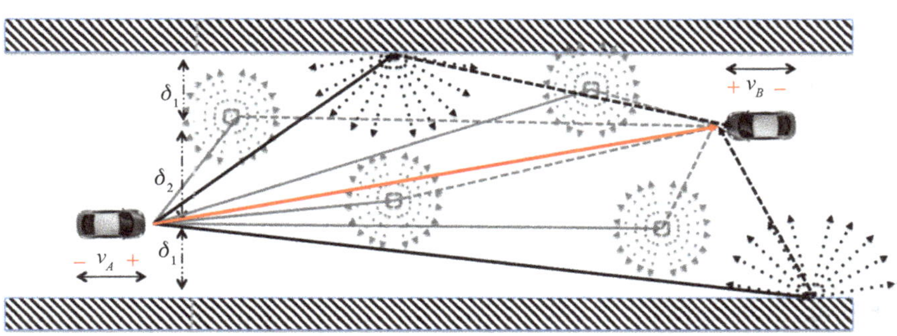

Figure 3.2 Scenario for Doppler with scatterers

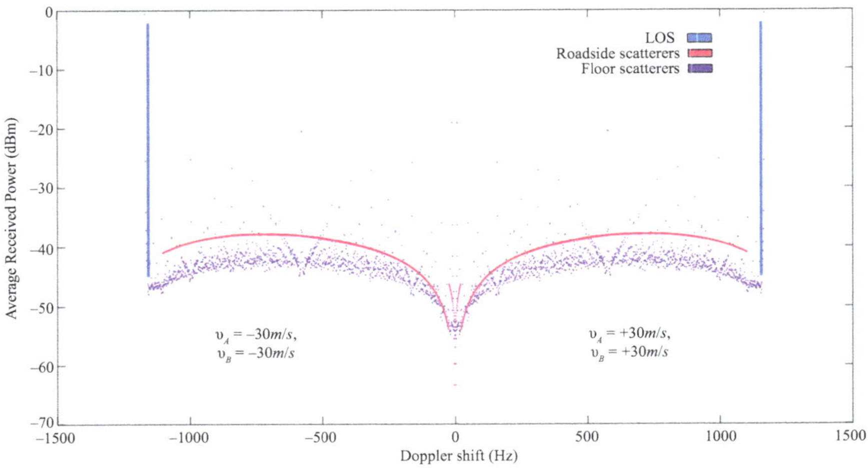

Figure 3.3 Average power received as function of Doppler shift

figure also shows the case when the vehicles have crossed each other and continue in the opposite direction. For this reason, the figure shows speeds with negative signs indicating direction. Results show the received power by the LOS component, the multipath components caused by the roadside scatterers and those caused by the floor scatterers. The maximum received power is by the LOS component, although it is at the maximum Doppler shift due to the speed and direction of the vehicles that coincides with the maximum separation where the movement of the vehicles is almost colinear.

3.1.2 Blockage

Blockage in vehicular scenarios is important due to the losses caused to signal propagation. Blockage in vehicle-to-everything (V2X) communications can be encountered on the road due to other vehicles and road geometry. Blockage of signal is the condition when the LOS component is not received, and the communication, in case of being established, is due to multipath components or propagation of the wave front such as diffraction.

Figure 3.4 shows two scenarios with blockage conditions. One is caused by neighbouring vehicles that do not let the LOS reach the receiver in a V2V communication, another one shown is the blockage caused by the same infrastructure that interrupts the signal in its transmission. A second scenario is shown when the infrastructure of a cell-type communication system is tried to be established and the LOS signal is interrupted by objects such as trees or it could be by buildings. In the figure, it is shown a possible solution to one of those blockage conditions that is the use of a multipath that needs to be strong enough so that the received signal has the sufficient power to reach the receiver as shown with the green collared multipath component.

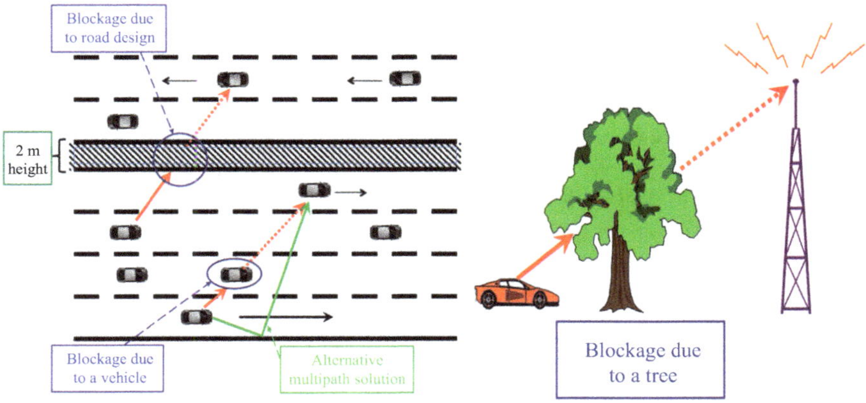

Figure 3.4 Blockage conditions on the road for V2X communications

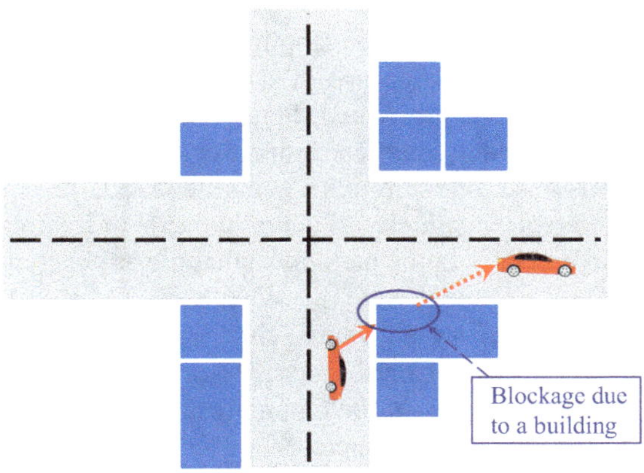

Figure 3.5 Blockage conditions on an intersection for V2X communications

Figure 3.5 shows a different scenario where blockage can be the result of those different situations as those shown in Figure 3.4, but the road geometry intensifies the blockage conditions due to the need to communicate with another vehicle or an element of the infrastructure that is right around the corner, hence buildings or trees might cause blocking conditions of the LOS component and a multipath might be the solution to communicate reliably with the receiver.

3.1.3 Platoons

Platoons in vehicular communications are more related to the scattering models and the network topology establishment. For networking, platoons can be used as clusters of nodes that create intra-vehicular communications and when information has a destination outside the cluster, then a cluster head will communicate with a different cluster (inter-cluster communication) in order to reach the final destination. This clustered network topology is also used for V2I communication when the cluster head communicates information from the cluster to a remote destination that is reachable through the infrastructure.

When talking about scattering models, then a cluster will define a set of scatterers surrounding a receiver and causing signal reflections that need to be characterized in order to deal with the problem of multipath effects which generate due to mobility, non-stationary random processes in the signals being received and the parameters of interest such as angles of departure and arrivals, delay spread, or Doppler shift, among other phenomena.

Platoons are also used for cooperative or collaborative decision making, so that vehicles can increase or decrease speed collectively, or decide to change lanes or decide how to let another vehicle merge into the line. Some negotiations are in place to let traffic run smoothly as decisions of which vehicle passes first in intersections is discussed among the vehicles in the platoon. The platoon leader can be the cluster head, and it can be changed for different circumstances as time evolves.

Platoons are naturally formed when stoplights are on the street, since vehicles can pass across an intersection in groups of size determined by the time taken by the stoplight to change from green to red. Platooning can also help with communication among vehicles under emergency situations, e.g., when emergency vehicles or first responders need to go through a congested road, the emergency vehicle can start a message to request vehicles to get on the side of the road. This message is propagated to other vehicles in the platoon so that the emergency vehicle can get through.

In Figure 3.6, one can see a road where vehicles are organized in platoons. The platoons can be organized by separation distance or by specific characteristics of the technology in use or interests in the communication needed. As shown in the figure, platoons form clusters of vehicles were communication can take place

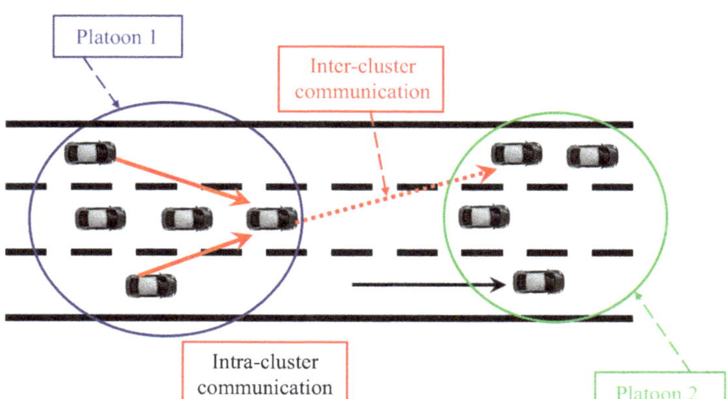

Figure 3.6 Use of platoons for V2X communications

inside the cluster and also links need to be established between neighbouring clusters or platoons. The use of platoons or clusters helps to organize the network, especially when there is a high density of vehicles in the area because the number of messages needed to be transmitted will be organized so that the number of links used in the network is decrease. Also, the use of communications through platoons can save energy and increase the network lifetime.

3.2 Modelling approaches

Channel modelling and simulations are important tools to characterize and describe signal propagation phenomena in vehicular communications scenarios. Modelling approaches can be grouped as follows:

(A) **Empirical methods:** models in this group need to collect data from real scenarios and then through approximations or regression methods a model can be derived

(B) **Stochastic channel models:** in this group, models are based on known probability distributions or the distribution of the channel behaviour is obtained. Issues such as stationarity and non-stationarity are also part of these models

(C) **GB stochastic models:** Although the models also contain probability distributions, the models are constructed based on geometrical descriptions of the environment.

(D) **Deterministic methods:** Although this group is also a GB modelling, signal propagation is studied through the generation of deterministic rays in a simulation scenario. The most used approach for the vehicular channel modelling is known as ray tracing (RT) and allows us to represent closer real scenarios.

In the following sections, we discussed some models that can be considered to be in one of the groups just mentioned.

3.2.1 Empirical methods

Empirical models are those that have been obtained by statistical analysis or regression methods based on sets of measurements. Once data of the signal that represents quality of reception are collected, generally a set of equations is obtained formulating propagation prediction for different scenarios. Most of these methods are concentrated on large-scale propagation to predict average received power as a function of distance or time. Also, the empirical modelling is used when simplified simulations and low computational complexity are desired.

Some of the most used empirical propagation models for outdoor environments are the Okumura–Hata model [4], the Longley–Rice model [5], the Walfisch–Bertoni model [6] and the COST 231 Walfisch–Ikegami model [7]. These methods are used to predict average power received in different situations and, at the same time, are used to determine initial coverage of the services. The models in general are computationally friendly providing prediction results rapidly, although the models require calibration when more data are available so that the models can characterize better different environments.

Models describe the received power as distance changes, or the power loss as distance changes, both in units of decibels. When new data are available, some new form of power loss is modelled, and the advantage of using expressions based on decibels is that the new model will only have new terms being summed to the previously used terms, and each term of the equation represents a loss of some kind. For example, the path losses predicted by the Okumura model denoted as $PL_{O,\mathrm{dB}}(d_{\mathrm{km}})$, see [5], considers the free space propagation loss, $PL_{\mathrm{dB}}(d_{\mathrm{km}})$, which is obtained by using the free space propagation model, see chapter 2, (2.9). Then, it adds some more terms to account for losses caused by the shape of the terrain, $A_{t,\mathrm{dB}}$, the base station (BS) and the mobile station height gains, G_T and G_R, respectively. The basic model is then

$$PL_{O,\mathrm{dB}}(d_{\mathrm{km}}) = PL_{\mathrm{dB}}(d_{\mathrm{km}}) + A_{t,\mathrm{dB}} + G_T + G_R \tag{3.1}$$

Usually, these methods consider random variables to model the changing behaviour of the signal propagation in the different environments. Some of the random variables used are the log-normal, the Rayleigh and the Ricean random variables that describe different fading and shadowing phenomena that the empirical model does not.

The Okumura–Hata model [4] is also similar to the Okumura model in the sense that terms are summed up and situations where the area of deployment is urban, a suburban or a rural environment is considered by some definition of terms. The model works for frequencies from 150 to 1,500 MHz. The Model is given by

$$\begin{aligned} PL_{H,\mathrm{dB}}(d_{\mathrm{km}}) &= 69.55 + 26.16\,\log f_c - 13.82\,\log h_T - A(h_R) \\ &\quad + (44.9 - 6.55\log h_T)\log d_{\mathrm{km}} \end{aligned} \tag{3.2}$$

where $PL_{H,\mathrm{dB}}(d_{\mathrm{km}})$ is the path loss of the Hata model, f_c is the carrier frequency, h_T is the transmission antenna height and $A(h_R)$ is the correcting factor that accounts for the different environments that depend on the characteristics of a small to medium size city. The COST 231 model is an extension of the Hata model to characterize environments with frequencies from 1,500 up to 2,000 MHz.

In [8], some empirical propagation models are introduced using simulation data of V2I communication at 28 GHz. The communication channel is characterized for an environment consisting of a highway with six lanes of traffic, three in one direction and the other three in the opposite direction. Free space path loss is used and some large-scale as well as small-scale parameters are calculated. Path loss exponent is obtained for LOS and non-line of sight scenarios under shadow fading conditions, and power delay profiles (PDP) are also calculated.

As an example of modelling, Figure 3.7 shows measurement data taken from a public square, transmitting with a frequency of 2.5 GHz. The models used to fit the measurement data were the Hata model given in (3.2), linear regression that also

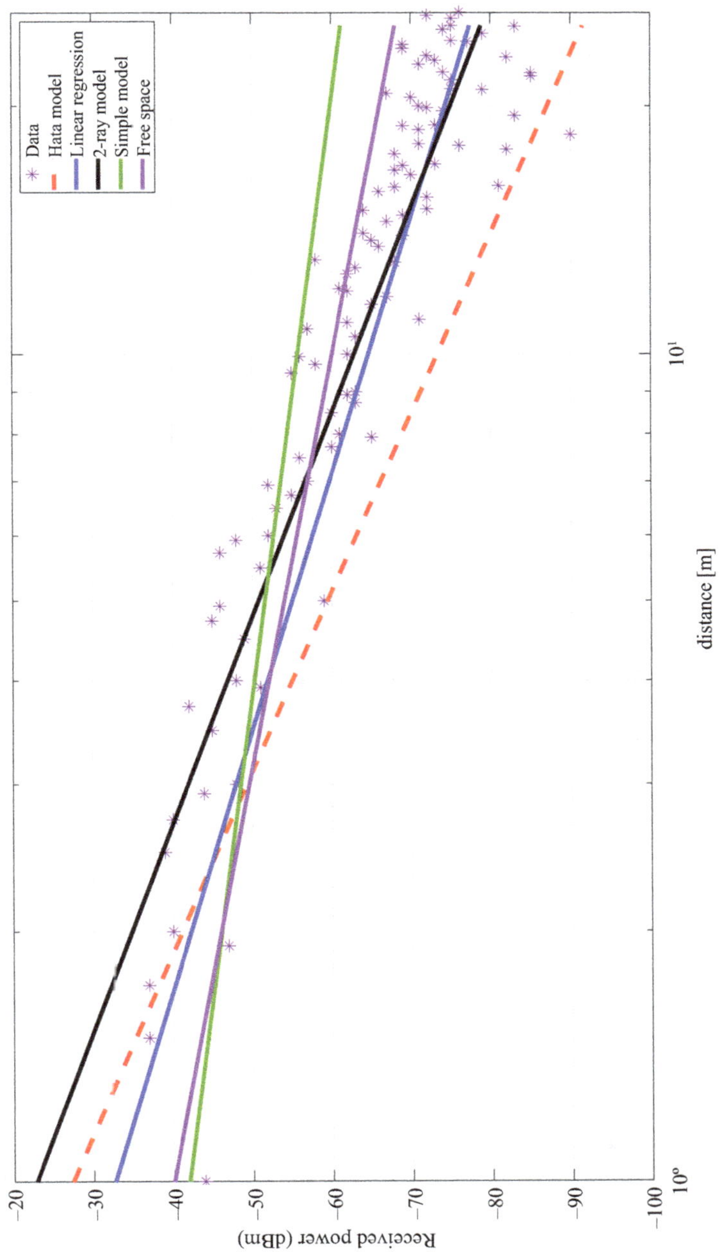

Figure 3.7 Statistical and empirical models for path loss

helps to estimate the path loss exponent, the two-ray model in (2.13), what we call the simple model given in (2.11) where the path loss exponent is obtained using the ML estimation in (2.28), and the free space path loss model in (2.12). One can see that the two-ray and the linear regression models are closer to the behaviour shown by the data. Also, the Hata model follows the same trend as that of the two-ray model, with only the difference of the starting point (y-intercept) of the straight line.

3.2.2 Stochastic channel models

In stochastic channel models, propagation loss is modelled as probability distributions. First, a statistical characterization of the propagation loss is obtained that includes path loss and fadings, and then a set of known distributions can be used to find out the closest to the real phenomena behaviour. Among the techniques used we find tools such as fit to different distributions, the Kolmogorov–Smirnov criterion, the Anderson–Darling and the Kullback–Leibler divergence, and several different options are chosen, one of them could be known distributions such as Rayleigh, Nakagami, log-normal or Gaussian mixture. The probability distribution chosen must be able to capture the effects of large- and small-scale propagation.

To introduce these probability distributions, it is usually common to use the probability density functions. For example, if we consider that the Rayleigh distribution is appropriate to model the channel fadings, then the envelope of the received signal, R, will have a probability density function given by

$$f_R(r) = \frac{r}{\sigma^2} e^{-r^2/2\sigma^2}, \; 0 \leq r < \infty \tag{3.3}$$

where the power of the received signal is σ^2, see [5]. As another example, if a random variable Y has a log-normal distribution [9], then it satisfies

$$X = \ln(Y) \sim \mathcal{N}(\mu, \sigma^2), \; 0 < Y < \infty \tag{3.4}$$

where $Y = e^X$, μ and σ are the mean and standard deviation in dB, and the probability density function of $Y_{dB} = 10\log_{10} Y$ is given by

$$f_{Y_{dB}}(y) = \frac{10/\ln 10}{y\sigma\sqrt{2\pi}} e^{-\frac{(10\log_{10} Y - \mu)^2}{2\sigma^2}}, \; 0 < y < \infty \tag{3.5}$$

Once a probability distribution model has been chosen, one can use it to characterize relevant quality indicators such as signal-to-noise ratio, outage probability or coverage area. In some cases, the channel is modelled by defining states of it, e.g., good and bad states, or outage and no outage state. For the case of multipath channels, one can have states defined by conditions of the existence of the LOS component or a non-LOS which represents a reception based on signal reflections.

Stochastic models are strongly used for vehicular scenarios due to the dynamic changes that the signal propagation undergoes, resulting in non-stationary channels that have time-varying statistical parameters that need to be characterized. Some models partition the scenario into smaller portions or observation times, where the channel parameters behave as stationary random processes, and then they put them together, but the dynamic conditions present in the changes of observation times are not captured by this technique.

To capture the effects of non-stationary phenomena, in the channel model, the gains of the signal need to depend on the time evolution for each path, so as discussed in chapter 2 in (2.32), we have impulse responses of the channel given by

$$h(t;\tau) = \sum_{i=-\infty}^{\infty} \sum_{k=0}^{\infty} \alpha_{i,k}(t)\delta(t - \tau_{i,k})e^{j(\theta_k + 2\pi f_{i,k} t)} \tag{3.6}$$

where the channel gains $\alpha_{i,k}(t)$ depend on time.

In [10], the V2V communications channel is modelled in a three-dimensional scenario where azimuth and elevation angles are considered to define distributions of clusters of multipath components. The models are based on measurements taken at 5.9 GHz in urban and suburban vehicular scenarios. The models deal with dynamic behaviour of multipath components and clusters of them through clustering and tracking algorithms. One of the conclusions of this modelling is that the probability distribution that describes the behaviour of the azimuth and elevation spreads is the log-normal distribution. For other parameters, such as power of multipath components, they concluded that the truncated Gaussian distribution characterizes that parameter, for the global cluster arrival angle the Laplace distribution has a good fit, and for the number of multipath components within the clusters the normal distribution is best. The model is validated by using the root mean square (rms) delay spread calculated from multiple sets of measurements.

Markov chains have been used to model wireless channels when representing the outage and non-outage state conditions of a channel by using a two-state Markov chain. In [11], authors present a model where Markov chains are used. Since they are using discrete-time Markov chains to represent shadowing effects, the range of values that shadowing takes is partitioned into a finite number of non-overlapping intervals and the transition probabilities are obtained by empirical methods through the measurements they have, or by considering that a log-normal model fits the shadowing effects of the signal. Although the application of the model is in a wireless system called WiMAX (IEEE802.16e standard), the modelling could be used in different systems and could also be extended to different frequencies and scenarios that could help the vehicular communications environment when modelling shadowing effects. The log-normal distribution is assumed to model received power signal variations, and it is used in conjunction with an exponential assumption in the autocorrelation function which results in a Markov process that can be characterized by an autoregressive (AR) model of first order, although the use of these two assumptions is not completely true since for some experiments the data is not described by a model with these two assumptions.

We just discussed how an AR model with exponential assumptions has been used to describe a channel model that characterizes shadowing effects. In [12], a model that describes multipath fading phenomena using an AR moving average (ARMA) model is introduced. It is also argued that the ARMA model is better to use than an AR model due to the low order linear system involved. Authors use two different ARMA models to characterize autocorrelation functions for the maximum Doppler. The parametrization of the ARMA model helps to avoid the need of prior knowledge as might be necessary with other methods such as those where machine learning is used.

In [13], we can see a method to generate non-stationary processes to represent fading channels. A Monte Carlo simulation method with windowing is used, and the Karhunen–Loeve expansion method is used to generate simulations that have high accuracy.

3.2.3 GB stochastic models

In vehicular propagation models, the GB channel models help to predict power losses in large-scale propagation. These GB models can be stochastic or deterministic. The deterministic models are discussed in section 3.2.4 and use ray-tracing techniques [14].

The GB models extract data from measurements or simulations to obtain channel statistics and use geometric descriptions of the environment. Different parameters are estimated and sets of them generated by evaluating in the environment different settings between transmitters and receivers, affecting LOS conditions. Some models have used signal components in their evaluations such as LOS, scattering and reflections from different objects including static or mobile objects. Others have used stochastic techniques with Markov chains where transition probabilities of the chain are obtained using probability distributions that describe the LOS conditions for different sets of settings, see [15].

The authors in [16] introduce a GB model for wireless channel emulation in real time. They define the hardware architecture needed in order to be a real-time model. The approach is to define the impulse response of the channel as the superposition of several multipath components, except that the channel is considered to be non-stationary. Thus, the non-stationary random process that defines the fading conditions is partitioned in portions that are stationary at least in the local sense. Therefore, the superposition of the multipath components will have attenuation factors that are time varying, Among the time-varying parameters, we have the delay, and the Doppler shift. A cross over is simulated and emulated using measurement techniques with field programmable gate array technology, and the PDP is compared showing a close match. They use the same methodology for delay spread and Doppler shift.

In [17], a GB model is also used to describe stochastically a non-stationary MIMO channel for vehicular communications. They use the autocorrelation function for this description and a space cross-correlation function. The model considers moving transmitters and receivers, and the scattering model used includes angles of arrival and departure that are time varying due to the mobility of the objects. The scatterers that surround the receiver are characterized by a ring pattern.

A model for V2V communication that includes the dynamism of the environment due to vehicular traffic is introduced in [18]. In V2V channel models, it has been known that propagation parameters for large and small scale have significant variations and changes due to objects large or small, static or mobile. The authors present a GB model that they call the GB efficient propagation model for V2V (GEMV^2). Links are classified as LOS and non-LOS, where the later are subdivided into non-LOS caused by surrounding moving vehicles or by static objects. The model is validated using measurements for different vehicular environments. Although GEMV^2 was started for V2V communications, nowadays it includes V2I communication as well.

Channel models for 5G technology are becoming more necessary to develop network planning and deployment solutions, that is why MIMO communications with V2V are needed to be characterized. In [19] a 3D GB non-stationary channel model in tunnel environments for MIMO V2V communications is presented. The authors obtain as characterization of the channel the time-varying spatial correlation functions, the frequency selectivity of the channel, the power spectral density based on Doppler, the PDP and the maximum duration of an observation time where the random process behaves at least as a wide sense stationary process.

3.2.4 Deterministic methods

The development of new channel models is needed as well as their statistical characterization so that one can obtain descriptions of the signal propagation behaviour. Such descriptions allow us to model among other things Doppler effects, path loss and fadings. Another objective of deterministic methods is the description of channel impulse responses that can be used for performance evaluation.

The application of deterministic methods in transportation systems and vehicular scenarios was discussed in [20], where three large groups of modelling approaches are presented that consist of empirical methods, stochastic channel models and GB stochastic models. One of the issues is that all the characteristics of the environment are not considered in these modelling approaches, and there is a need of a fourth group which is the one about deterministic approaches. The most common approach among deterministic methods for vehicular channel modelling is RT techniques where approximations of Maxwell equations are solved, with high precision but also high computing complexity.

Previously, RL and RT were both classified as ray-tracing methods, although more recently both methods are distinguished. The differences are mainly due to different approaches. The RL technique principle is that the transmitter launches thousands of test rays in a solid angle and the true path is determined by looking for the rays arriving at the receiver, whereas in classical RT methods the reflected paths by walls and furniture are found by computing the image of the transmitter or of the receiver. These methods are precise but are time-consuming due to inherent computational complexity.

RT and ray launching (RL) are an essential part of these deterministic methods. Previously, they were both classified as RT methods, although more recently both methods are distinguished. They consist basically of simulation scenarios where transmitters are positioned in a 3D environment and rays are launched from those. The number of reflections and the number of rays with their direction of departure are defined so that when a ray encounters an object within the simulation scenario, depending on the material the object is made of, absorption, reflected ray or new reflections and diffractions will be generated. All these rays are traced so that their paths are followed from the transmitter to the receiver throughout LOS rays, reflection of rays or diffracted rays. The environment has detailed models of objects that affect signal propagation such as trees, buildings, vehicles, etc. The differences between RT and RL are mainly due to different approaches. The RL technique principle is that the transmitter launches thousands of test rays in a solid angle and the true path is determined by looking for the rays arriving at the receiver, whereas in classical RT methods the reflected paths by the scatterers are found by computing the image of the transmitter or of the receiver. These methods are precise but are time-consuming due to inherent computational complexity.

There are also models that combine the GB methodology and the RL techniques to obtain time-varying descriptions of the channel impulse response and frequency response. In [21], a hybrid modelling technique is used to describe the time-varying nature of the V2V communication channel in a tunnel that contains scattering elements. Non-stationary fading processes are modelled, PDP are obtained using RL techniques. From these, delay spread and path loss can be calculated. The model is based on the time variation as well as discrete components in the frequency domain based on a carrier frequency f_c and a 'sampling' frequency f_s. With n denoting the discrete time, m discrete frequency and t_s sampling time, the channel frequency response would be considered a function of these parameters as $H(nt_s, f_c + mf_s)$. This channel response would consist of two parts, one calculated through data obtained from a RL simulation and a second part of diffuse scattering that is obtained from graph specific situations where small objects are considered.

The use of paths in the simulation and the way those are traced provide a platform where realistic scenario simulations can be carried out. Figure 3.8 shows a simplification of a 3D simulation environment where RL is used. The 3D environment is divided into cubes of some specific size that help control the heights at which propagation can be predicted with the models. One can see a BS antenna radiating three rays that are have a different number of reflections in objects (not shown) that are also specified within the environment. Parameters are evaluated at each of these cubes such as power received, angle of arrival of each ray, etc. The cubes can be of smaller sizes, but the simulation time and processing become difficult at sometimes.

3.2.4.1 Ray-tracing techniques

As an example, deterministic techniques based on RL approaches are commonly used for complex environments. In these methods, 3D scenarios can be created considering all the obstacles, walls, transceivers and the whole components of the environment. The principle is that a grid is defined in the space to save the parameters of each ray. Accordingly, the environment is divided into a number of

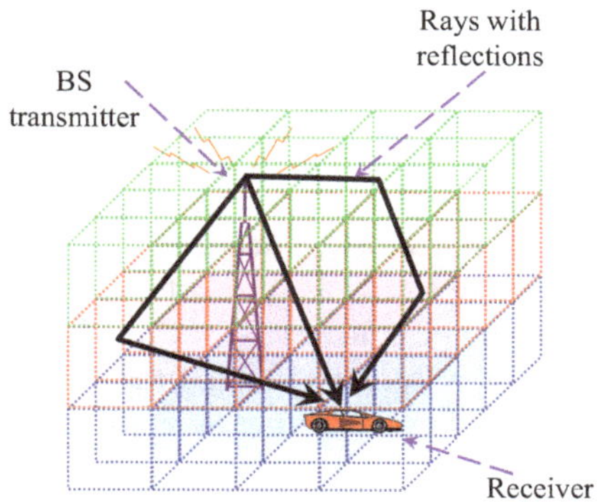

Figure 3.8 Concept of 3D ray launching. Space is divided in cuboids

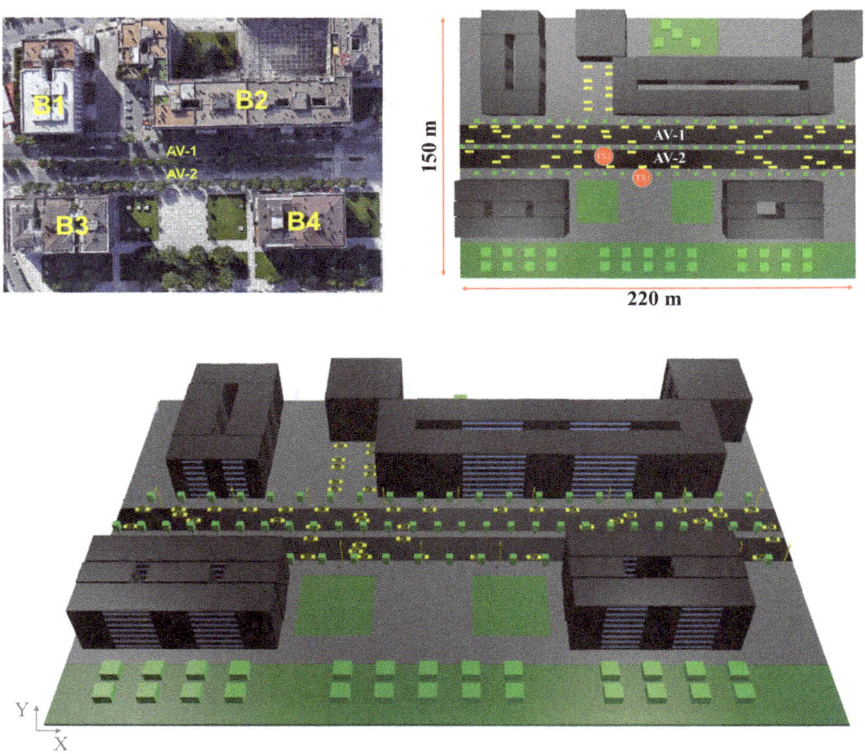

Figure 3.9 Urban scenario: Google map aerial view (left top); 2D-map schematic aerial view (right top); 3D-map frontal rendered frontal view (bottom)

cuboids of a fixed size. When a ray enters a specific hexahedron, its parameters are saved in a matrix. Parameters such as frequency of operation, radiation patterns of the antennas, number of multipath reflections and separation angle between rays and cuboid dimension can be modified as input parameters. A set of rays are launched and propagated along the space interacting with the obstacles, causing physical phenomena such as reflection, refraction, diffraction and scattering. The parameters of these rays can be stored as they enter to each hexahedral. Then, by using the stored parameters and considering the predefined transceiver's characteristics, such as radiated power, the emitting antenna's directivity, wave polarization and the carrier frequency, it is possible to easily derive the relevant parameters for channel modelling.

In chapter 5, for inter-vehicle communications, some urban scenarios that were taken from real cities were specified in a simulator where 3D RL was used. Figure 3.9 (also Figure 5.1 in chapter 5) shows an aerial photograph of a city and the corresponding top and panoramic views of the scenario created for a ray-tracing simulation.

Figure 3.10 shows received power for the 3D urban scenario simulated. See that even vehicles can be included in the simulator and different frequencies at which signals are transmitted can be varied. Also, the results shown are for a 1.5 m height, which is defined by the plane of cuboids that were discussed

These and many different results obtained by RL simulations for intra-vehicle and inter-vehicle scenarios will be discussed in chapters 4 and 5.

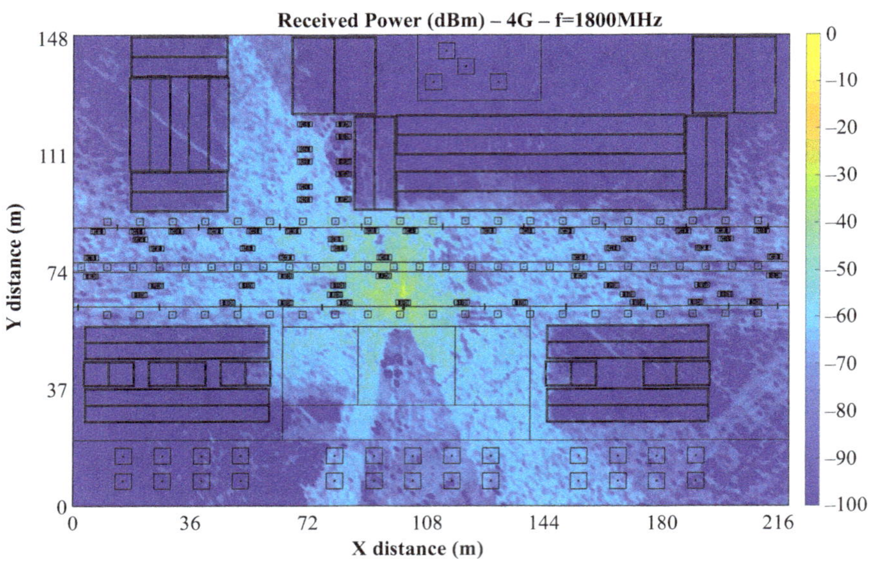

Figure 3.10 Received power (dBm) for the planes of 1.5 m height for the 1,800 MHz frequency band

References

[1] Amaro-Ramos, A., 'Caracterización del Efecto Doppler en Canales de Comunicación Vehiculares en Ambientes ITS', *M.Sc. Thesis*, Tecnologico de Monterrey, May 2013.

[2] Amaro-Ramos, A., Munoz-Rodriguez, D., and Vargas-Rosales, C., 'Geometrical channel model for vehicle-to-vehicle systems', *AUE-International Journal of Electronics and Communications*, 2014, **68**(8): 779–82.

[3] Liang, X., Cao, W., and Zhao, X., 'Doppler power spectra for 3D vehicle-to-vehicle channels with moving scatterers', *IEEE Access*, 2018, **6**: 42822–28.

[4] Hata, M., 'Empirical formula for propagation loss in land mobile radio services', *IEEE Transactions on Vehicular Technology*, 2008, **VT-29**(3): 317–25.

[5] Rappaport, T., *Wireless Communications: Principles and Practice*, Prentice Hall Communications Engineering and Emerging Technologies Series, 2nd ed., Prentice-Hall PTR, Upper Saddle River, NJ, USA, 2002.

[6] Bertoni, H.L., *Radio Propagation for Modern Wireless Systems*, Prentice-Hall PTR, Upper Saddle River, NJ, USA, 2000.

[7] Damosso, E., and De Brito, G., 'COST231 achievements as a support to the development of UMTS: a look into the future', *IEEE Communications Magazine*, 1996, 34(2): 90–96.

[8] Wang, L., Ai, B., He, D., *et al.*, 'Vehicle-to-infrastructure channel characterization in urban environment at 28 GHz', *China Communications*, 2019, **16**(2): 36–48.

[9] Goldsmith, A., *Wireless Communications*, 1st edn, Cambridge University Press, Cambridge, 2005.

[10] Yang, M., Ai, B., He, R., *et al.*, 'A cluster-based three-dimensional channel model for vehicle-to-vehicle communications', *IEEE Transactions on Vehicular Technology*, 2019, **68**(6): 5208–20.

[11] Seetharam, A., Kurose, J., and Goeckel, D., 'A Markovian model for coarse-timescale channel variation in wireless networks', *IEEE Transactions on Vehicular Technology*, 2016, **65**(3): 1701–10.

[12] Mera, D., Tavares, G., and Ortigueira, M., 'An alternative method for ARMA (p, q) model characterization of multipath fading channels', *Proceedings of the 81st Vehicular Technology Conference (VTC Spring)*, Glasgow, UK, 2015.

[13] Parra-Michel, R., Vazquez Castillo, J., Vela-Garcia, L.R., Kontorovich, V., and Pena-Campos, F., 'A channel model and simulation technique for reproducing channel realizations with predefined stationary or non-stationary PSD', *IEEE Transactions on Wireless Communications*, 2018, **17**(8): 5409–24.

[14] Viriyasitavat, W., Boban, M., Tsai, H.-M., and Vasilakos, A.V., 'Vehicular communications, survey and challenges of channel and propagation models', *IEEE Vehicular Technology Magazine*, 2015, **10**(2): 55–66.

[15] T. Abbas, K. Sjöberg, J. Karedal, and F. Tufvesson, 'A measurement-based shadow fading model for vehicle-to-vehicle network simulations', arXiv preprint arXiv:1203.3370v2, 2012.

[16] Hofer, M., Xu, Z., Vlastaras, D., *et al.*, 'Real-time geometry-based wireless channel emulation', *IEEE Transactions on Vehicular Technology*, 2019, **68** (2): 1631–45.

[17] Ma, Y., Yang, L., and Zheng, X., 'A geometry-based non-stationary MIMO channel model for vehicular communications', *China Communications*, 2019, **15**(7): 30–38.

[18] Boban, M., Barros, J., and Tonguz, O.K., 'Geometry-based vehicle-to-vehicle channel modeling for large-scale simulation', *IEEE Transactions on Vehicular Technology*, 2019, **63**(9): 4146–64.

[19] Jiang, H., Zhang, Z., Dang, J., and Wu, L., 'A novel 3-D massive MIMO channel model for vehicle-to-vehicle communication environments', *IEEE Transactions on Communications*, 2018, **66**(1): 79–90.

[20] Azpilicueta, L., Vargas-Rosales, C., and Falcone, F., 'Intelligent vehicle communication: deterministic propagation prediction in transportation systems', *IEEE Vehicular Technology Magazine*, 2016, **11**(3): 29–37.

[21] Gan, M., Steinbock, G., Xu, Z., Pedersen, T., and Zemen, T., 'A hybrid ray and graph model for simulating vehicle-to-vehicle channels in tunnels', *IEEE Transactions on Vehicular Technology*, 2018, **67**(9): 7955–68.

Chapter 4
Intra-vehicle short-range wireless channel characterization

In this chapter, an analysis of radio wave propagation in different intra-vehicle complex scenarios is introduced. We present different intra-vehicle scenarios, and the most important characteristics of the electromagnetic channel in these environments are assessed with the aid of a 3D ray-launching algorithm. Several case studies are presented, and the differences among them are explained. Two specific case studies to analyse the electromagnetic propagation properties for intra-vehicle communications in intelligent transportation systems are presented, being the first one fast-moving vehicles in underground tunnels and the second one, slow-moving vehicles in underground mines and cave scenarios. At the end of the chapter, the optimal position of the transceivers is proposed for different environments based on simulation, as well as some empirical modelling to be used for this type of wireless communications systems in intra-vehicle scenarios.

4.1 Intra-vehicle scenarios radio wave propagation analysis

In a typical indoor scenario, the surrounding objects and obstacles around transmitters and receivers in a wireless communication system have a great impact on the mobile propagation channel. The way in which the signal propagates along a specific path is determined by these obstructions. Specifically, vehicular scenarios have stronger effects as they have a higher density of clutter in a small area than typical indoor scenarios. Thus, it is very important to identify the characteristics of such obstructions and their distribution in vehicular scenarios in order to account for these changes and variations in field strength.

Intra-vehicle environments are complex indoor environments where the particular signal propagation behaviour should not be underestimated. These are environments with a high density of objects that become obstacles with different material properties and sizes, where metallic objects are the most popular within the vehicles. The dimensions of the vehicles vary, ranging from small dimensions, such as a car, to wider environments, such as a bus, a railway passenger vehicle or an aircraft. All of these vehicles have in common a large number of obstacles within them, such as seats, handholds, metallic walls, windows and people, which can range from 5 people in a typical car to a maximum of 56 people in a bus or 74 people in a wagon train car, approximately.

When analysing the electromagnetic propagation in these environments, it is important to consider all the aforementioned factors for radio planning purposes, since the diversity of obstacles and different types of materials within the vehicles, as well as the presence of people, are important elements that can affect the proper functioning of wireless communication systems.

The propagation mechanisms involved in indoor propagation environments are considered much more difficult and challenging than outdoor radio propagation environments [1]. In fact, intra-vehicle propagation can be considered as a specific indoor propagation case, in which a high density of clutter, specifically a large number of metallic obstacles as well as different densities of people, can be encountered in a relatively small area. This is precisely one of the reasons why the propagation mechanisms in these specific environments need to be fully examined.

Thus, in order to comprehend how the specific morphology and topology of vehicular scenarios affect radio wave propagation, we need to understand several propagation mechanisms that are relevant to indoor wireless communication systems within vehicular environments. These propagation mechanisms have been explained in chapter 2, section 2.2. It is important to consider that propagation mechanisms inside vehicles are more complex than typical indoor scenarios, as vehicle geometries, shapes, roughness, metallic obstacles, people, etc., confined in a relatively small area, have an effect on the resulting field at a receiver. Accordingly, intra-vehicle radio wave propagation analysis needs a good amount of understanding and effort.

This chapter describes different case studies of electromagnetic propagation analysis in a general context for different intra-vehicular environments, with the aid of a 3D ray-launching algorithm. Different modelling approaches can be considered for this analysis with their advantages and drawbacks being explained in chapter 3, section 3.2. Specifically, the 3D ray-launching technique is a deterministic method based on ray-tracing approaches. The complete details of this algorithm can be found in [2]. This technique has been validated in different environments with a root mean square (rms) error of approximately 3–5 dB and a standard deviation of 5–7 dB [3–6]. The main advantage of the 3D simulation tool is the high accuracy, but the drawback is the high computational cost required, which can be unaffordable for large scenarios. Because of that, hybrid techniques in which the 3D ray-launching algorithm can be used combined with neural networks [7], diffusion equations [8] or collaborative filtering [9] can be found in the literature. These hybrid techniques reduce the computational cost with acceptable accuracy for engineering purposes.

The different intra-vehicle scenarios considered for the radio wave propagation analysis in this chapter have consisted of a typical car, two different types of urban transportation buses and a train wagon car. The analysis has been done with a different number of people to understand their influence and impact in wave propagation within the vehicles. The specific characteristics of each vehicle are presented in each subsection, yet here we present the most significant differences that can affect the electromagnetic propagation inside the vehicles:

- Vehicle height: the car has smaller dimensions than the rest of the considered vehicles, in which people within the vehicle can only be seated. In addition, the distance between the transceivers will be shorter in this case than in the rest of the vehicles.
- People density within the vehicles: in the case of the car, the number of people will be lower than in the case of the bus and the train wagon car, because of the dimensions of the vehicles.
- Obstacles and materials: in any of the considered vehicles, the amount of clutter within them is significant, as well as the quantity of metal objects. This led to the meaningfulness of analysing multipath propagation inside the vehicles. In the case of the bus and the train wagon car, the number of obstacles and clutter within the vehicles is greater, since there are bars, handholds and in some cases glass sheets to separate environments, as is the case of the bus when separating the environment of the driver with the rest of the passengers. In addition, windows on the walls of larger vehicles, such as the buses or the train, have bigger dimensions than in the car, which should be considered when assessing wave signal propagation.

In the following subsections, we present the electromagnetic propagation signal analysis within different types of vehicles in terms of the main propagation mechanisms in these specific scenarios. For that purpose, a wireless personal area network (WPAN) based on ZigBee and Bluetooth technology has been considered in different vehicles. These technologies have been selected to present a typical case currently considered for an intra-vehicular system. In addition, in order to be able to compare the signal behaviour in the different vehicles, the same analysis have been carried out for all the considered vehicles. Simulation parameters considered in the 3D ray-launching algorithm as well as the parameters for the considered technologies are presented in Table 4.1.

The considered vehicles are furnished with different types of obstacles and clutter. The electromagnetic material properties of these obstacles have been considered in the 3D ray-launching algorithm by means of the conductivity and dielectric permittivity of the materials at the frequency of the system under analysis. Table 4.2 presents the material parameters used in the simulation models [10]. The differences between the vehicles are explained in detail in the following subsections.

For every in-vehicle scenario, radio wave propagation has been analysed by means of large- and small-scale propagation parameters, which is explained in chapter 2, section 2.2.

4.1.1 Within a car

A typical sedan car model based on real dimensions (4.57 m $\times$ 1.6 m $\times$ 1.46 m) has been implemented in the 3D ray-launching tool for the simulations. A detailed view of the modelled car can be seen in Figure 4.1, with the transmitter position. To perform more realistic simulations, the car model has been populated with 2 and 4 people, and simulations have been carried out for both cases.

Table 4.1 3D ray-launching simulation parameters

System	Parameters	Values
Bluetooth Class 2	Transmitter power	4 dBm
	Frequency/bit rate	2.4 GHz/1 Mbps
	Antenna radiation pattern (RX, TX)/gain	Omnidirectional/0 dB
	3D ray launching: angular resolution/reflections/diffraction/cuboids size	1 degree/6/yes/0.1 m
	Receiver sensitivity	−82 dBm
ZigBee 868	Transmitter power	13 dBm
	Frequency/bit rate	868 MHz/80 kbps
	Antenna radiation pattern (RX, TX)/gain	Omnidirectional/0 dB
	3D ray launching: angular resolution/reflections/diffraction/cuboids size	1 degree/6/yes/0.1 m
	Receiver sensitivity	−106 dBm
ZigBee 915	Transmitter power	13 dBm
	Frequency/bit rate	915 MHz/250 kbps
	Antenna radiation pattern (RX, TX)/gain	Omnidirectional/0 dB
	3D ray launching: angular resolution/reflections/diffraction/cuboids size	1 degree/6/yes/0.1 m
	Receiver sensitivity	−103 dBm
ZigBee 2400	Transmitter power	8 dBm
	Frequency/bit rate	2.4 GHz/250 kbps
	Antenna radiation pattern (RX, TX)/gain	Omnidirectional/0 dB
	3D ray launching: angular resolution/reflections/diffraction/cuboids size	1 degree/6/yes/0.1 m
	Receiver sensitivity	−100 dBm

The constitutive materials of the different parts of the vehicle, such as dashboard, seats, windows, engine, the car body and people seated within it, have been defined, taking into account their real size and dielectric characteristics, as presented in Table 4.2. For the simulations, an antenna has been placed within the car, just over the dashboard, as can be seen in Figure 4.1, considering that position to be a place where a real device could be situated. Simulation parameters can be seen in Table 4.1.

4.1.1.1 Received signal strength indicator

The received signal strength indicator (RSSI) or received power represents the signal level that reaches each spatial point in the three-dimensional scenario considered. In Figure 4.2, we can see the power cuts received at a height of 1 m, which is the same height where the transmitter has been placed on the dashboard. In the simulations, results are obtained for all spatial points, but only a single height (1 m)

Table 4.2 Electromagnetic material properties used for the selected frequencies

Material	Parameters			
	Frequency (MHz)			
	868–915		2,400	
	Permittivity (ε_r)	Conductivity (σ)	Permittivity (ε_r)	Conductivity (σ)
Air	1	0	1	0
Polycarbonate	3.2	0.2	3	0.2
Aluminium	4.5	4×10^7	4.5	4×10^7
Glass	5.5	10^{-12}	6.06	10^{-12}
People	[11]	[11]	[11]	[11]

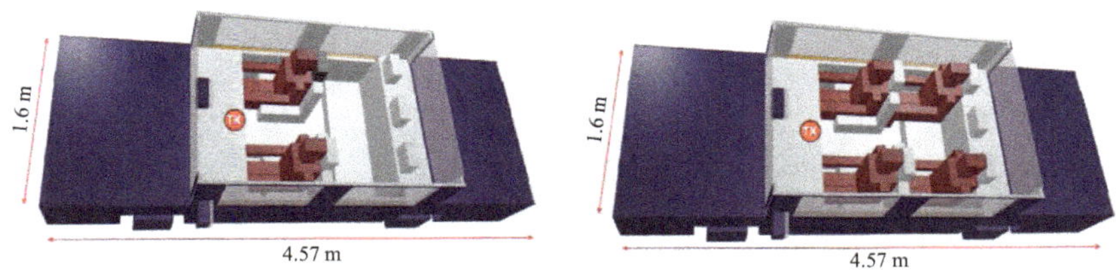

Figure 4.1 Car modelled used in the 3D ray-launching tool: 2 people (left) and 4 people (right)

is shown here for terms of comparison. The following points can be noticed from Figure 4.2:

- The received power levels vary considerably within the scenario, with the highest values close to the transmitting antenna for all cases.
- The influence that obstacles (such as seats, dashboard, windows, engine and people) have on radio wave propagation inside the car can be appreciated, as they absorb and reflect part of the emitted power, creating areas with lower received power levels behind them.
- The metallic body of the car makes most of the power remain inside the car, due to the reflective properties of metals, yet a significant amount of energy is leaked through the windows. Low received power areas can also be found in the engine area, as well as the car's trunk area, once again due to the metallic body of the car.
- Fast fading or multipath propagation is visible within the car, as there are some areas within the car where the received power level is low (blue-green areas). Multipath components appear due to the high reflectivity (great amount of metallic content) and diffraction (many obstacles).
- As the frequency is increased, the signal attenuation is bigger, as is observed from the 2.4 GHz frequency case (Bluetooth and ZigBee 2.4 GHz). Yet, it must

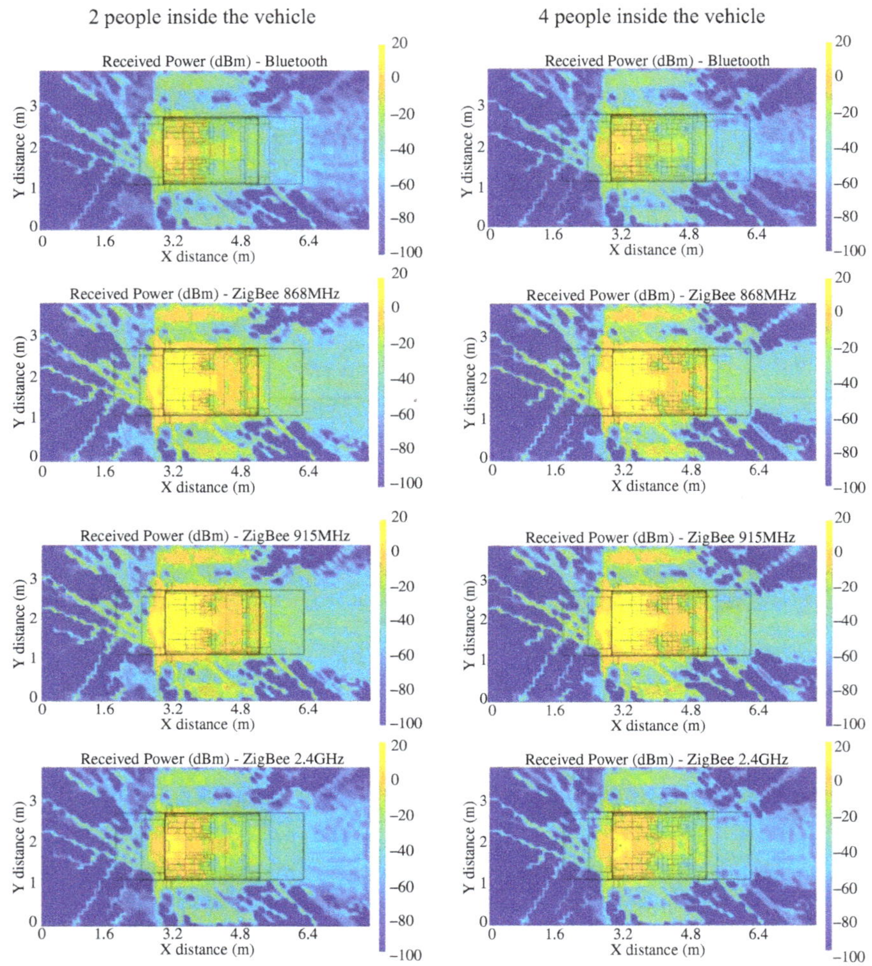

Figure 4.2 Comparison of received power (dBm) for the bi-dimensional planes of 1 m height for the different considered systems within the car: two people (left) and four people (right)

be considered that different transmitter power has been considered for each case, as is presented in Table 4.1.

- The number of people within the car also affects radio wave propagation, inducing more multipath propagation inside the vehicle, as well as scattering and absorption, when there are more people. To have insight into this phenomenon, Table 4.3 shows the additional attenuation encountered in a specific point within the vehicle, which corresponds to the place of a passenger in the back of the vehicle, as is presented in Figure 4.3. It can also be observed that

Table 4.3 Additional attenuation with four people within the car

System	Additional attenuation point (4.9 m, 2.9 m, 1 m)
Bluetooth Class 2	7.55 dB
ZigBee 868	9.45 dB
ZigBee 915	8.47 dB
ZigBee 2400	7.55 dB

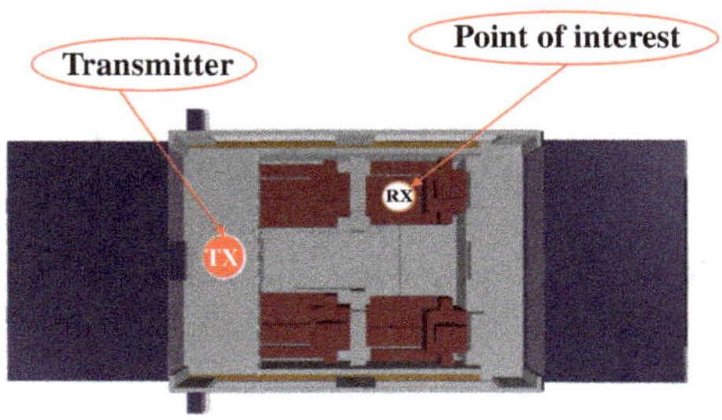

Figure 4.3 Transmitter and receiver position within the car for the calculus of additional attenuation in Table 4.3

the presence of two more people in the back seat of the car leads to an additional signal attenuation in this point which ranges from 7 to 9 dB depending on the considered technology.

As can be seen in the estimations of received power levels (Figure 4.2), both the topology and morphology of the considered scenario affect considerably the expected received power values, which depends on the location within the scenario. These RSSI variations are given by multipath propagation or fast fading phenomena, which should not be underestimated in these highly reflective environments. In order to show better the multipath propagation phenomena, power delay estimations are presented in the following subsection.

4.1.1.2 Multipath propagation

Multipath propagation is usually described by fast signal variations on the scale of half-wavelength and commonly introduces fluctuations as large as 35–40 dB for typical indoor environments [1]. It results from the constructive and destructive interference between multiple waves reaching the receiver from the transmitter. Multipath propagation is also known as multipath fading or fast fading.

In order to have insight into the multipath propagation phenomena within the car, the worst case has been considered, which corresponds to four people inside the car, as losses due to absorption and scattering will be larger in this case. The specific case of ZigBee 2.4 GHz has been analysed.

In Figure 4.4, a power delay profile for a spatial point behind the passenger seat is presented, exhibiting echoes in a time span of approximately 0–30 ns. The power delay profile shows the power level and arrival time of each ray that crosses a single point in the scenario. As can be seen, the number of components that arrives at different instants of time is significantly high, although the considered point is non-line-of-sight (nLoS) with the transmitter, which is placed on the dashboard. This confirm the high reflectivity of the environment, where a high number of reflections are produced due to the metallic parts of the car body. This quantity of reflections increases the probability of rays crossing the same observation point.

Time dispersion varies widely in a mobile radio channel depending on the geometrical position relationships among transmitters, receivers and the surrounding physical environment. Due to this fact, another parameter that can grossly quantify the multipath propagation is the rms delay spread, which shows the effects of dispersion and is determined by the power delay profile. It must be pointed out that the calculation of the rms delay spread depends on the choice of the noise threshold to process the received power at each delay. The noise threshold used to analyse the rms delay spread within the car is−120 dBm, chosen to emulate the typical sensitivity of a wireless sensor network (WSN), which can be deployed within a car.

The rms delay spread gives us information about multipath propagation for all spatial points of the considered scenario. Figure 4.5 shows a bi-dimensional plane for 1 m height of rms delay spread within the car. The strong dependence between

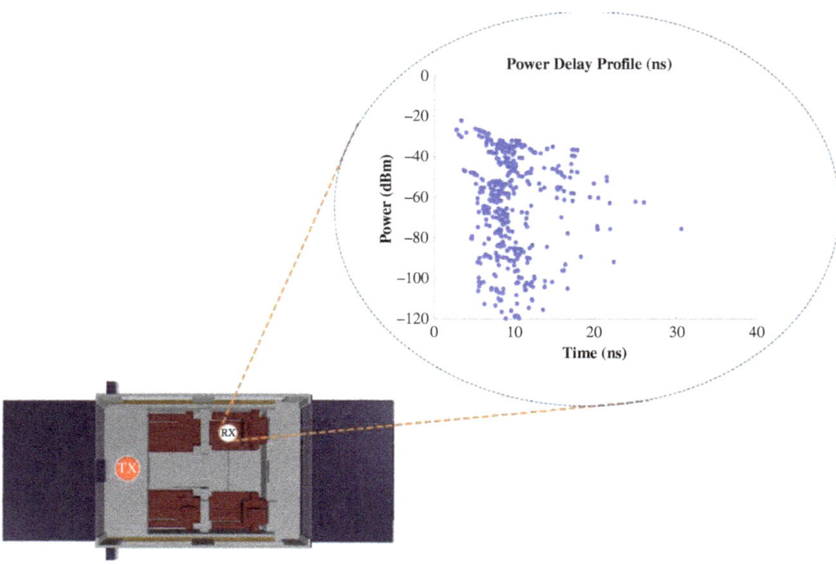

Figure 4.4 Power delay profile behind the passenger seat

the complex morphology of the car and multipath components at each spatial point can be seen from the figure. Specifically, the effect of the metallic car body and the highly reflectivity can be clearly seen, since the rms delay spread values within the car body are much higher than the values just outside, due as stated before to the highly reflective environment, which make most of the signal to be confined inside the car. Nevertheless, there are also leakages from the windows that should not be underestimated. It can also be observed that losses are greater when the distance from the transmitter to the receiver increases, as is shown from a distance outside the car, where no rays arrive.

4.1.1.3 Frequency domain channel description

To understand better the frequency domain channel behaviour, we characterized the coherence bandwidth from the rms delay spread within the car. The coherence bandwidth is the range of frequencies over which two frequency components have a strong potential for amplitude correlation [12]. The frequency correlation function has been considered above a value of 0.9 for this analysis. Figure 4.6 shows a bi-dimensional plane of coherence bandwidth within the car at 1 m height, showing values that vary from 0 to 20 MHz. It can be seen that there is a great variability in the scenario, with lower values within the car body. The number of people also affects coherence bandwidth values, as it can be seen that when people are placed inside the car, coherence bandwidth is bigger. The areas with bigger values of coherence bandwidth mean that in those areas, the range of frequencies that include components with strong potential for amplitude correlation is higher. Therefore, an intended receiver in those areas potentially can experience stronger multipath interference. From the point of view of network planning, those areas of high

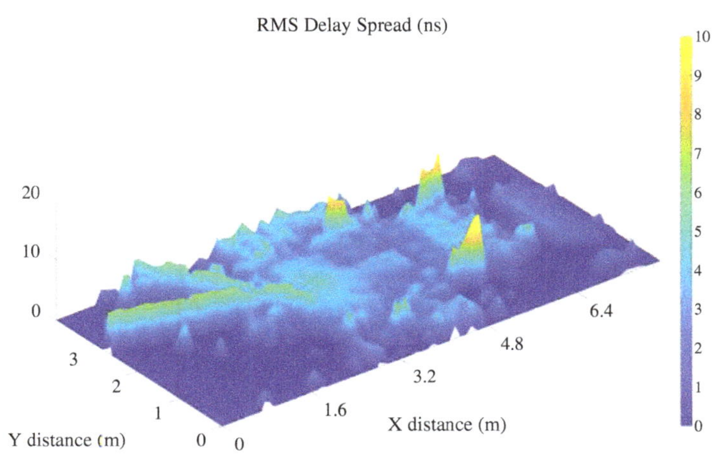

Figure 4.5 RMS delay spread within the car at 1 m height for ZigBee 2.4 GHz frequency

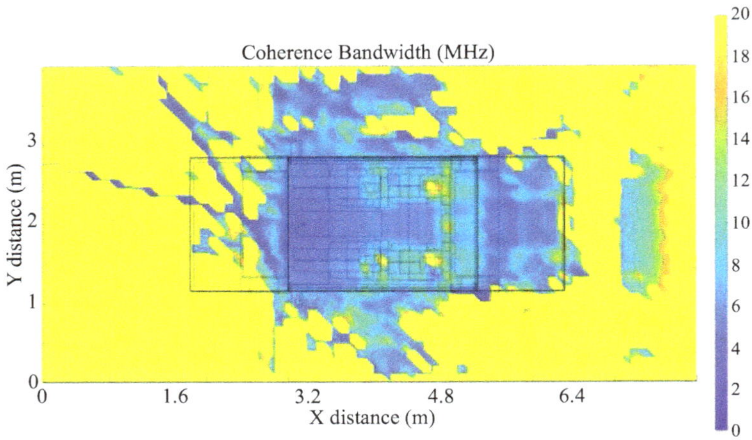

Figure 4.6 Coherence bandwidth within the car at 1 m height for ZigBee 2.4 GHz frequency

values of coherence bandwidth define regions where the same frequency can be reused installing another transmitter–receiver pair.

4.1.2 Within a bus

In this subsection, two different types of urban buses have been assessed: an articulated and a rigid bus. The main differences between them are their length, their internal organization and the bouncy metallic structure in the centre of the articulated bus. The dimensions are (2.7 m × 17.05 m × 2.45 m) and (2.7 m × 11.7 m × 2.45 m) for the articulated and rigid bus, respectively. Figure 4.7 shows both types of buses (left) and presents the internal distribution of the articulated bus, with seats and handhold allocation (right). Two different types of seats can be observed: the red ones, reserved for seniors, pregnant women, adults with children or individuals with disabilities, which have different dimensions or are prepared to engage wheelchairs or prams, and the turquoise blue seats, for the rest of the passengers.

The same scenarios of Figure 4.7 have been implemented in the ray-launching tool, considering in both cases all furnishings and the dielectric properties of the materials employed, as is presented in Table 4.2. As we know that the main goal of urban buses is the transportation of people, simplified human body models have been randomly introduced within the vehicles, with the aim of analysing the dispersive characteristics of the human body. Besides, not only the location of people has been varied, but also their density has been ranged to evaluate their impact in radio wave propagation according to their location and proximity to the transmitting antenna.

Figure 4.8 depicts the four scenarios implemented along with the different people densities considered for simulation. Low- and high-density conditions have

Figure 4.7 Real transportations buses considered for simulations (left) and internal distribution of the articulated bus (right)

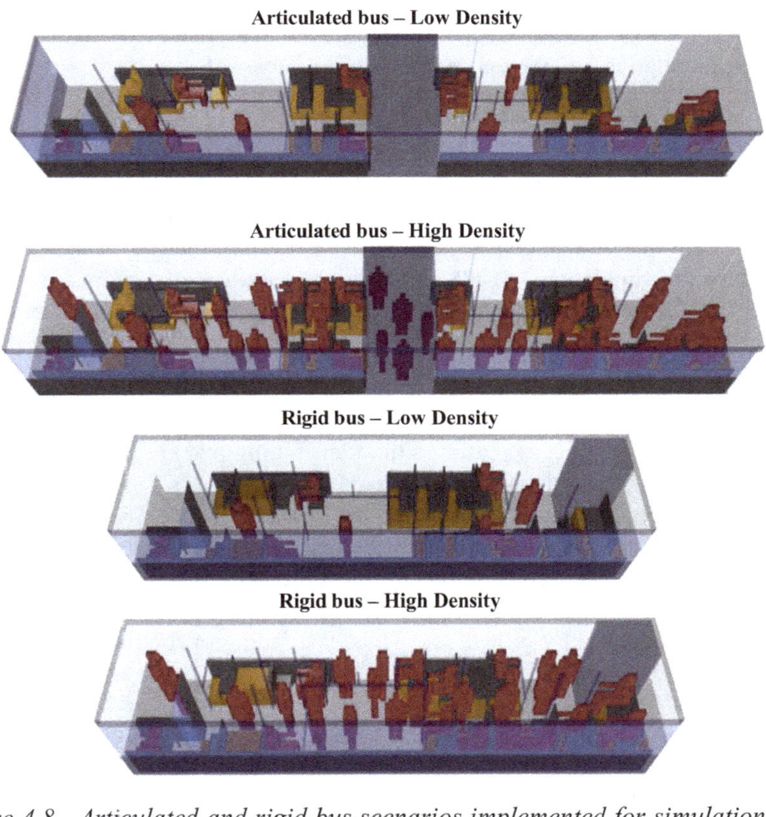

Figure 4.8 Articulated and rigid bus scenarios implemented for simulation with different indoor designs and passenger densities

been considered for both types of buses, which represent the 25% and 75% of the maximum capacity of passengers of the buses, respectively. The maximum capacity of passengers in the articulated bus has been considered to be 56 passengers and in the rigid bus 39 passengers.

The main differences that can be encountered comparing with the previous case of the car can be described as follows:

- The buses are larger vehicles than the car. They have more obstacles within them, and they can be more densely populated.
- In comparison to the car, although the buses have more obstacles and people within them, they also have more open spaces as the dimensions (height and length) are bigger.
- All these factors could lead to different fast fading behaviour than other vehicles. Thus, radio wave propagation within these vehicles should be fully examined and should not be underestimated.

For the simulations, a transmitter antenna has been placed within the two types of buses at the same position ($X = 3.2$ m, $Y = 2$ m, $Z = 1.1$ m), which correspond to a position at the front of the vehicles, emulating a wireless device of a WSN within the buses. Simulation parameters can be seen in Table 4.1.

4.1.2.1 Received signal strength indicator

The RSSI or received power has been calculated for each spatial point in the three-dimensional considered scenario. In Figure 4.9, we can see the power cuts received in the rigid bus at 1.1 m height, which is the same height where the transmitter is placed on both buses. The comparison between different people densities can be analysed. In the simulations, results are obtained for all spatial points, but only a single height (1.1 m) is shown here for comparison. The following points can be noticed from Figure 4.9:

- As in the car, the received power levels vary considerably within the rigid bus scenario, with the highest values close to the transmitting antenna. In all cases, a hot spot area is observed around the transmitting antenna, which, as can be seen, is at the front part of the bus.
- The influence that obstacles (such as the different kinds of seats, handholds, windows, metallic bouncy structures and people) has on radio wave propagation inside the buses can be appreciated, as they absorb and reflect part of the emitted power, creating areas with lower received power levels and areas with more power caused by reflections. Thus, fast fading or multipath propagation is clearly visible within the rigid bus.
- If we compare between the different technologies, we can see that, for the case of Zigbee 868, we have the highest received power values along the whole plane, since the frequency in this case is the lowest, and therefore, the signal suffers less absorption by obstacles. In the case of ZigBee 915, results are very similar, and no significant differences can be appreciated. However, in the case of Bluetooth and ZigBee 2.4, it can be seen that the received power levels are

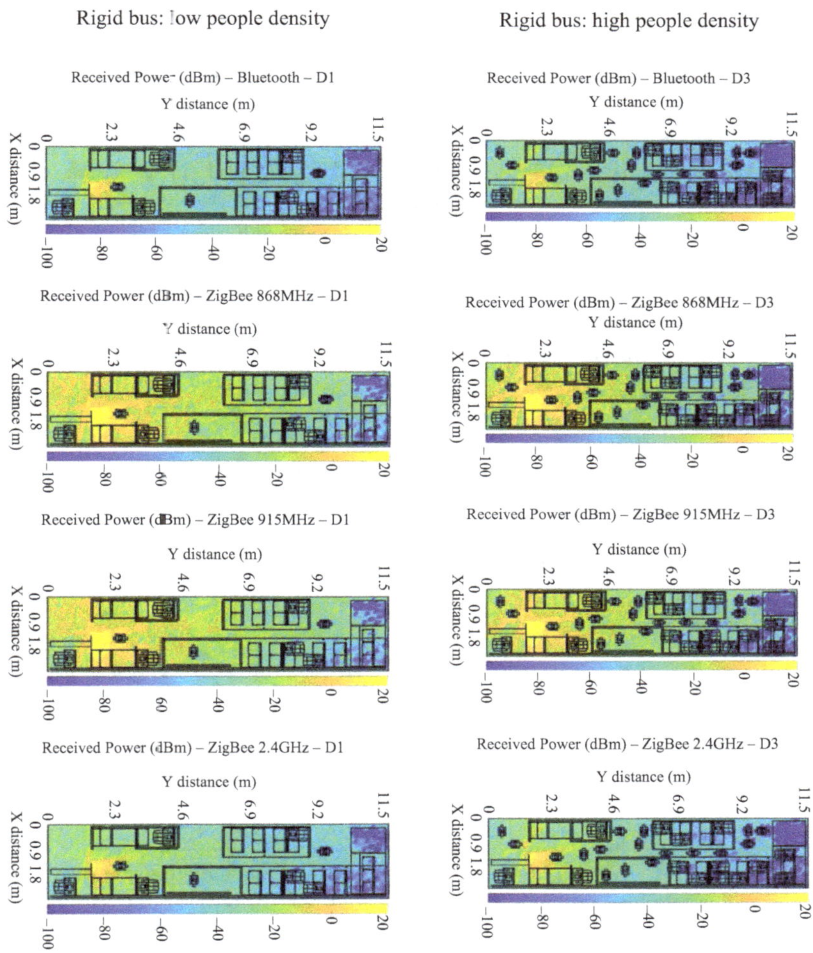

Figure 4.9 Comparison of received power (dBm) for the bi-dimensional planes of 1.1 m height for the different considered systems within the rigid bus: low people density (left) and high people density (right)

lower throughout the entire bi-dimensional plane mainly due to two factors: the use of a higher frequency and the lower transmission power used in these systems.

- For all technologies, it is observed that the number of people affects electromagnetic wave propagation, being the worst case when a high density is considered (D3), where very low power levels arrive to the back part of the vehicle, from $Y = 7$ m approximately. In this case (D3), there are some points where the signal fades completely after this distance. This is due to the fact that a large number of obstacles combined with the high quantity of metallic parts in the vehicle provokes a tremendously reflective environment, which makes the signal arrive less distance than in an open environment.

- To have better insight into the impact of people within the rigid bus, Table 4.4 shows the additional attenuation encountered in a specific point within the vehicle, which corresponds to the location of a passenger in the back part of the vehicle, represented in Figure 4.10. It can be seen that the presence of 50% more people randomly placed within the rigid bus leads to an additional attenuation in this point ranging from 32 to 39 dB depending on the considered technology. It can be appreciated that the additional attenuation is significantly bigger than in the case of the car. This is due to the fact that in the case of the rigid bus, there are more obstacles and people, combined with a larger distance, between the transmitter and receiver, which can make the signal have more losses. These attenuation values are in accordance with typical indoor environments, as is presented in [1].

The same results of received power levels have been obtained for the articulated bus, which are represented in Figure 4.11. The transmitter point has been considered the same as the rigid bus, as can be seen in Figure 4.12. The following points can be noticed from Figure 4.11:

- In this case, unlike the rigid bus, the bus length is bigger, which makes that the area where received power levels are very low is bigger. Approximately, from the same distance as the rigid bus ($Y = 7$ m), signal received levels are below the thresholds for almost all the points.

Table 4.4 Additional attenuation with high people density within the rigid bus

System	Additional attenuation point (2 m, 7.2 m, 1.1 m)
Bluetooth Class 2	39.81 dB
ZigBee 868	32.28 dB
ZigBee 915	35.39 dB
ZigBee 2400	39.80 dB

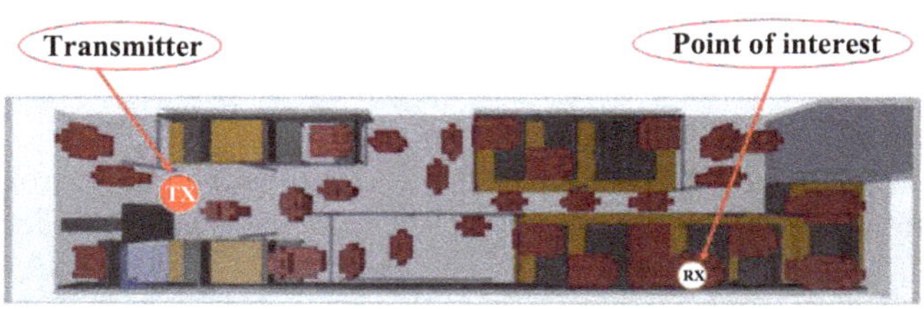

Figure 4.10 Transmitter and receiver position within the rigid bus for the calculus of additional attenuation in Table 4.4

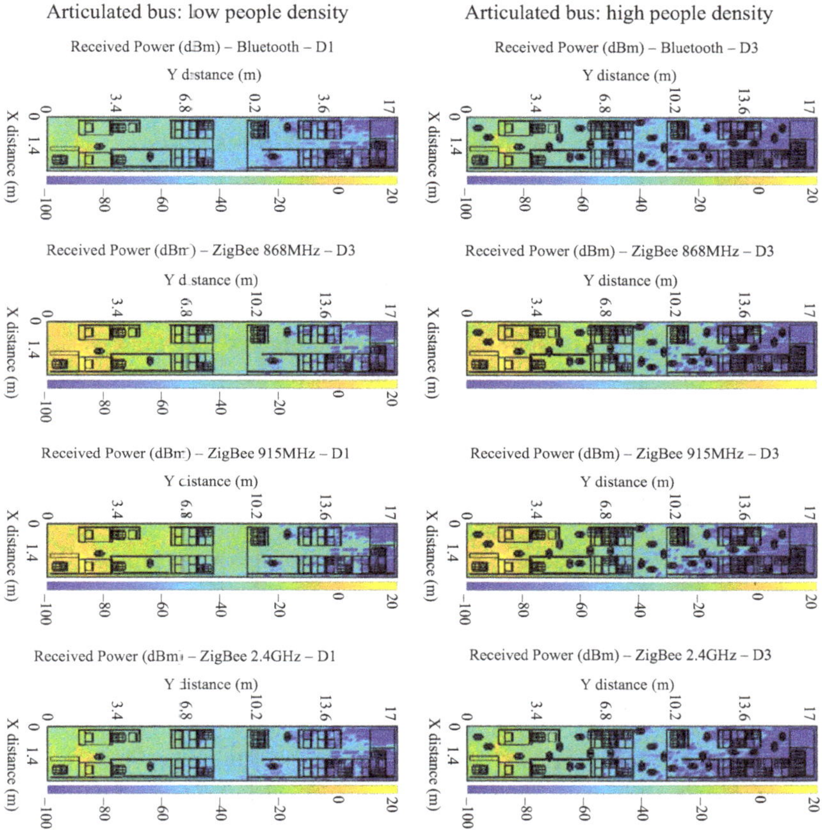

Figure 4.11 Comparison of received power (dBm) for the bi-dimensional planes of 1.1 m height for the different considered systems within the articulated bus: low people density (left) and high people density (right)

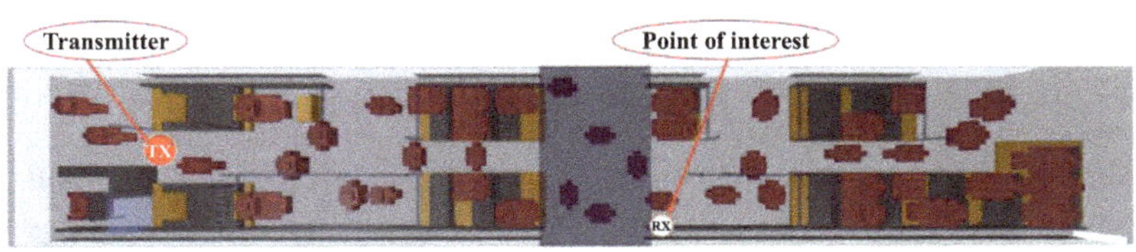

Figure 4.12 Transmitter and receiver position within the articulated bus for the calculus of additional attenuation in Table 4.5

Table 4.5 *Additional attenuation with high people density within the articulated bus*

System	Additional attenuation point (2 m, 7.2 m, 1.1 m)
Bluetooth Class 2	25.67 dB
ZigBee 868	35.75 dB
ZigBee 915	34.44 dB
ZigBee 2400	25.66 dB

- In the same way as the rigid bus, the large number of metallic obstacles provokes the effects of fast fading and multipath propagation, which are clearly visible within the articulated bus.
- The impact of people within the articulated bus is also observable in this case. In Table 4.5, the additional attenuation is observed for the same point as in the previous case, which is located in this case, just after the metal bouncy structure located in the centre of the vehicle, as is represented in Figure 4.12. It can be seen that levels of additional attenuation are also very high in this case, approximately the same values as the rigid bus, varying between 25 and 35 dB. This is due, once again, to the increase in the number of obstacles and people, combined with the larger distance between the transmitter and the receiver, when compared with the car previously analysed case.

Thereby, we can conclude in this subsection with the statement that the morphology and topology of the articulated and rigid bus have a considerable impact in received power levels within the scenarios. In comparison with the car, the larger dimensions of the buses, along with the higher number of obstacles and people that the signal can be encountered within the buses, provoke larger propagation losses, due to absorption from obstacles, as well as reflection, diffraction from edges and scattering from non-homogeneous surfaces such as the human body. As stated before, these RSSI variations are given by multipath propagation or fast fading phenomena, which should not be underestimated in these highly reflective environments. In order to better show the multipath propagation phenomena within the buses, we present power delay estimations in the following subsection.

4.1.2.2 Multipath propagation

In order to have insight into the multipath propagation phenomena within the urban transportation buses, we have considered the worst case, which corresponds with high people density within the buses, as losses due to absorption and scattering will be larger in this case. The specific case of ZigBee 2.4 GHz has been analysed.

In Figure 4.13, a power delay profile for a spatial point placed in the back part of the buses is presented (same point for both buses), exhibiting echoes in a time span of approximately 0–50 ns. As stated before, the power delay profile

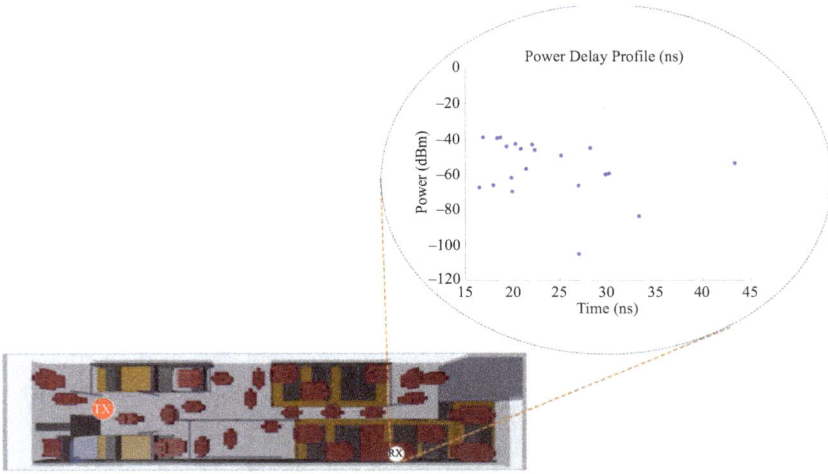

(a)

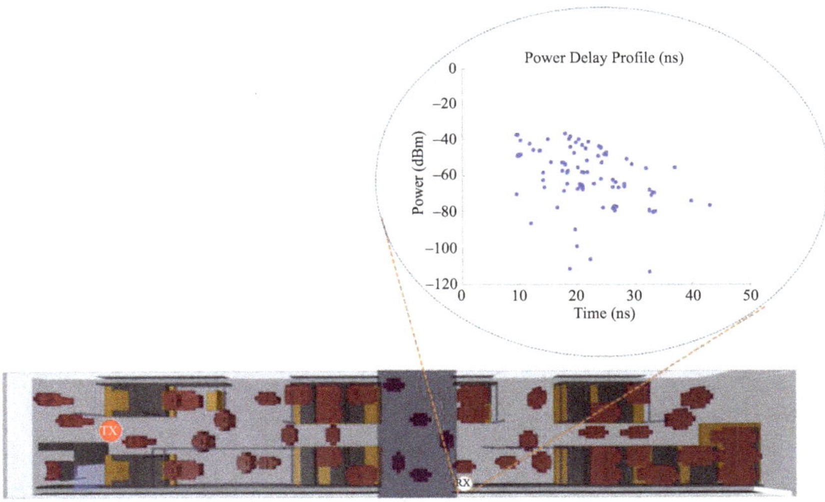

(b)

Figure 4.13 Power delay profile for high people density (a) rigid bus and (b) articulated bus

shows the power level and arrival time of each ray that crosses a single point in the scenario. As can be seen, the number of components that arrives at different instants of time is significantly high. For the articulated bus, the number of multipath components is bigger than for the rigid bus in this specific analysed spatial point. This is because the articulated bus is more densely populated than the rigid bus for high people density. Thus, reflections due to obstacles are higher, and the number of multipath components is also higher. These results confirm that the morphology and topology of the considered scenario, as well as the number of obstacles within it, has a significant impact in multipath propagation, provoking that different multipath components arrive to a specific point depending on the surrounding environment.

As stated before, the rms delay spread gives us information about multipath propagation for all spatial points of the considered scenario. Figure 4.14 shows the bi-dimensional planes for 1.1 m height of rms delay spread within the rigid and articulated buses. From the figure, a strong dependence between the complex morphology of the buses and multipath components at each spatial point can be seen. In the articulated bus, higher values of rms delay spread appear in some spatial points, since the number of obstacles and people is higher because of its bigger dimensions. It is observed that in the final part of the vehicles, no signal levels arrive, being the largest area where there is no signal in the articulated bus.

4.1.2.3 Frequency domain channel description

To understand the frequency domain channel behaviour, we characterized coherence bandwidth from the rms delay spread within both types of buses [12]. Figure 4.15 shows the bi-dimensional planes of coherence bandwidth within the

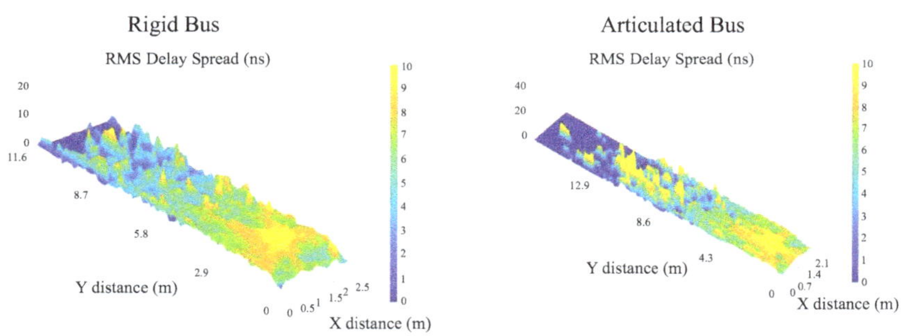

Figure 4.14 RMS delay spread for the rigid bus (left) and articulated bus (right) for high people density

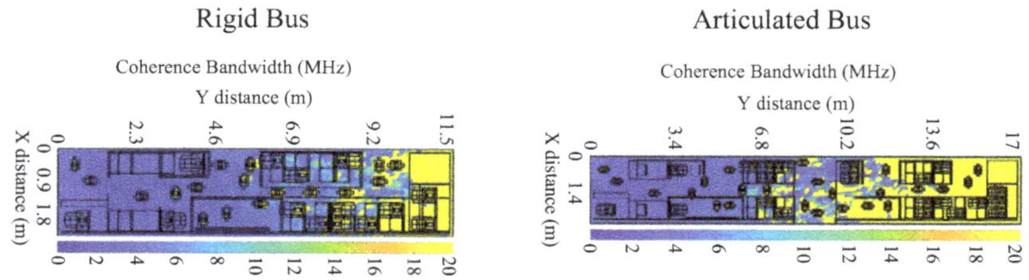

Figure 4.15 Coherence bandwidth for the rigid bus (left) and articulated bus (right) for high people density

buses for 1.1 m height, showing values that vary from 0 to 20 MHz. It can be seen that bigger values are encountered in the back part of the buses, as there is more distance with the transmitter. These high values are an indicator of channel availability for other transmitters at these positions given that according to Figure 4.11 the RSSI is below the sensitivity and the communication with the considered transmitter is unfeasible, leaving the channel available for its use at that zone. The area with higher values of coherence bandwidth is greater for the articulated bus because of its dimensions (bigger) and the fact that is more densely populated than the rigid bus.

4.1.3 Within a train

In this subsection, a real passenger wagon train car has been implemented in the ray-launching tool to analyse signal propagation within it. The main differences that can be encountered comparing with the case of the buses and the car previously analysed, can be described as follows:

- The train environment is the longest vehicle, approximately 16 m in length.
- The number of obstacles, specifically metallic obstacles within the vehicle, is also significant in this case, which can cause once again high multipath propagation.
- The main difference with the urban buses is that the obstacles distribution within the vehicle is different, having more seats, distributed uniformly along all the length of the wagon train car. The number of bars and handholds is lower in this case, and there are not glass walls within the vehicle as in the buses.
- There is not a metal bouncy structure in the central part of the vehicle as it was the case of the articulated bus.
- In comparison with the car, although the buses have more obstacles and people within them, they also have more open spaces as the dimensions (height and length) are bigger.
- All these factors could lead to different fast fading behaviour than other vehicles. Thus, radio wave propagation within these vehicles should also be fully examined.

The dimensions of the wagon train car are (15.46 m × 2.48 m × 2.18 m). Figure 4.16 represents the two scenarios implemented for simulation along with the people densities considered within them. Like in the buses, a low- and high-density conditions have been considered in the wagon train car, which represent the 25% and 75% of the maximum capacity of passengers of the vehicle. The main difference with the buses is that, in this case, only seated people have been considered, as this type of vehicle is not designed to transport stand people. The maximum capacity of seated people within the wagon train is 74 passengers.

For the simulations, a transmitter antenna has been placed in the central part of the vehicle ($X = 3$ m, $Y = 1.24$ m, $Z = 1.1$ m). In this case, this position has been selected to analyse the influence of the transmitter antenna location and

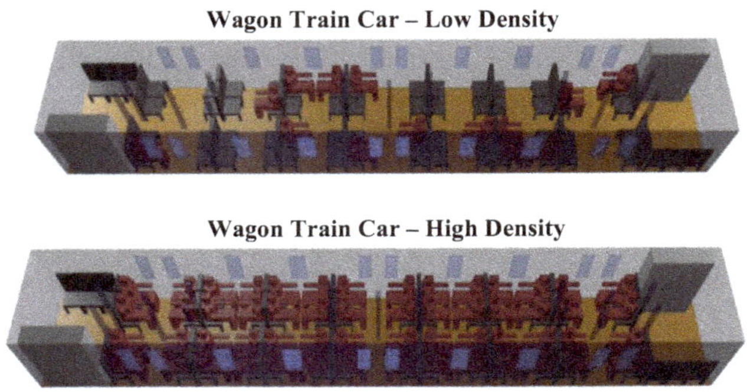

Figure 4.16 Wagon train car scenario implemented for simulation with different passenger densities

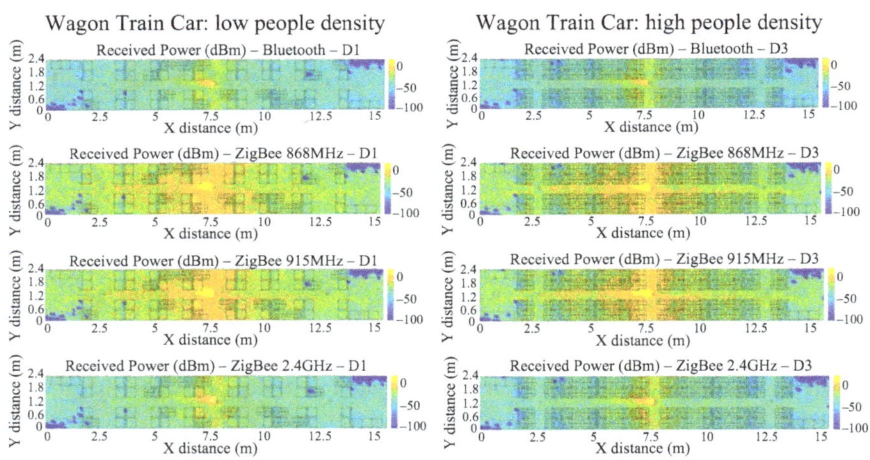

Figure 4.17 Comparison of received power (dBm) for the bi-dimensional planes of 1.1 m height for the different considered systems within the wagon train car: low people density (left) and high people density (right)

the impact in the received signal. Simulation parameters can be seen in Table 4.1.

4.1.3.1 Received signal strength indicator

In Figure 4.17, we can see the power cuts received in the wagon train car at 1.1 m height, which is the same height where the transmitter has been placed. The comparison between different human beings' densities can be analysed. In the simulations, results are obtained for all spatial points within the scenario, but only a single height (1.1 m) is shown here for terms of comparison. The following points can be noticed from Figure 4.17:

- As in the previous cases of the car and the urban buses, the received power levels vary considerably within the wagon train car scenario, with the highest values close to the transmitting antenna for all analysed systems. In this specific case of the wagon train car scenario, it can be seen that received signal levels are greater than the receiver sensitivity for the entire train car, unlike the previous cases of the buses, in which the back part of the vehicles does not reach signal levels. This is because, in this case, the transmitter has been placed in the central part of the wagon train car, which means less distance to arrive to the extremes of the vehicle. In addition, the bi-dimensional planes shown are for 1.1 m height (same as the transmitter position), which corresponds to the head of the seated people. As in this case, we have not considered stand people within the vehicle, absorption due to the human body has less influence at this represented height. Because of that, signal received power levels are distributed more uniformly throughout the vehicle for this specific height. At the extremes of the wagon train car, the presence of metal cabinets can be observed, where very low signal power levels arrive.
- As in previous cases, the influence that obstacles have on radio wave propagation inside the wagon train car can be appreciated, as they absorb and reflect part of the emitted power, creating areas with lower received power levels and areas with more power levels caused by reflections. Thus, fast fading or multipath propagation is clearly visible within the wagon train car.
- If we compare the different technologies, we can see that for the case of ZigBee 868 we have the highest received power values along the whole plane, since the frequency in this case is the lowest, and therefore, the signal suffers less absorption by obstacles. In the case of ZigBee 915, results are very similar, and no significant differences can be appreciated. However, in the case of Bluetooth and ZigBee 2.4, it can be seen that the received power levels are lower throughout the entire bi-dimensional plane mainly due to two factors: the use of a higher frequency and the lower transmission power used in these systems.
- In this case, the differences that can be appreciated between the different people densities are not as significant as in the previously analysed cases (urban transportation buses). This is because the analysed height has not as much as obstacles as the buses, as it has less handholds, the same type of seats and only seated people within the wagon train car.

4.1.3.2 Multipath propagation

In order to gain insight into the multipath propagation phenomena within the wagon train car, we have considered the worst case, which corresponds to a high people density within the passenger wagon. As stated before, the main difference with the previously analysed cases of the urban buses is that in the wagon train car, all people considered is seated. The specific case of ZigBee 2.4 GHz has been analysed.

In Figure 4.18, a power delay profile for a spatial point placed in the back part of the wagon train is presented exhibiting echoes in a time span of approximately 0–80 ns. As can be seen, the number of components that arrives at different instants of time is significantly high. The main difference that we can appreciate if we compare with the urban transportation buses is that in this case the delay for some components is bigger (in order of 30 ns more). This can be explained because for the considered receiver point the distance from the transmitter is bigger than in the buses. Besides, the high number of obstacles embedded in the vehicle makes the signal to have multiple reflections, but not having standing people within it (only seated people) leads to less signal absorption at the analysed height (1.1 m), which makes that more multipath components achieve the receiver at this specific considered point.

Once again, these results confirm that the morphology and topology of the considered scenario, as well as the number of obstacles and people within it, have a significant impact in multipath propagation, provoking that different multipath components arrive to a specific point depending on the surrounding environment.

Figure 4.19 shows the rms delay spread for the wagon train car for high people density. With the rms delay spread, we can see how the multipath propagation affects the bi-dimensional plane at height 1.1 m in the train wagon car. It is observed that in the vicinity of the transmitter (centre of the wagon train car), the rms delay spread is lower because the signal does not impact with so many obstacles as the transmitter is placed in an open area just in the middle of the vehicle (see Figure 4.16). However, as the signal propagates throughout the wagon

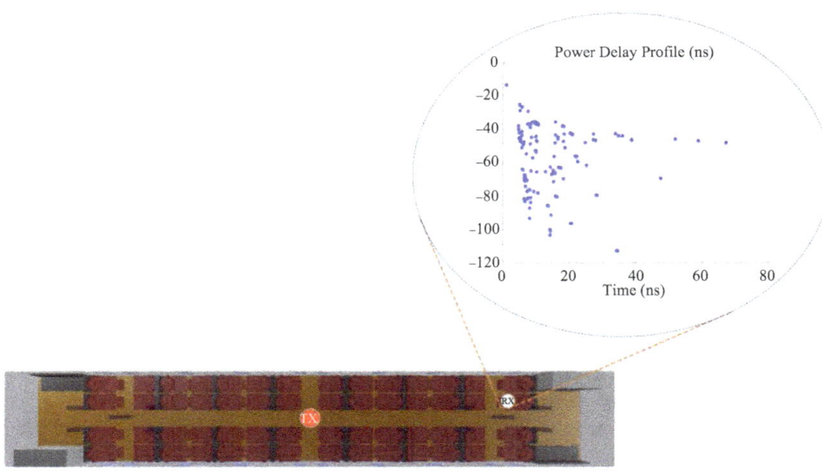

Figure 4.18 Power delay profile for the wagon train car for high people density

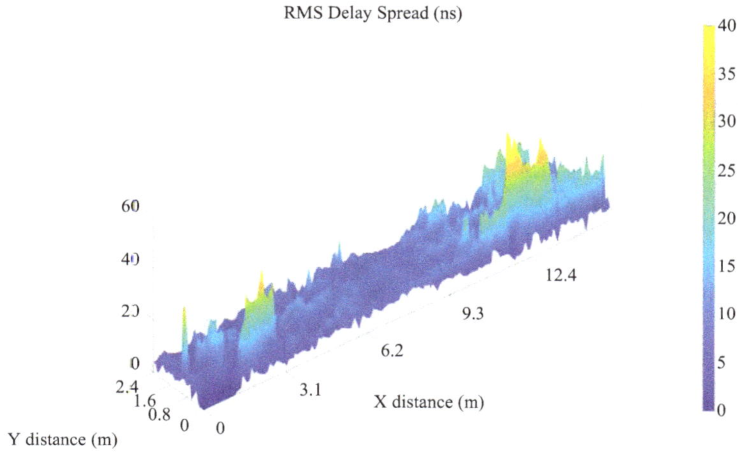

Figure 4.19 RMS delay spread for the wagon train car for high people density

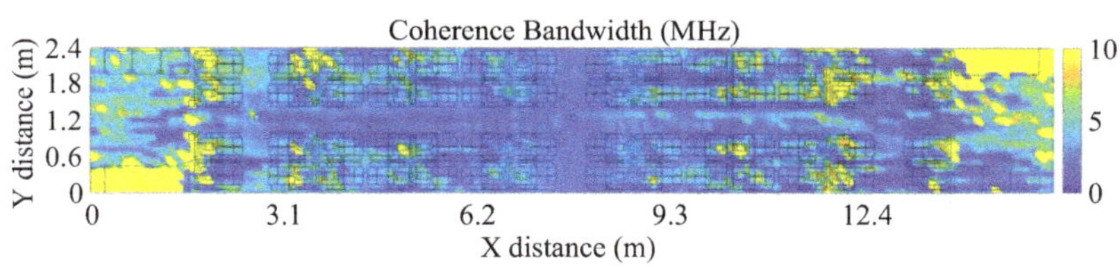

Figure 4.20 Coherence bandwidth for the wagon train car for high people density

car train, the signal impacts with more obstacles and people making that higher values of rms delay spread can be encountered at some points.

4.1.3.3 Frequency domain channel description

At this point, the coherence bandwidth has been analysed within the wagon train car. As stated before, coherence bandwidth is the range of frequencies over which two frequency components have a strong potential for amplitude correlation, and it is a defined relation derived from the rms delay spread [12]. The frequency correlation function has been considered above 0.9 for the analysis of coherence bandwidth. Figure 4.20 depicts the coherence bandwidth values and shows its inverse relationship with the rms delay spread, showing higher values in the extremes of the vehicle and lower values in the central and middle part of the vehicle, mainly due to the strong multipath environment which means high channel occupancy by the transmitter. Coherence bandwidth analysis is complementary with the RSSI previous analysis in order to contrast the channel availability with the received power.

4.2 Case studies for intra-vehicle electromagnetic properties in ITS

In this subsection, two different case studies for intra-vehicle communication systems within an underground tunnel has been assessed: slow-moving vehicles and fast-moving vehicles. Figure 4.21 presents the considered modelled scenario, which is an underground tunnel of 100 m length, 9 m wide, and 13 m height, where two lanes of cars can circulate at the same time. The car model presented in Section 4.1.1 has been embedded within the tunnel, analysing a random distribution of cars within the tunnel. The car model with four people inside has been considered to assess a more realistic scenario. The tunnel walls, as well as the floor scenario, have been considered as a concrete material, taking into account the different material properties. Table 4.6 presents the material parameters used in the simulation models [10].

The 3D ray-launching tool has been used for the analysis of the time-varying nature of the channel in this scenario caused by either relative motion between the transceivers or the movement of objects in the channel (other vehicles). The most important aspect of this analysis is that the vehicles are inside a tunnel, which can be considered as a unique indoor environment, with more challenging propagation characterization, which need to be fully examined. The non-stationarity of the channel has been assessed considering different velocities for the mobiles, leading to Doppler shift and Doppler spread results, providing insight into the performance of the wireless communication system. In chapter 2, section 2.2, propagation phenomena associated with the non-stationarity of the channel are presented.

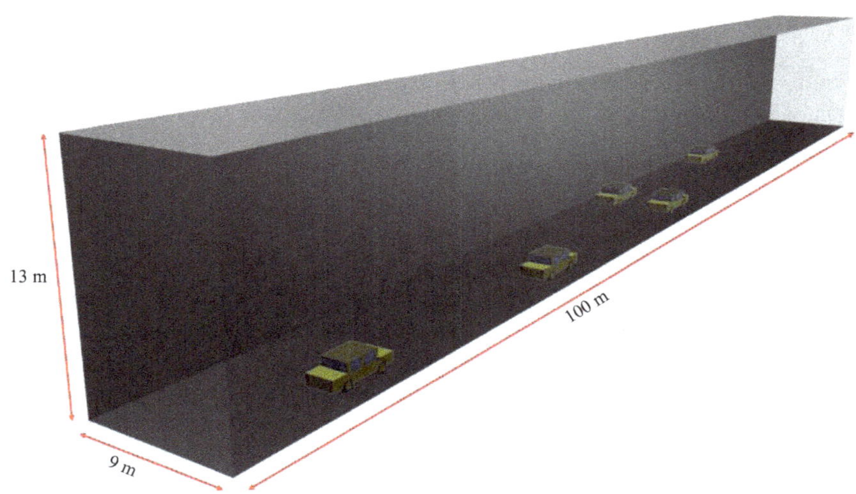

Figure 4.21 Tunnel scenario with vehicles implemented in the 3D ray-launching tool

Table 4.6 Electromagnetic material properties used for the selected frequencies

Material	Parameters			
	Frequency (MHz)			
	868–915		2,400	
	Permittivity (ε_r)	Conductivity (σ)	Permittivity (ε_r)	Conductivity (σ)
Air	1	0	1	0
Polycarbonate	3.2	0.2	3	0.2
Aluminium	4.5	4×10^7	4.5	4×10^7
Glass	5.5	10^{-12}	6.06	10^{-12}
Concrete	8.1	0.02	8.1	0.02

In the following subsections, we present the electromagnetic propagation signal analysis associated with the non-stationarity of the channel in terms of Doppler shift and Doppler spread for different velocities of the vehicles. For that purpose, the same WPAN as in previous sections based on ZigBee and Bluetooth technology has been considered. These technologies have been selected to present a typical case currently considered for an intra-vehicular system. Simulation parameters considered in the 3D ray-launching algorithm, as well as the parameters for the considered technologies, are presented in Table 4.1.

4.2.1 Fast-moving vehicles (in underground tunnels)

As a starting step, several simulations of the considered scenario have been performed considering different instants of time denoted as t_i, $i=1, \ldots, 5$, for the selected vehicles position. As an example, Figure 4.21 presents the instant of time t_2. In this case, two different velocities have been considered for the vehicles: 22.22 m/s (80 km/h) and 27.77 m/s (100 km/h). The transmitter has been placed within the car just on the dashboard in the front part of the vehicle.

First, we present the received power for a bi-dimensional plane of 1 m for two instants of time for the considered scenario, as is presented in Figures 4.22 and 4.23 for t_2 and t_4, respectively. The influence of the movement of the vehicle in signal fading and changes in the received power in the considered environment are shown. Nevertheless, in this case of intra-vehicle communications, it is observed that received power levels are above the sensitivity levels within the vehicle, and the surrounding vehicles received few signal levels unless the distance with the transmitter systems is very small. Besides, other vehicles and the tunnel walls are going to affect signal propagation, but the impact is higher when analysing inter-vehicle communications, which is presented in chapter 5, section 5.2.

To get knowledge about the time-varying nature of the channel caused by the different velocities of the vehicles considered, the Doppler shift (f_d) at the receiver can be obtained by

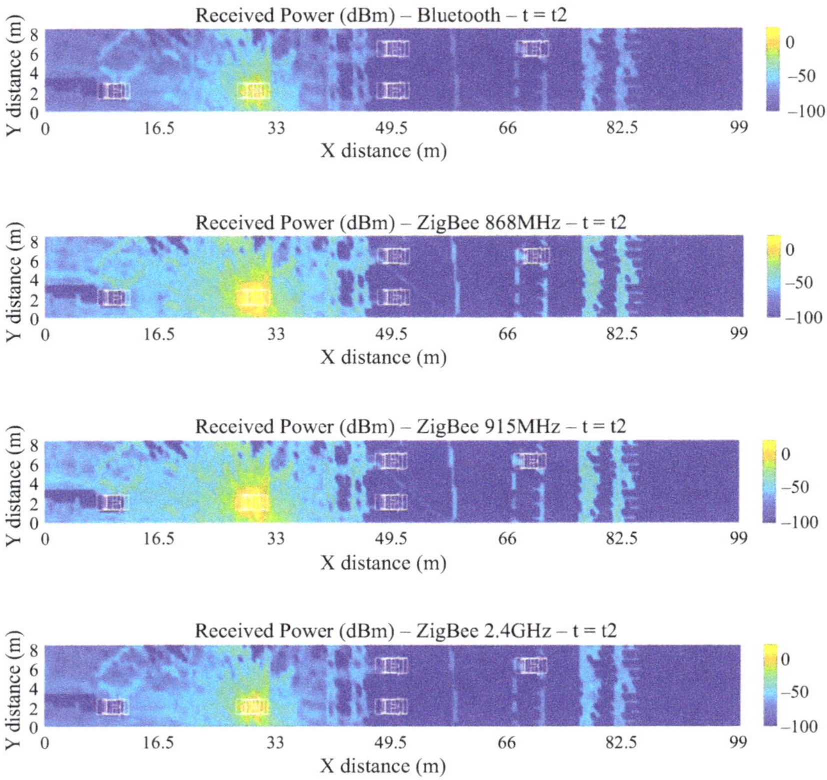

Figure 4.22 Comparison of received power (dBm) for the bi-dimensional planes of 1 m height for the different considered systems within the tunnel at the instant of time t_2

$$f_d = \frac{v}{\lambda} \cos(\psi) \tag{4.1}$$

where ψ represents the angle between the direction of arrival of the specific examined propagation path and the receiver's movement, v is the speed of the vehicle and λ is the wavelength of the signal.

Figure 4.24 presents the Doppler shift for a specific point behind the passenger seat within the vehicle for the considered scenario for two different vehicle velocities: 22.22 m/s (80 km/h) and 27.77 m/s (100 km/h) for the instant of time t_2. It can be seen that for 22.22 m/s velocity, we can observe that the maximum Doppler shift is 150 Hz for ZigBee 2.4 GHz and Bluetooth systems, which use a higher frequency band than the other considered systems. This means that systems working in 2.4 GHz frequency bands have more impact on the Doppler shift than systems in lower frequency bands (868 and 915 MHz).

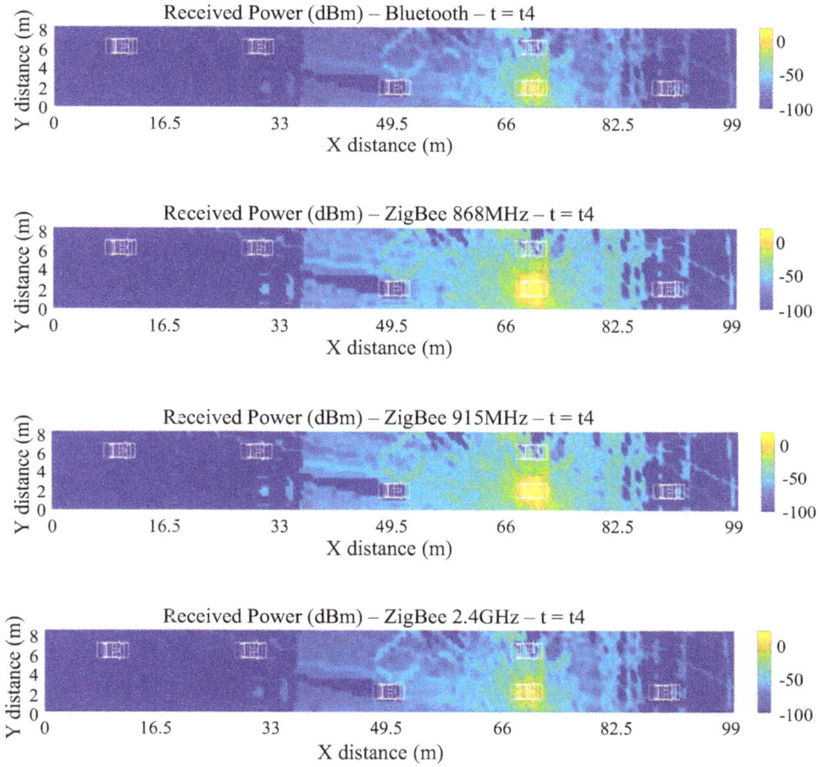

Figure 4.23 Comparison of received power (dBm) for the bi-dimensional planes of 1 m height for the different considered systems within the tunnel at the instant of time t_4

We can also observe that if we increase the speed to 27.77 m/s (100 km/h), the Doppler shift also increases, being the maximum for the 2.4 GHz frequency band in approximately 180 Hz.

To get an insight of the impact of the Doppler shift in the considered scenario, the Doppler spread has been calculated for all the spatial samples of the scenario and it is shown in Figure 4.25. The Doppler spread has been defined for each spatial sample as the difference of the maximum and the minimum frequency shifts. The entire tunnel has been considered for the analysis, but we have previously seen that the transmitted signal only arrives in the vicinity of the vehicle, causing a frequency shift at receiver points inside the car, and in nearby vehicles, but from a certain distance, there is no frequency shift since there is no signal level.

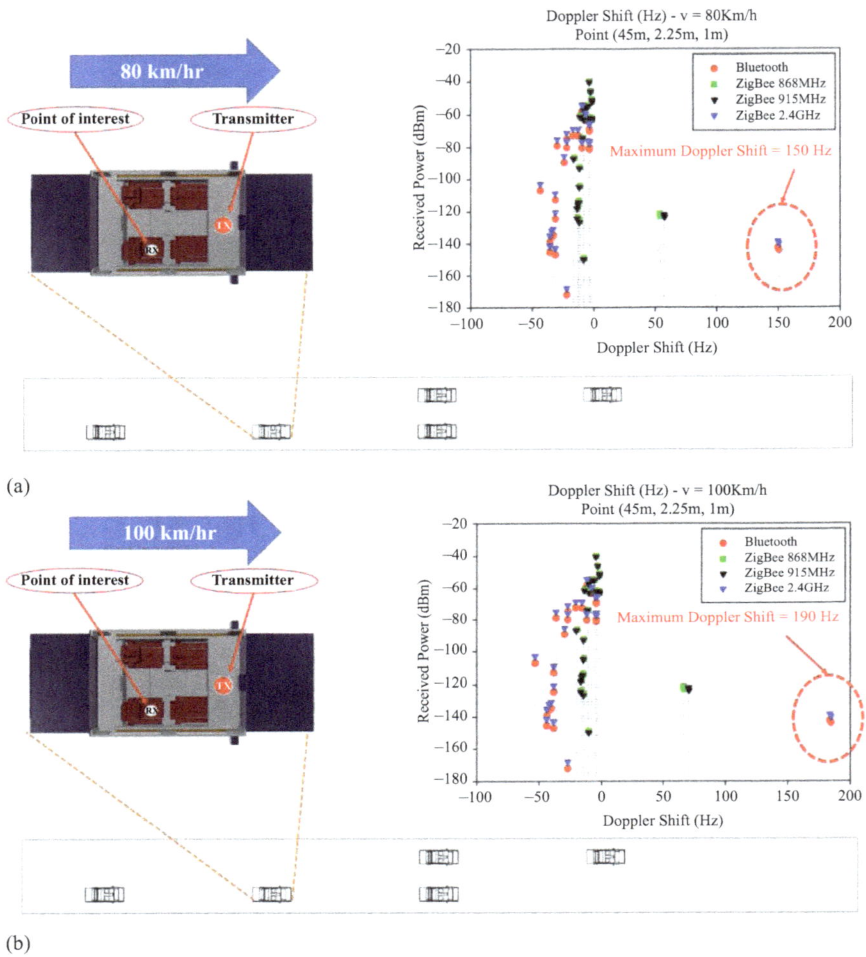

Figure 4.24 Doppler shift (Hz) for fast-moving vehicles within the tunnel at an instant of time t_2; (a) 80 km/h and (b) 100 km/h

For the case of 27.77 m/s (100 km/h), we can see that values of Doppler spread are higher than for the speed of 22.22 m/s (80 km/h), which verify that an increase in the velocity of the vehicle will imply a higher frequency variation.

It can also be observed that for the technologies with higher frequency (2.4 GHz), Doppler spread levels are greater than for ZigBee 868 and Zigbee 915. This frequency shift can affect the reception of the signal for underground environments, depending on the speed of the vehicle.

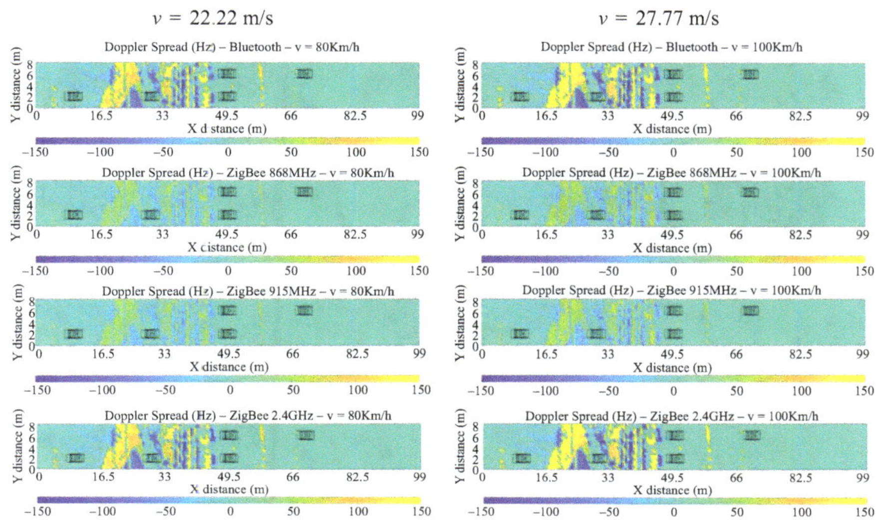

Figure 4.25 Doppler spread (Hz) for fast-moving vehicles within the tunnel at the instant of time t_2

4.2.2 Slow-moving vehicles (in underground tunnels)

In order to compare with different velocities, Doppler shift and Doppler spread have been analysed for the same specific point within the vehicle for the considered scenario for slow-moving vehicles: 5.55 m/s (20 km/h) and 11.11 m/s (40 km/h). Figure 4.26 presents the Doppler shift for the same point as Figure 4.24 but for slow-moving vehicles for the instant of time t_2. It can be seen that for 5.55 m/s velocity, we can observe that the maximum Doppler shift is 38 Hz for ZigBee 2.4 GHz and Bluetooth systems, which use a higher frequency band than the other considered systems. This means that systems working in the 2.4 GHz frequency band have more impact in the Doppler shift than systems in lower frequency bands (868 and 915 MHz). We can also observe that if we increase the speed to 11.11 m/s (40 km/h), the Doppler shift also increases, being the maximum for 2.4 GHz frequency band in approximately 78 Hz.

It can be seen that for slow-moving vehicles, Doppler shift is significantly lower than that of fast-moving vehicles. Thus, it is highly important to consider the non-stationarity of the channel in the design of the transceiver when it is going to be subjected to high speeds.

To get an insight of the impact of the Doppler shift in the considered scenario, the Doppler spread has also been calculated for all the spatial samples of the scenario for slow-moving vehicles and is shown in Figure 4.27.

As in the previous case of fast-moving vehicles, we can see an increase of the Doppler spread when higher velocities are considered. Nevertheless, in this case, as

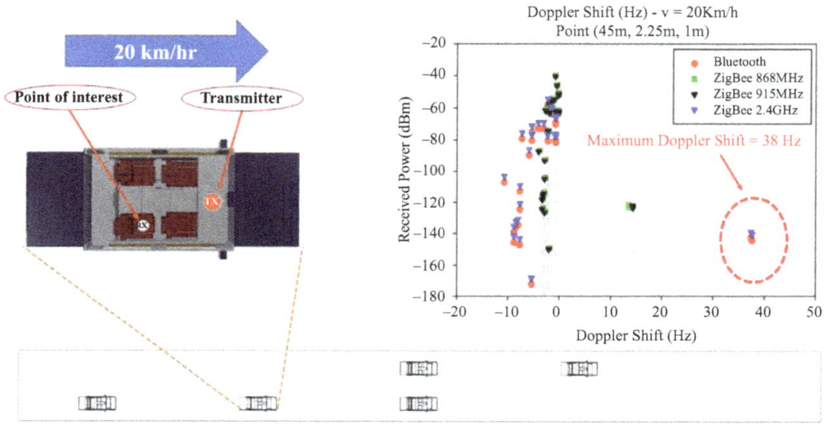

(a)

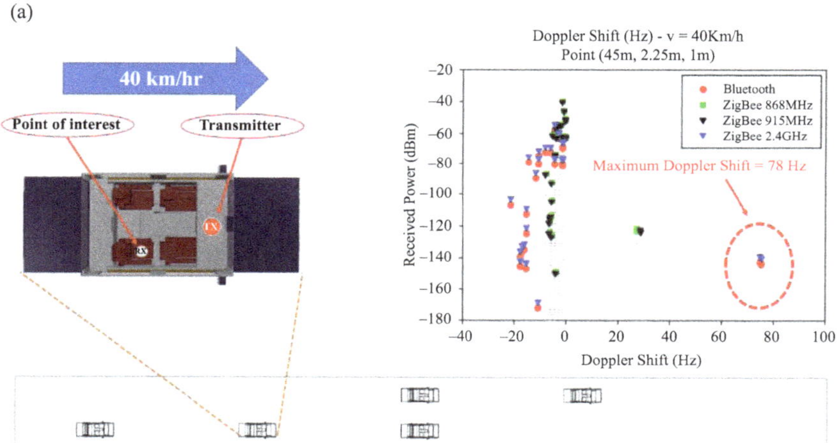

(b)

Figure 4.26 Doppler shift (Hz) for slow-moving vehicles within the tunnel at the instant of time t_2: (a) 20 km/h and (b) 40 km/h

the velocity is lower, Doppler spread values also have lower values than fast-moving vehicles case, in the order of approximately 100 Hz. Note that the scale in Figure 4.27 is different from Figure 4.25. For higher velocities, which is the case of Figure 4.25, the presented Doppler spread ranges from −150 to 150 Hz, while for lower velocities, which is the case of Figure 4.27, the presented Doppler spread ranges from −100 to 100 Hz.

In order to have a better knowledge about the comparison between the different considered systems and different velocities within the tunnel, Figure 4.28 shows the Doppler shift comparison for the different communication systems considered and different velocities within the tunnel at the instant of time t_2.

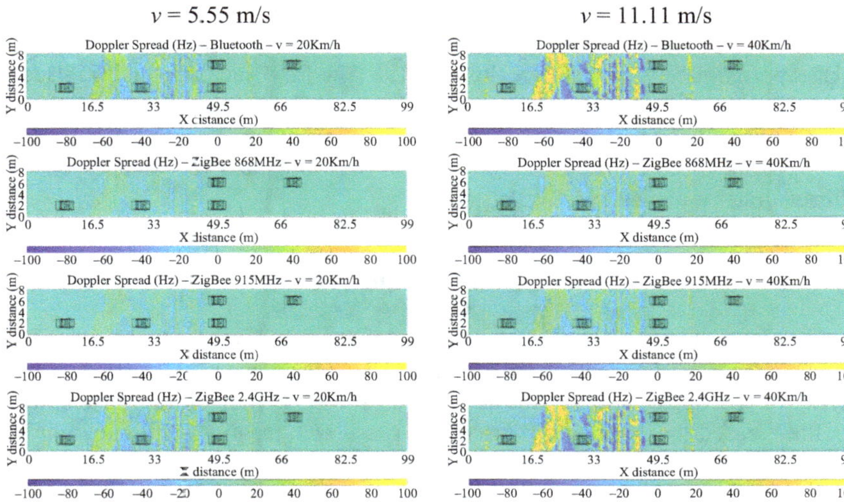

Figure 4.27 Doppler spread (Hz) for slow-moving vehicles within the tunnel at the instant of time t_2

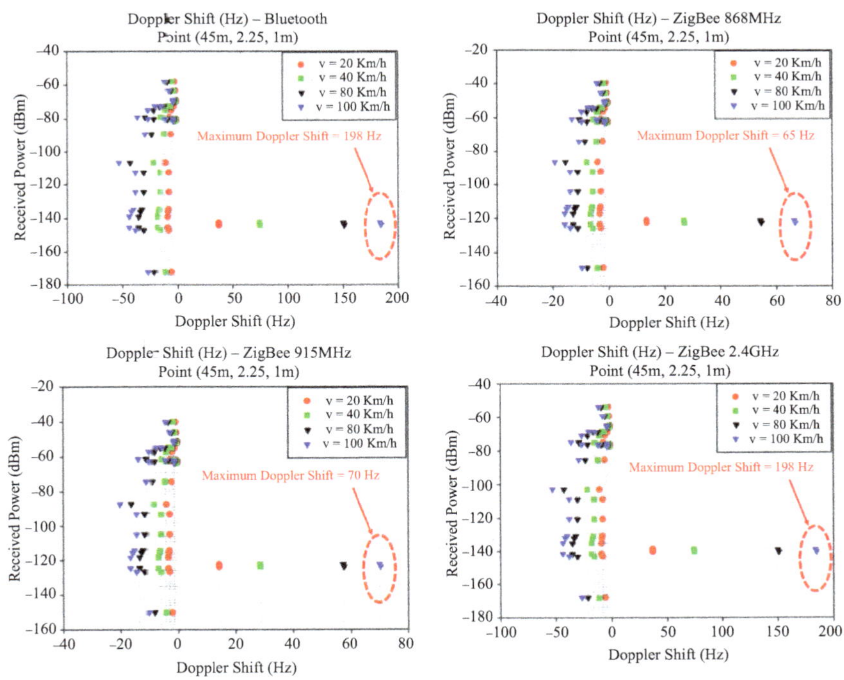

Figure 4.28 Doppler shift (Hz) comparison for different communication systems and different velocities within the tunnel at the instant of time t_2

It is observed that for all the systems analysed, the highest Doppler shift is always when the highest speed is considered: 27.77 m/s (100 km/h), varying from 70 Hz approximately for ZigBee 868 to 198 Hz for Bluetooth and ZigBee at 2.4 GHz. The lowest Doppler shift is when we consider the lowest speed: 5.55 m/s (20 km/h) varying from 18 Hz for ZigBee 868 to 40 Hz for Bluetooth and ZigBee. It is important to note that these values have been obtained for vehicles moving within a tunnel with concrete walls.

4.3 Intra-vehicle radio planning analysis

In this section, radio planning analysis within the different vehicles presented before has been performed in terms of potential communication transceiver placement for the deployment of the considered systems. In addition, an empirical modelling based on simulation results is presented for the different analysed vehicles. The worst-case condition has been considered for this analysis for all the vehicles, in terms that four people have been considered within the car, and high people density for the buses and for the wagon train car have been considered.

4.3.1 Empirical modelling

The path loss (PL) have been analysed and modelled under the different considered systems presented in Table 4.1 for the different vehicles (car, buses and wagon train car). PL is defined as the ratio of the effective transmitted power to the received power, calibrating out cable losses and antenna gains. The modelling method used for PL, taking into account the complex environment of the vehicles, is given by the empirical model of log-distance [12] expressed as

$$\mathrm{PL}(d) = \mathrm{PL}_0 + 10n\log_{10}(d) + X_\sigma \tag{4.2}$$

where PL_0 is the intercept, d is the transmitter–receiver separation distance in meters and n is the PL exponent dependent on the specific propagation environment indicating the rate at which PL increases with distance. In free-space propagation, the exponent n is 2. X_σ denotes shadow fading with a zero-mean normally distributed random variable with standard deviation σ (dB). The values of PL_0 and n are extracted by using the linear regression method. Estimated PL values have been calculated within the vehicles and compared with the lognormal distribution. From the results of the received power stated previously, PL can be predicted for each spatial point of the environment applying:

$$\mathrm{PL}(d)(\mathrm{dBm}) = P_t(\mathrm{dBm}) - P_r(\mathrm{dBm}) \tag{4.3}$$

With the antenna gains included in P_t(dBm) and P_r(dBm).

4.3.1.1 Within a car

The modelling results for the car for the Bluetooth Class 2 system case are shown in Figure 4.29 for a height of 1.1 m with respect to the transmitter–receiver

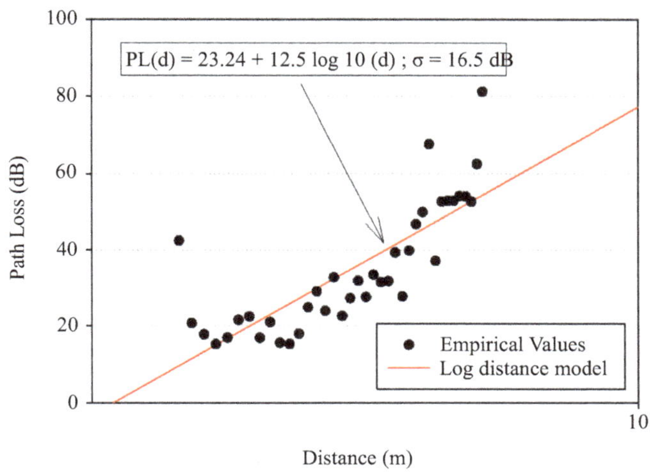

Figure 4.29 Path loss versus transmitter–receiver separation with linear regression fit for Bluetooth Class 2 system

Table 4.7 Empirical modelling within a car

System	**PLo (dB)**	***n***	***σ* (dB)**
Bluetooth Class 2	23.24	1.25	16.55
ZigBee 868	43.60	1.30	17.14
ZigBee 915	42.46	1.29	17.04
ZigBee 2400	27.25	1.25	16.55

separation for a 2.4 GHz frequency of operation. Additionally, the linear regression line, resulting from a minimum mean square error analysis, is shown, which is expressed as

$$\mathrm{PL}(d) = 23.24 + 12.5\log_{10}(d); \sigma = 16.5\ \mathrm{dB} \tag{4.4}$$

The corresponding path loss exponent is $n = 1.25$ with a standard deviation $\sigma = 16.5$ dB. Table 4.7 presents the different values of PL_0, n and σ for the different systems analysed. It is noteworthy that n values are below the free-space path loss exponent. This indicates that a larger amount of energy is available due to multipath propagation in addition to the line-of-sight (LoS) component, leading to a smaller PL than for free-space propagation conditions.

In Figure 4.30, an aerial view of the car is presented with dashed lines along the *X*-axis for different values of *Y*. These dashed lines represent the lines that correspond to the linear distribution lines of received power, which are represented in Figure 4.31, for 1 m height. It can be seen that there are locations where signal

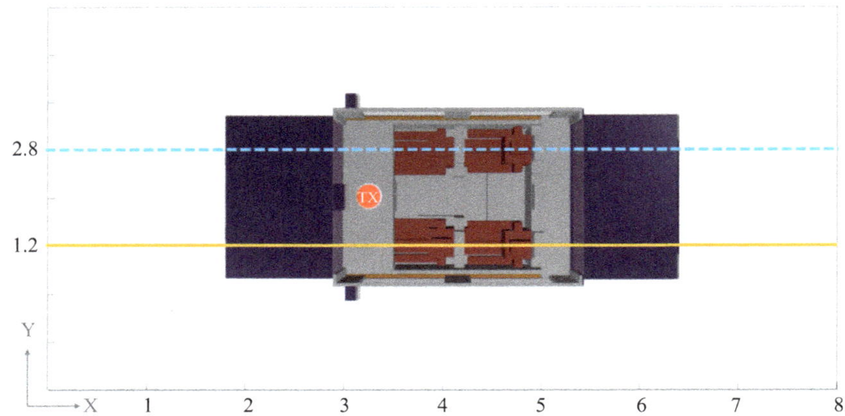

Figure 4.30 Radial lines considered for the representation of linear distribution of received power levels in Figure 4.28

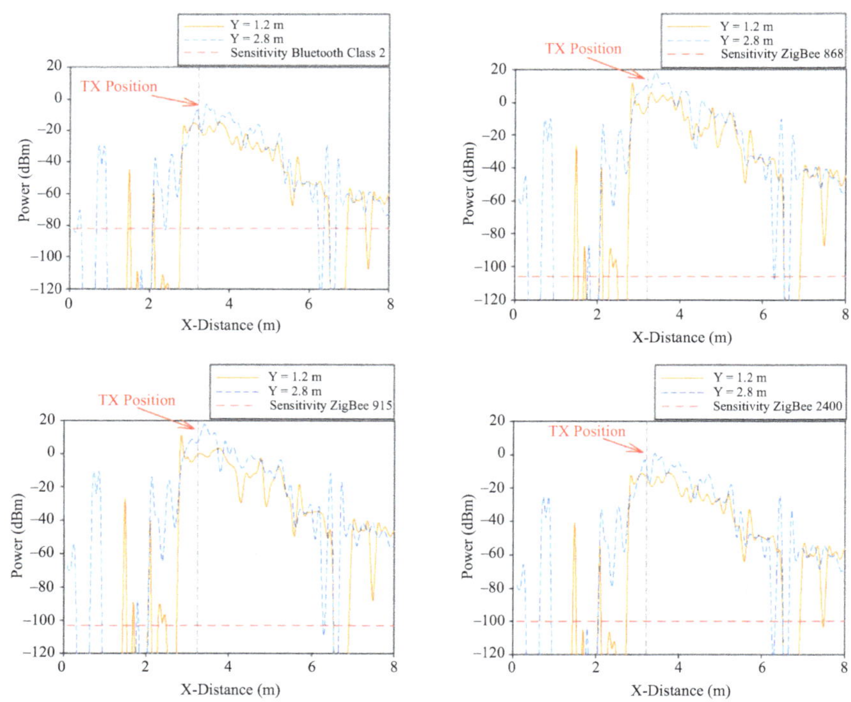

Figure 4.31 Comparison of linear distribution of received power levels for the different considered systems within the car

levels decrease below the sensitivity level, but these locations correspond to the trunk of the car and the engine. For all cases, all received power levels for the spatial points within the body car are above the sensitivity level, which means that with one transmitter coverage will be achieved within all the body car for all considered systems.

4.3.1.2 Within a bus

The PL modelling results for the rigid and articulated buses for the different considered systems are presented in Tables 4.8 and 4.9, respectively. These models can be used for radio planning purposes before the deployment of a wireless communication system within the buses. As in the previous case of the car, the resulting PL exponent for both types of buses is below the free-space PL exponent, showing once again the considerable impact of multipath propagation in these indoor vehicular environments. It can also be appreciated that the standard deviation for the shadow fading phenomena is 10 dB approximately bigger for the case of the articulated bus, due mainly to a higher number of obstacles and people within it.

In Figure 4.32, an aerial view of the rigid and articulated buses is presented with dashed lines along the *X*-axis for different values of *Y*. These dashed lines represent the lines that correspond to the linear distribution lines of received power which are represented in Figures 4.33 and 4.34, for 1.1 m height. It can be seen that there are locations after a certain distance where signal levels decrease below the sensitivity level. To achieve coverage in these points, additional transmitter will be needed within both types of buses for all the considered systems.

Table 4.8 Empirical modelling within the rigid bus

System	PLo (dB)	n	σ (dB)
Bluetooth Class 2	3.35	1.13	17.41
ZigBee 868	28.26	1.27	17.49
ZigBee 915	26.29	1.25	17.87
ZigBee 2400	8.06	1.15	17.40

Table 4.9 Empirical modelling within the articulated bus

System	PLo (dB)	n	σ (dB)
Bluetooth Class 2	11.67	1.07	27.24
ZigBee 868	33.48	1.16	26.11
ZigBee 915	33.04	1.15	26.21
ZigBee 2400	16.54	1.09	27.22

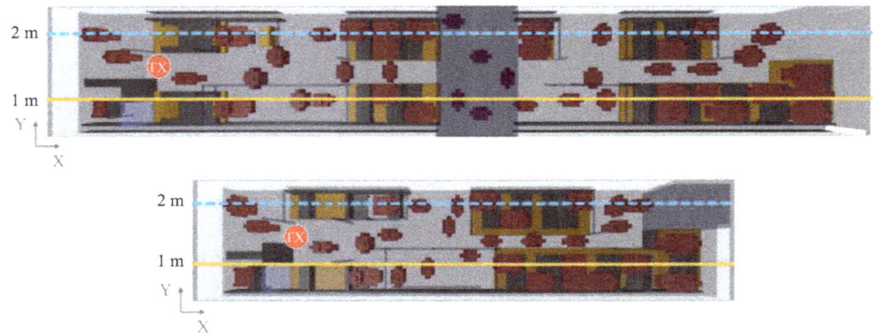

Figure 4.32 Radial lines considered for the representation of linear distribution of received power levels in Figure 4.30 (rigid bus) and Figure 4.31 (articulated bus)

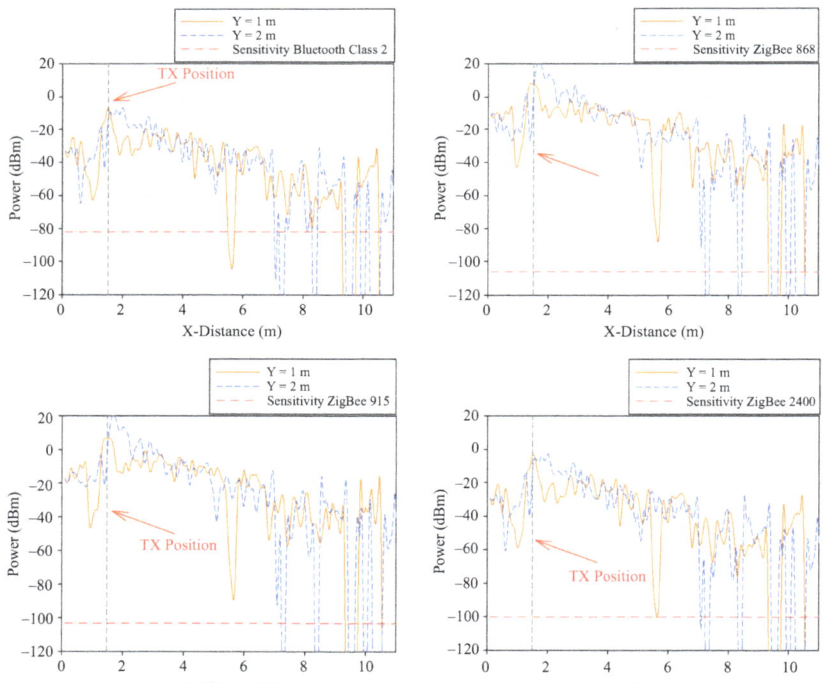

Figure 4.33 Comparison of linear distribution of received power levels for the different considered systems within the rigid bus

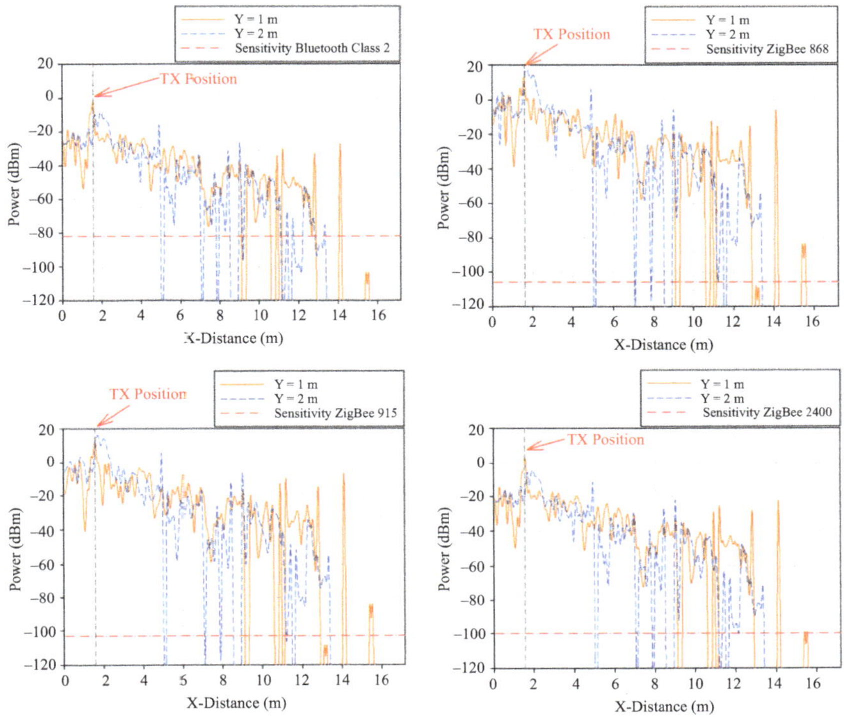

Figure 4.34 Comparison of linear distribution of received power levels for the different considered systems within the articulated bus

Table 4.10 Empirical modelling within the train

System	PLo (dB)	n	σ (dB)
Bluetooth Class 2	27.76	1.49	11.09
ZigBee 868	7.17	1.84	12.01
ZigBee 915	9.96	1.64	11.60
ZigBee 2400	23.69	1.51	11.11

4.3.1.3 Within a train

The PL modelling results for the wagon train car for the different considered systems are presented in Table 4.10. As in the previous cases of the car and the buses, the resulting PL exponent for both types of buses is below the free-space PL exponent, showing once again the considerable impact of multipath propagation in these indoor vehicular environments. In this event, it must be considered that the transmitter antenna has been placed at a central point of the wagon train car, and people within the wagon train are considered only seated. Because of that, in this case the standard deviation for the shadow fading phenomena is lower than in the other cases due to the signal impact with a smaller number of obstacles at the considered height.

In Figure 4.35, an aerial view of the wagon train car is presented with dashed lines along the *X*-axis for different values of *Y*. These dashed lines represent the lines that correspond to the linear distribution lines of received power, which are represented in Figure 4.36, for 1.1 m height. It can be seen that

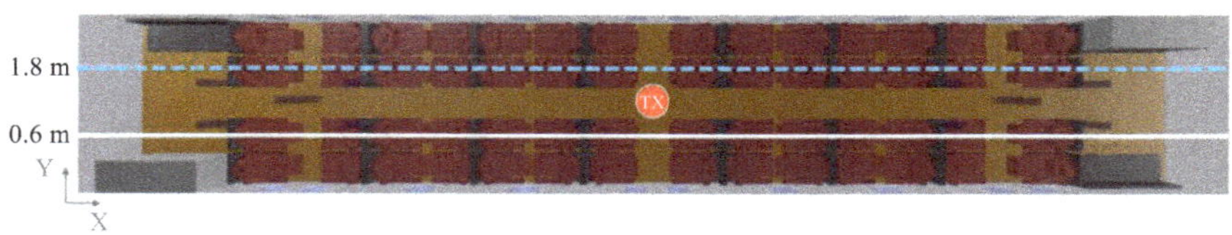

Figure 4.35 Radial lines considered for the representation of linear distribution of received power levels in Figure 4.33

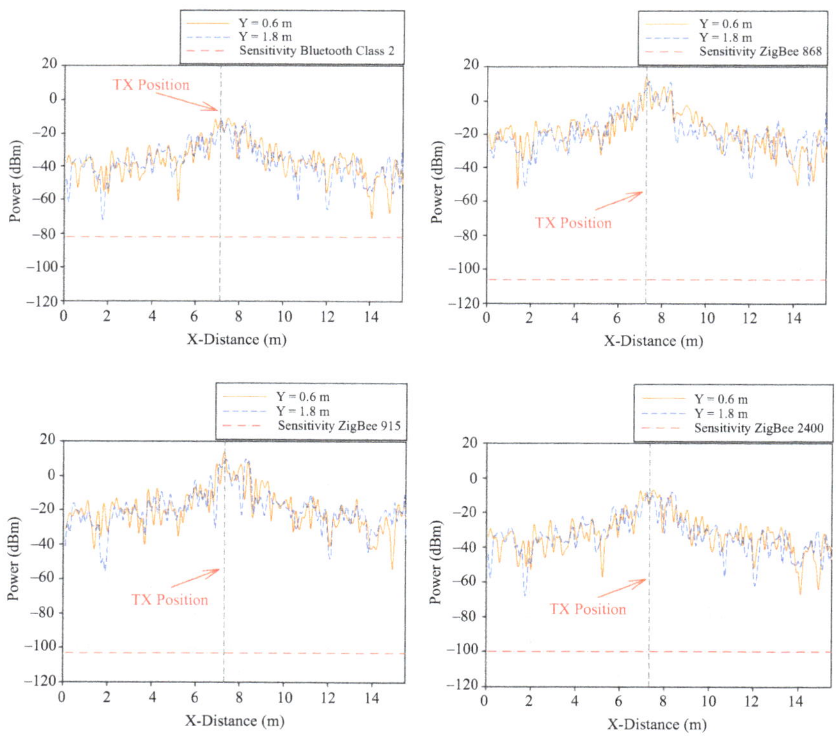

Figure 4.36 Comparison of linear distribution of received power levels for the different considered systems within the wagon train car

in this case received power levels are above sensitivity levels for all the considered systems. This is because of two factors: the location of the transmitter antenna, which has been placed at the centre of the wagon train car, and the considered height, which corresponds to the head of the seated people, which cause less absorption than the previous cases, where stand people have been considered.

References

[1] Aragón-Zavala, A., *Indoor Wireless Communications: From Theory to Implementation*, Wiley, New York, NY, USA, 2017.

[2] Azpilicueta, L., Rawat, M., Rawat, K., Ghannouchi F., and Falcone, F., 'Convergence analysis in deterministic 3D ray launching radio channel estimation in complex environments', *ACES Journal*, 2014, 29(4): 256–71.

[3] Granda, F., Azpilicueta, L., Vargas-Rosales, *et al.*, 'Deterministic propagation modeling for intelligent vehicle communication in smart cities', *Sensors* 2018, 18, 2133.

[4] Azpilicueta, L., Lopez-Iturri, P., Aguirre, E., Vargas-Rosales, C., León, A., and Falcone, F., 'Influence of meshing adaption in convergence performance of deterministic ray launching estimation in indoor scenarios', *Journal of Electromagnetic Waves and Applications*, 2017, 31(5): 544–59.

[5] Aguirre, E., Lopez-Iturri, P., Azpilicueta, L., *et al.*, 'Design and implementation of context aware applications with wireless sensor network support in urban train transportation environments', *IEEE Sensors Journal*, 2017, 17(1): 169–178.

[6] Azpilicueta, L., Vargas-Rosales C., and Falcone, F., 'Deterministic propagation prediction in vehicular environments', *IEEE Vehicular Technology Magazine*, 2016, 11(3): 29–37.

[7] Azpilicueta, L., Rawat, M., Rawat, K., Ghannouchi, F., and Falcone, F., 'A ray launching-neural network approach for radio wave propagation analysis in complex indoor environments', *IEEE Transactions on Antennas and Propagation*, 2014, 62(5): 2777–86.

[8] Azpilicueta, L., Falcone, F., and Janaswamy, R., 'A hybrid ray launching-diffusion equation approach for propagation prediction in complex indoor environments', *IEEE Antennas and Wireless Propagation Letters*, 2017, 16: 214–17.

[9] Casino, F., Azpilicueta, L., López-Iturri, P., Aguirre, E., Falcone, F., and Solanas, A., 'Optimised wireless channel characterisation in large complex environments by hybrid ray launching-collaborative filtering approach', *IEEE Antennas and Wireless Propagation Letters*, 2017, 16: 780–83.

[10] Balanis, C.A., *Advanced Engineering Electromagnetics*, Wiley, New York, NY, USA, 1989, 205.

[11] Aguirre, E., Arpón, J., Azpilicueta, L., De Miguel, S., Ramos, V., and Falcone, F., 'Evaluation of electromagnetic dosimetry of wireless systems in complex indoor scenarios with human body interaction', *PIER B*, 2012, 43: 189–209.

[12] Rappaport, T., *Wireless Communications: Principles and Practice*, Prentice Hall Communications Engineering and Emerging Technologies Series, 2nd edn, Prentice-Hall PTR, Upper Saddle River, NJ, USA, 2002.

Chapter 5
Inter-vehicle short-range wireless channel characterization

In this chapter, an analysis of radio wave propagation in different inter-vehicle complex scenarios is presented. The inter-vehicle scenarios that have been analysed are both urban and rural environments. The main electromagnetic phenomena that occur in this type of scenarios are assessed, including the non-stationarity of the channel, which is an important factor in these types of environments. Two specific case studies to analyse the electromagnetic propagation properties for inter-vehicle communications in intelligent transportation systems are presented, the first one being fast-moving vehicles in underground tunnels and the second one slow-moving vehicles in underground mines and caves scenarios. At the end of the chapter, the optimal position of the transceivers is proposed for the different environments based on simulation, as well as some empirical modelling to be used for this type of wireless communications systems in inter-vehicle scenarios.

5.1 Inter-vehicle scenarios radio wave propagation analysis

One of the key aspects of Intelligent Transportation Systems (ITS) is the wireless communications between cars, typically called vehicle-to-vehicle communication (V2V), or between cars and a fixed infrastructure, referred to as vehicle-to-infrastructure (V2I) communication. The general term to refer to these communications is V2X communications.

In the last decade, V2I applications such as toll-booth payments have been implemented mostly in the unlicensed band of 915 MHz. Nevertheless, there are additional applications that can be supported with V2I communications, related to safety and service or infotainment, such as road-side units (RSUs) in intersections that could alert a driver as to other cars are approaching the intersection, although they may not be visible to the driver.

V2V communications also allow several desirable possible applications. For example, sudden braking caused by an accident or road hazard could automatically initiate a wireless alert message that could be relayed down the highway to warn approaching drivers of the need to slow and approach with caution. Drivers could also leave the highway and find alternative routes as they would have the information about major congestion or delays caused by accidents or construction, which could be relayed kilometres down the highway.

Hitherto, V2V communications have only been implemented in small-scale experimental configurations. The most important problem for this deployment is the necessity of a significant fraction of the cars on the road equipped with the wireless capability to offer useful functions. The fact that there are few equipped cars on the roadway influence negatively in the motivation for early adopters to purchase wireless capability. Nevertheless, if a government agency or service provider take the first step and install RSU infrastructure, then wireless capability in the car would provide immediate benefits. V2V communication would then be an arising capacity as the installed base of wireless-enabled cars increases.

The V2V and V2I radio channel links are characterized by being highly time-varying and are described through a non-stationary fading process. Propagation characteristics in these environments are very peculiar, caused by different factors such as the large number of scattering objects in the surrounding environment, the high velocities of the transmitter and receiver or the low antenna position in the car. As a result, the deployment of a trustworthy and highly reliable system for vehicular communications is a challenging task. A deep understanding of the underlying radio propagation channel is needed for setting up an efficient vehicular communication system [1].

In this section, radio wave propagation for I2V and V2V communication links have been assessed for two different types of scenarios: urban and rural environment, by means of a ray-launching propagation model, which has been presented in chapter 3, section 3.2.

For the I2V link, the following assumptions have been considered:

- RSU placed at a streetlight at a height of 3.5 m.
- Omnidirectional antenna for transmitter and receiver.
- The cellular V2X (C-V2X) communication link has been analysed considering different frequency bands for 4G (800/1,800/2,600 MHz) and 3G (900/2,100 MHz).
- The IEEE 802.11p standard [2], which is the standard allocated for vehicular communications nowadays, has also been analysed considering its allocated frequency band of 5.9 GHz.

For the V2V link, the following assumptions have been considered:

- Transmitter antenna placed at the rooftop of a vehicle.
- Omnidirectional antenna for transmitter and receiver.
- The IEEE 802.11p standard has been analysed considering its allocated frequency band of 5.9 GHz.

The properties of the V2X channel are different in several significant ways to the cellular communication channel. First, the communication range needed is much shorter, typically hundreds of metres rather than kilometres. Second, a LoS path is present the majority of the time. Third, if the 5.9 GHz frequency band is used, channel behaviour is very different due to differences in diffraction, absorption by foliage, and scattering from fixed-side objects. Finally, the scattering objects are not randomly distributed in an omnidirectional manner, but instead tend to be aligned along the roadway.

For these reasons, it is important to analyse V2X communication links and characterize channel properties in representative environments [3–5].

For the inter-vehicle wireless channel analysis presented in this chapter, two different scenarios have been considered: an urban and a rural environment. The main difference between them is the existence of buildings in the first one. Both scenarios have a main road with four lanes for each side. The specific details of each scenario are presented in the following subsections. All the obstacles and details of the environment have been considered for the simulation, such as cars embedded in the scenario, buildings, roads, streetlights, trees and vegetation. The material properties for all the obstacles have been considered by means of the relative permittivity and conductivity at the frequency of operation. These parameters can be found in [6,7].

For the simulations, different I2V and V2V links have been analysed, as it has been previously explained. Tables 5.1 and 5.2 present simulation parameters considered in the 3D ray-launching algorithm for the I2V and V2V links, respectively.

The receiver sensitivity has been selected according to the minimum sensitivity information reported for some V2X radio-communications commercial products [8–10].

Table 5.1 3D ray-launching simulation parameters for the I2V link

I2V link	Parameters	Values
4G	Transmitter power/position	0 dBm/streetlight
	Frequency/bit rate	800/1,800/2,600 MHz/ 25 Mbps
	Antenna radiation pattern (RX, TX)/gain	Omnidirectional/0 dB
	3D ray launching: angular resolution/reflections/diffraction/cuboids size	1 degree/6/yes/1 m
	Receiver sensitivity	−100 dBm
3G	Transmitter power/position	0 dBm/streetlight
	Frequency/bit rate	900/2,100 MHz/1 Mbps
	Antenna radiation pattern (RX, TX)/gain	Omnidirectional/0 dB
	3D ray launching: angular resolution/reflections/diffraction/cuboid size	1 degree/6/yes/1 m
	Receiver sensitivity	−100 dBm
802.11p	Transmitter power/position	0 dBm/streetlight
	Frequency/bit rate	5,900 MHz/27 Mbps
	Antenna radiation pattern (RX, TX)/gain	Omnidirectional/0 dB
	3D ray launching: angular resolution/reflections/diffraction/cuboid size	1 degree/6/yes/1 m
	Receiver sensitivity	−100 dBm

Table 5.2 3D ray-launching simulation parameters for the V2V link

V2V link	Parameters	Values
802.11p	Transmitter power/position	0 dBm/rooftop vehicle
	Frequency/bit rate	5,900 MHz/27 Mbps
	Antenna radiation pattern (RX, TX)/gain	Omnidirectional/0 dB
	3D ray launching: angular resolution/reflections/ diffraction/cuboid size	1 degree/6/yes/1 m
	Receiver sensitivity	−100 dBm

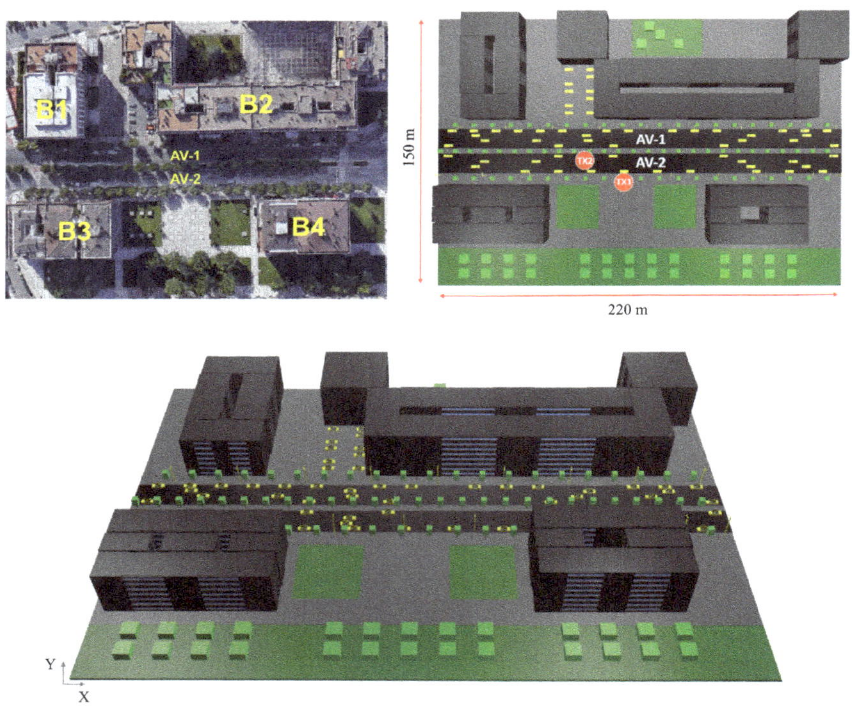

Figure 5.1 Urban scenario: Google map aerial view (left top); 2D-map schematic aerial view (right top); 3D-map frontal rendered frontal view (bottom)

5.1.1 Urban environment

The modelled and simulated urban scenario are represented in Figure 5.1, with the scenario aerial view, the 2D schematic aerial view map and the 3D frontal-view map used in the simulation. Table 5.3 identifies the Cartesian coordinates (x, y, z) of points of interest according to their relative position from the origin (0, 0, 0).

Table 5.3 Scenario reference

Reference	Abbreviation	Coordinates (*x*, *y*, *z*) m
Main avenues	AV1/AV2	(*x*, 83, 0)/(*x*, 69, 0)
Streetlight transmitter	TX1	(101.5, 63, 3.5)
Car transmitter	TX2	(80.5, 67.25, 1.5)
Receiver antenna	RX	(*x*, *y*, 1.5)
Buildings	B1, B2, B3, B4	Not applicable

The selected urban scenario corresponds to a central avenue of the city of Pamplona, Spain. The scenario dimensions are 220 m × 150 m × 22 m and includes the typical elements of an urban environment, such as buildings, trees, cars, streetlights and pedestrians. The scenario has one main avenue surrounded by buildings with different geometries. The main avenue has two sides, referred to as Avenue 1 (AV1) and Avenue 2 (AV2) in Figure 5.1, with three lanes of vehicles each one of them, and parallel parked vehicles on both sides. The parked vehicles coexist with metal and fibreglass garbage containers. For the simulations, two different antenna positions have been considered, the transmitter antenna placed in a streetlight on the roadway and the second one just over the rooftop on a vehicle, as is presented in Figure 5.1. Simulation parameters can be seen in Tables 5.1 and 5.2.

5.1.1.1 Received signal strength indicator

The received signal strength indicator (RSSI) or received power represents the signal level that reaches each spatial point in the 3D scenario considered. First, the RSSI levels for the C-V2X communication link has been assessed for the urban scenario with the transmitter antenna placed at a streetlight at 3.5 m height. Figures 5.2 and 5.3 present the power cuts received at 1.5 m height, which is the height of the rooftop of the cars, for different 4G and 3G frequency bands, respectively. The following points can be noticed from these figures:

- The great influence that the different elements of the scenario have on signal propagation, especially the buildings, which absorb a large part of the signal and consequently, inside the buildings received power levels are very low (around −90/−100 dBm).
- For lower frequencies, such as 4G 800 MHz and 3G 900 MHz, signal levels along AV1 and AV2 are higher than in the rest of the cases, however, from a certain distance along the highway, there are certain points with very low received power levels. The high impact of multipath propagation is observed along the lanes, caused by the scatterers inherently present in the scenario.
- The placement of the transmitting antenna on the streetlight facing the road causes a cone of lower received power levels in the back of the streetlight, not affecting the road. Thus, the location of the transmitting antenna, which in this case is in close contact with metallic elements, such as the lamp post structure, can modify partially antenna matching conditions.

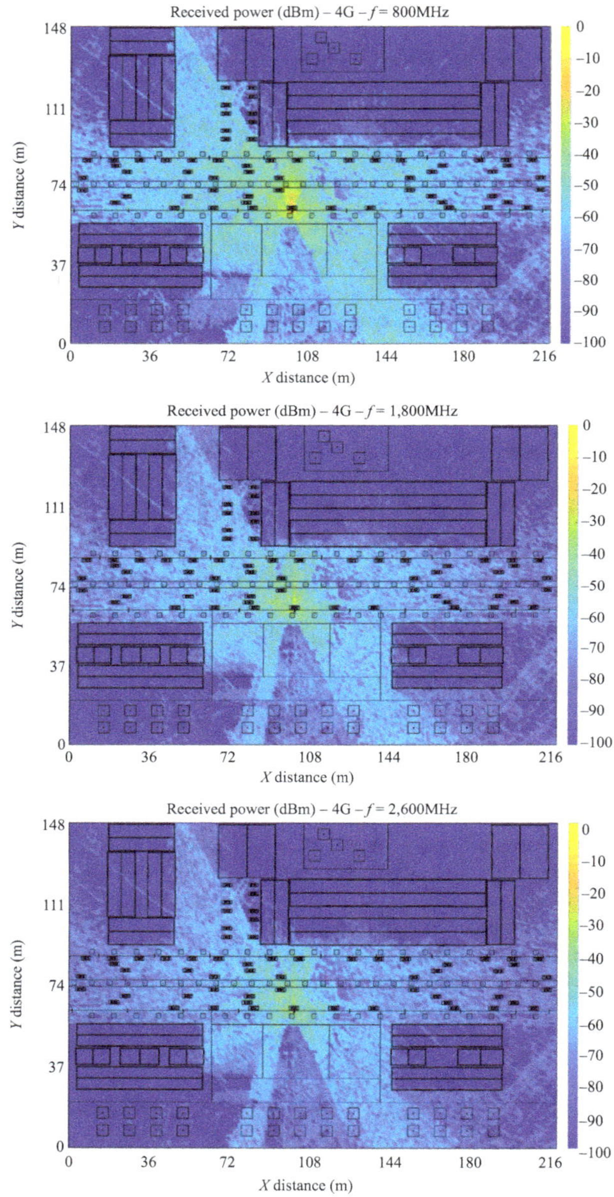

Figure 5.2 Comparison of received power (dBm) for the bi-dimensional planes of 1.5 m height for the different 4G frequency bands

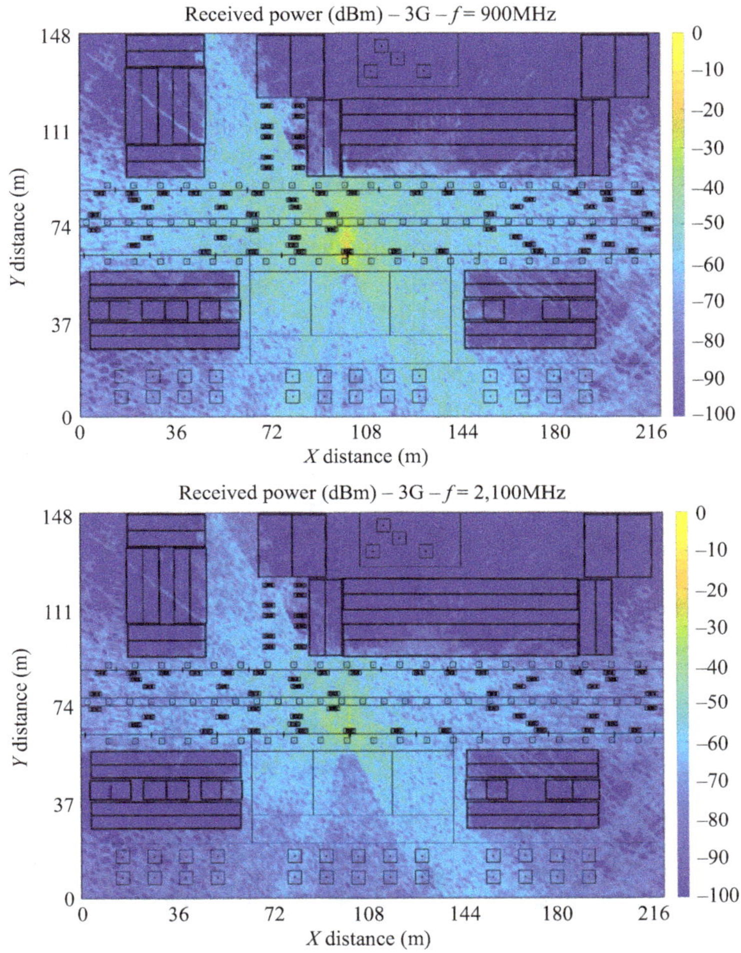

Figure 5.3 Comparison of received power (dBm) for the bi-dimensional planes of 1.5 m height for the different 3G frequency bands

- Signal fast fading is also an important factor in these environments as the signal impacts with many obstacles in its path. In the I2V communication link analysed, we must consider that the transmitter is placed at higher height than the receivers, which cause that most of the spatial receiver points are in LoS. However, the surrounding scatterers such as buildings, cars, trees and street-lights make the rise of multipath components, causing fast fading.
- The diffraction phenomenon is also important in an urban environment, since the corners and edges of buildings cause the appearance of diffracted field on the scenario.

- Scattering can also provoke signal losses in this type of scenario depending on the building's walls roughness, as well as the different vegetation environments. For these simulations, buildings' walls have been considered flat, and trees have been considered with homogeneous materials.

In order to have insight into radio wave propagation considering the standard allocated for vehicular communications, IEEE 802.11p at the 5.9 GHz frequency band [2], the I2V and V2V communication links have been analysed and are represented in Figure 5.4. It is noteworthy that the transmitter antenna has been placed at different positions for the analysis, as presented in Table 5.3. The following points can be noticed from the figure:

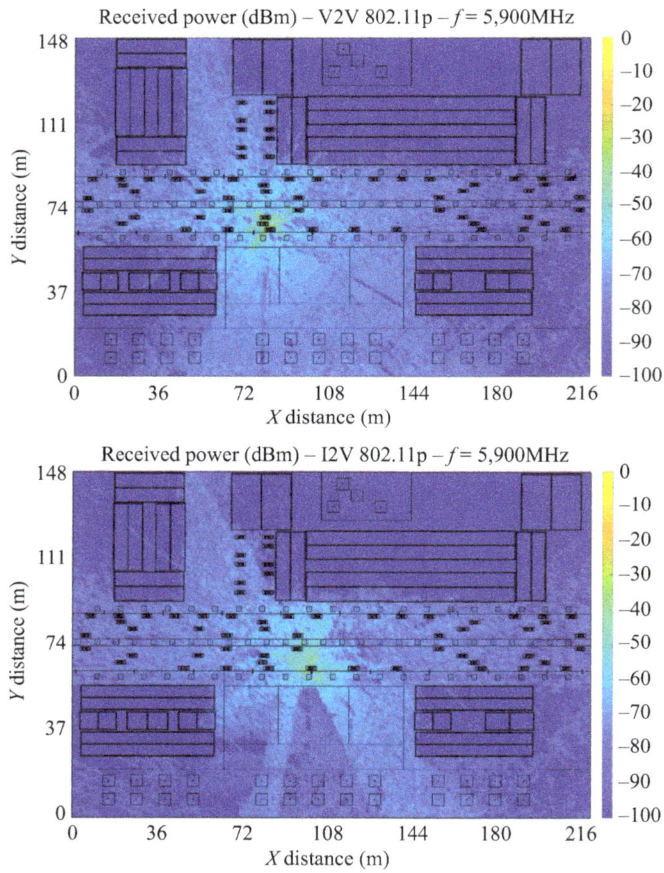

Figure 5.4 Comparison of received power (dBm) for the bi-dimensional planes of 1.5 m height for the 802.11p standard at the 5.9 GHz frequency band

- For the 5.9 GHz frequency band, signal received levels are lower, due to the higher absorption of obstacles at these frequencies, as well as differences in diffraction and scattering from fixed-size objects.
- It is observed that for the V2V communications link, higher power levels are reached for greater distances. This is because the bi-dimensional plane that has been represented is at the same height as the transmitter, unlike the case I2V in which the transmitter is at a height higher than the receivers, which are the cars.

5.1.1.2 Multipath propagation

As stated in chapter 4, multipath propagation is one of the most important phenomenon in cluttered environments, such as an urban environment with buildings, streetlights, people, vegetation and trees, among others. For the analysis of multipath propagation, the V2V communication link has been considered, with the transmitter antenna placed at the rooftop of a vehicle operating at the 5.9 GHz frequency band. We know that time delays between the multipath components arrivals result in different echoes that can cause errors, especially if the delays are large compared with a symbol period. Figure 5.5 presents the power delay profile for a spatial point at the rooftop of a vehicle located in a centre of the scenario at the same height of the transmitter, showing echoes in a time span of approximately 0–1,500 ns. As can be seen, time span is significantly higher than in-vehicle environments analysed in the previous chapter, as the signal travels greater distances in the outdoor environment. The limited distance to buildings in the urban environment results in a narrower range of echo times comparing with the rural environment.

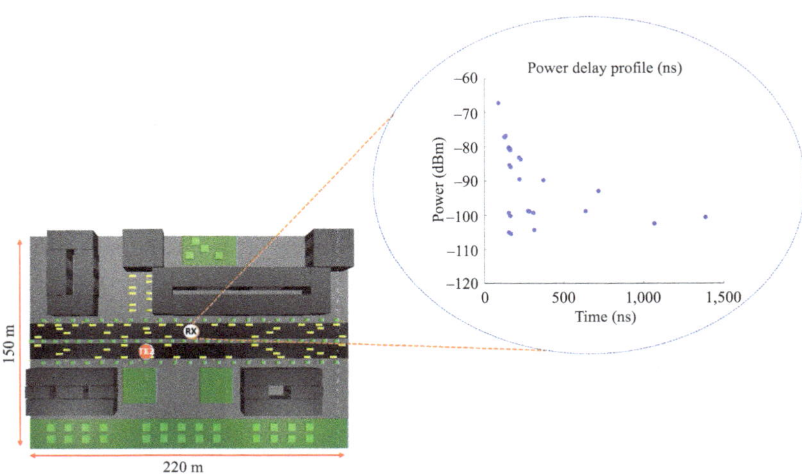

Figure 5.5 Power delay profile at the rooftop of a vehicle located in the centre of the scenario in the urban environment

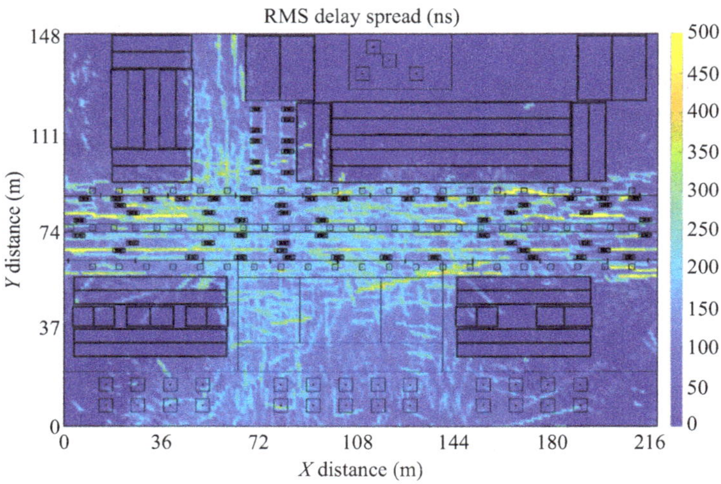

Figure 5.6 RMS delay spread within the urban environment at 1.5 m height for 5.9 GHz frequency

In time domain, the multipath components are commonly characterized by the rms delay spread, which is the square root of the second central moment of the measured power delay profile and discussed by Rappaport [11]. Figure 5.6 shows the rms delay spread for the bi-dimensional plane of 1.5 m height for the selected urban environment. It is observed the high impact of the buildings in the scenario in terms of signal absorption. In addition, the presence of closer objects (other cars) near the transmitter location results in the appearance of high variability of echoes along the road.

5.1.1.3 Frequency domain channel description

In the frequency domain, the echoes add in phase at some frequencies and out of phase at others, resulting in constructive or destructive frequency selective fades. As stated before, the coherence bandwidth is a measure of the range of frequencies over which the channel can be considered to be approximately constant. Figure 5.7 depicts the coherence bandwidth values and shows its inverse relationship with the rms delay spread, showing higher values inside the buildings and at a certain distance from the transmitter and lower values along the road. The high values within the buildings are an indicator of channel availability for other transmitters at these positions in accord with Figure 5.4. The RSSI is below the sensitivity and the communication with the considered transmitter is unfeasible, leaving the channel available for its use at that zone. The lower values within the road means that communication with other vehicles will be feasible up to a certain distance (approximately 100 m along the road).

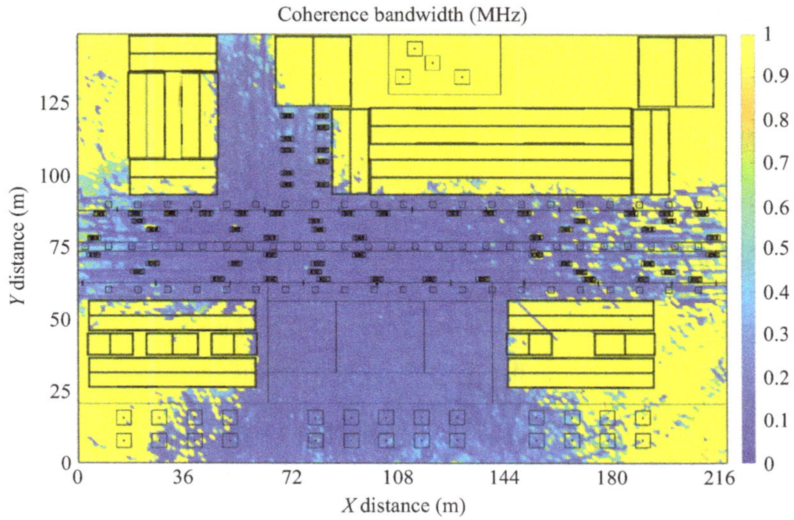

Figure 5.7 Coherence bandwidth within the urban environment at 1.5 m height for 5.9 GHz frequency

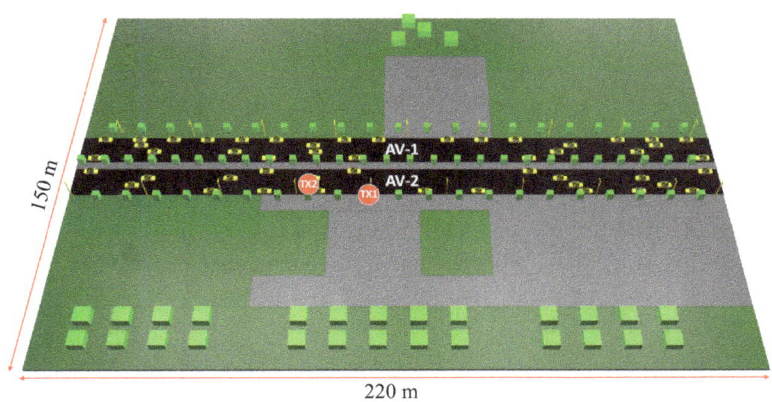

Figure 5.8 Rural scenario: 3D-map frontal rendered frontal view

5.1.2 Rural environment

The modelled and simulated rural scenario selected for radio wave propagation characterization has been the same scenario considered for the urban environment but in this case without buildings. Figure 5.8 presents the 3D frontal-view map used in the simulation. The points of interest according to its relative position are the same as the urban environment and are presented in Table 5.3. For the simulations, the same antenna positions as the urban environment has been considered: the first one being the transmitter antenna placed in a streetlight on the roadway and the second one just over the rooftop on a vehicle, as is presented in Figure 5.8. Simulation parameters can be seen in Tables 5.1 and 5.2.

5.1.2.1 Received signal strength indicator

Similar simulations as the urban scenario case has been performed in this subsection for the rural scenario. First, the RSSI levels for the C-V2X communication link has been assessed with the transmitter antenna placed at a streetlight at 3.5 m height. Figures 5.9 and 5.10 present the power cuts received at 1.5 m height, which is the height of the rooftop of the cars, for different 4G and 3G frequency bands, respectively. The following points can be noticed from these figures:

- The high impact of buildings in radio wave propagation for all the frequencies considered. It is observed that in the rural environment signal propagates larger distances because buildings are not obstructing the signal.
- For lower frequencies, such as 4G 800 MHz and 3G 900 MHz, signal levels along AV1 and AV2 are higher than in the rest of the cases. Besides, if we compare with the urban environment, we can observe that in the rural scenario case, received power levels are above the sensitivity in all the road, in contrast with the urban environment that from a certain distance along the highway, there are certain points with values below sensitivity levels (-100dBm). In this case, it is also observed the high impact of multipath propagation along the lanes, caused by the scatterers inherently present in the scenario (other cars, streetlights, trees, etc.).
- The placement of the transmitting antenna on the streetlight facing the road causes a cone of lower received power levels in the back of the streetlight, not affecting the road. Thus, the location of the transmitting antenna, which in this case is in close contact with metallic elements, such as the lamp post structure, can modify partially antenna matching conditions.
- In the same way that the urban environment, signal fast fading is also an important factor in rural environments as the signal impacts with many obstacles in its path. In the I2V communication link analysed, we must consider that the transmitter is placed at higher height than the receivers, which cause that most of the spatial receiver points are in LoS. However, the surrounding scatterers such as other cars, trees and streetlights make the rise of multipath components, causing fast fading.
- The diffraction phenomenon is less important in these environments than in an urban environment, as there are not edges of buildings than can cause the appearance of diffracted field in the scenario. Other objects with edges and corners can also cause diffracted field if there is NLoS between the transmitter and the receiver.
- Scattering can also provoke signal losses in this type of scenarios depending on the different vegetation environments within the rural scenario. For these simulations, the trees and the grass on the ground have been considered with homogeneous materials.

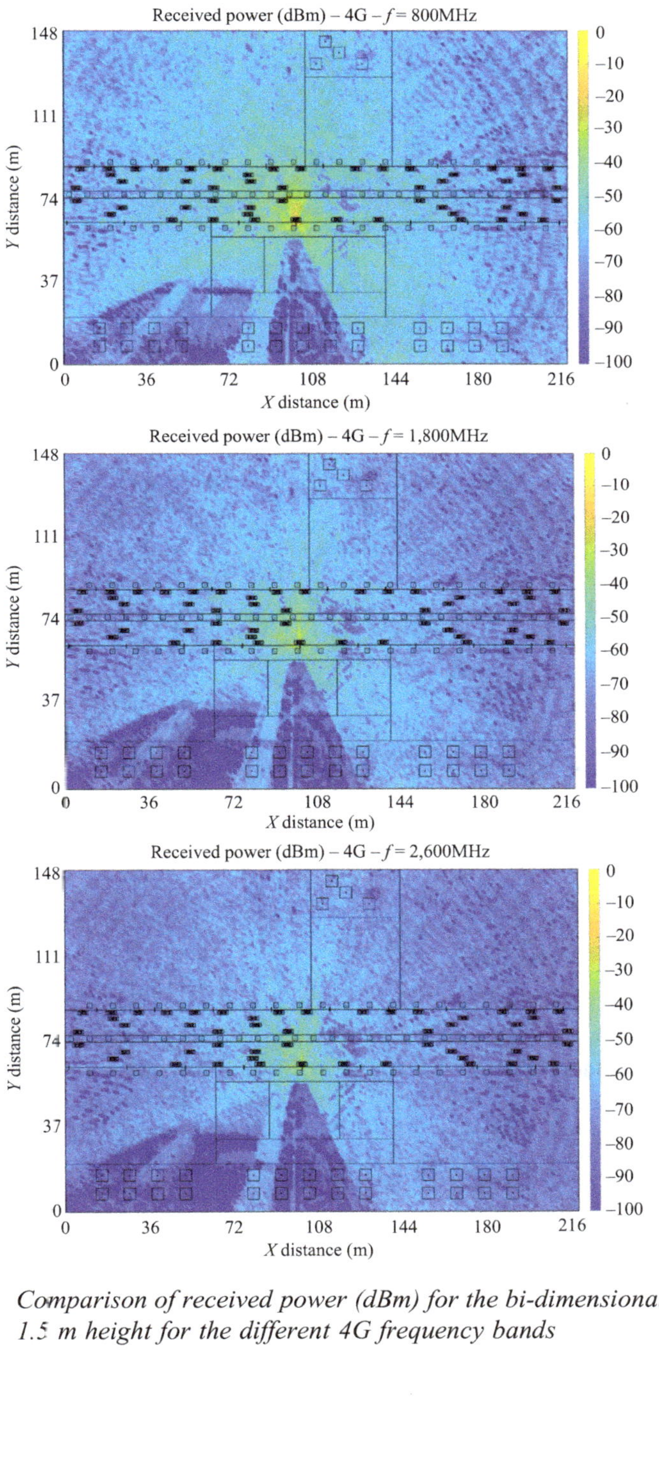

Figure 5.9 Comparison of received power (dBm) for the bi-dimensional planes of 1.5 m height for the different 4G frequency bands

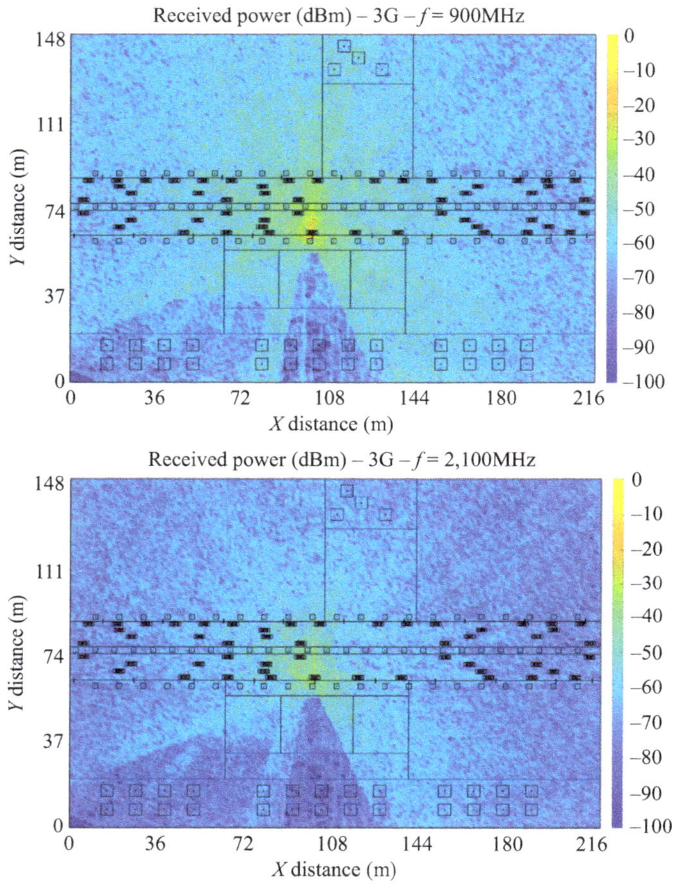

Figure 5.10 Comparison of received power (dBm) for the bi-dimensional planes of 1.5 m height for the different 3G frequency bands

In order to have insight into radio wave propagation considering the standard allocated for vehicular communications, IEEE 802.11p at the 5.9 GHz frequency band, the I2V and V2V communication links have been also analysed for the rural scenario and are represented in Figure 5.11. It is noteworthy that the transmitter antenna has been placed at different positions for the analysis, as presented in Table 5.3. The following points can be noticed from the figure:

- For the 5.9 GHz frequency band, signal received levels are lower, due to the higher absorption of obstacles at these frequencies, as well as differences in diffraction and scattering from fixed-size objects.
- The main difference with the urban environment is that in this case, there are not absorption by the buildings, so signal can be propagated farther distances.

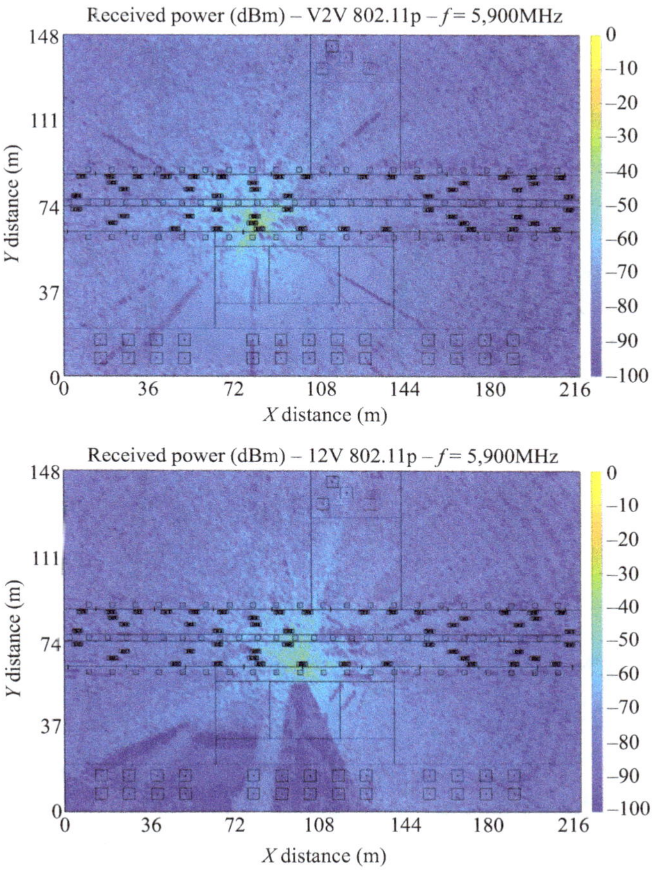

Figure 5.11 Comparison of received power (dBm) for the bi-dimensional planes of 1.5 m height for the 802.11p standard at the 5.9 GHz frequency band

- As in the case of urban environment, it is observed that for V2V communications link, higher power levels are reached for greater distances. This is because the bi-dimensional plane that has been represented is at the same height as the transmitter, unlike the case I2V in which the transmitter is at a height higher than the receivers, which are the cars.

5.1.2.2 Multipath propagation

Figure 5.12 presents the power delay profile for a spatial point at the rooftop of a vehicle located in a centre of the scenario at the same height of the transmitter, showing echoes in a time span of approximately 0–1500 ns. As it can be seen time span is in the same range as the urban environment, but in this case, the presence of larger open areas results in less and more separated echoes.

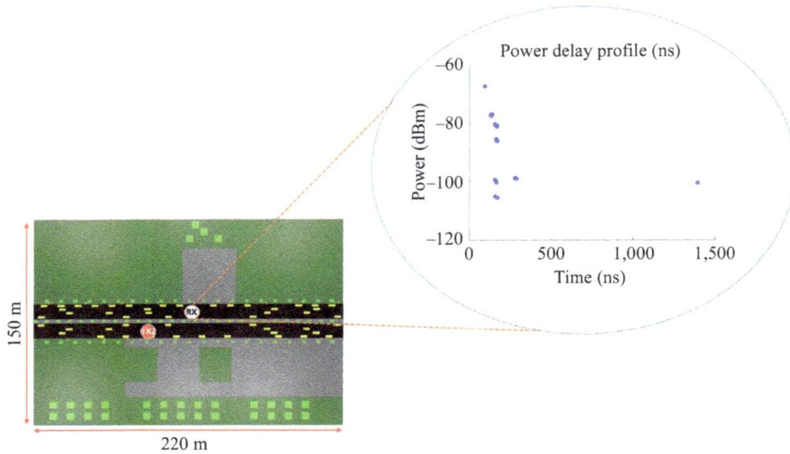

Figure 5.12 Power delay profile at the rooftop of a vehicle located in the centre of the scenario in the rural environment

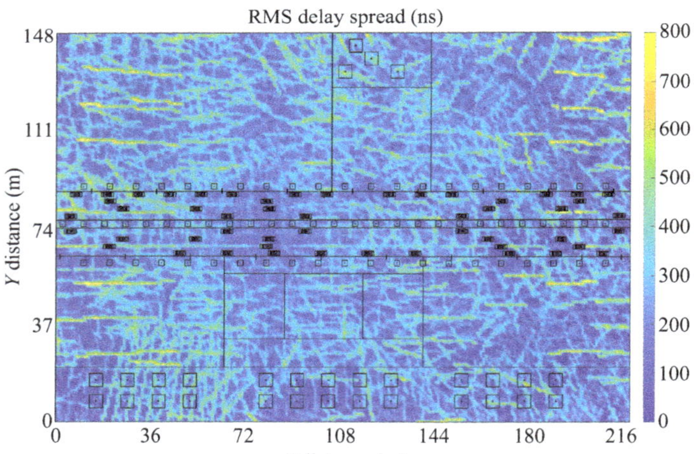

Figure 5.13 RMS delay spread within the rural environment at 1.5 m height for 5.9 GHz frequency

Figure 5.13 shows the rms delay spread for the bi-dimensional plane of 1.5 m height for the selected rural environment. It is observed, once again, that signal propagates along all the scenario, as the buildings are not obstructing the signal. In addition, as in the previous case, the presence of closer objects (other cars) near the transmitter location results in the appearance of high variability of echoes along the road.

5.1.2.3 Frequency domain channel description

Figure 5.14 depicts the coherence bandwidth values simulated and shows its inverse relationship with the rms delay spread. In this case, almost all the area of the scenario has low values of coherence bandwidth, in contrast with the previous analysed case of the urban environment. These low values of coherence bandwidth mean that communication with other vehicles or infrastructure at those spatial points will be feasible.

5.2 Case studies for inter-vehicle electromagnetic properties in ITS

In this subsection, two different case studies for inter-vehicle communication systems within an underground tunnel has been assessed: slow-moving vehicles and fast-moving vehicles. The selected scenario has been previously presented in chapter 4, which is an underground tunnel of 100 m length, 9 m wide and 13 m height, where two lanes of cars can circulate at the same time. The car model presented in chapter 4, section 4.1.1 has been embedded within the tunnel, analysing a random distribution of cars within the tunnel. The car model with four people inside has been considered to assess a more realistic scenario. The tunnel walls, as well as the floor scenario, have been considered in this case taking into account the different material properties, as explained in chapter 4, section 4.2. Figure 5.15 presents the tunnel scenario with the position of the transmitter and the V2V communication link that has been analysed in the following subsections.

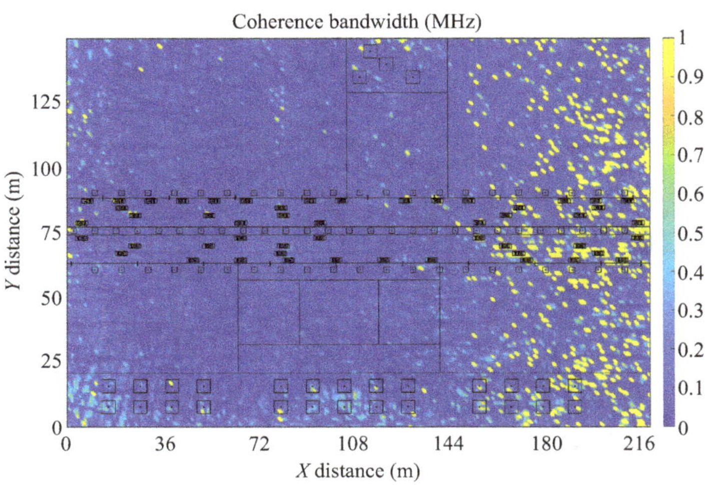

Figure 5.14 Coherence bandwidth within the rural environment at 1.5 m height for 5.9 GHz frequency

The 3D ray-launching tool has been used for the analysis of the time-varying nature of the channel in this scenario caused by either relative motion between the transceivers or the movement of objects in the channel (other vehicles). The most important aspect of this analysis is that the vehicles are inside a tunnel, which can be considered as an indoor environment and propagation characterization in these environments is more challenging. The non-stationarity of the channel has been assessed considering different velocities for the mobiles, leading to Doppler shift and Doppler spread results, providing insight into the performance of the wireless communication system. In chapter 2, section 2.2, propagation phenomena associated with the non-stationarity of the channel have been presented.

In the following subsections, we present the electromagnetic propagation signal analysis associated with the non-stationarity of the channel in terms of Doppler shift and Doppler spread for different velocities of the vehicles for V2V communication link. For that purpose, the same C-V2X and IEEE 802.11p standard previously used for the analysis of V2X communications has been considered in this case. Simulation parameters considered in the 3D ray-launching algorithm, as well as the parameters for the V2V links, are presented in Tables 5.1 and 5.2. It is noteworthy that the transmitter antenna has been placed at the rooftop of the car in this case, as is presented in Figure 5.15, to analyse inter-vehicle communications.

5.2.1 Fast-moving vehicles (in underground tunnels)

As a starting step, several simulations of the considered scenario have been performed considering different instants of time denotes as t_i, $i=1, \ldots, 5$, for the selected vehicle position as shown in Figure 5.15 for the instant of time t_2. In this case, two different velocities have been considered for the vehicles: 22.22 m/s (80 km/h) and 27.77 m/s (100 km/h).

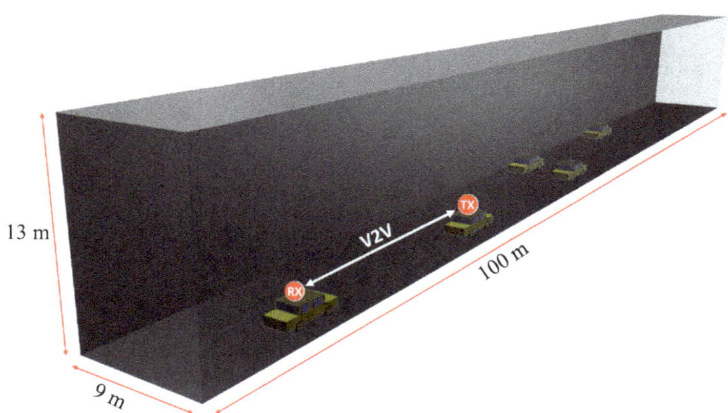

Figure 5.15 Tunnel scenario with vehicles implemented in the 3D ray-launching tool

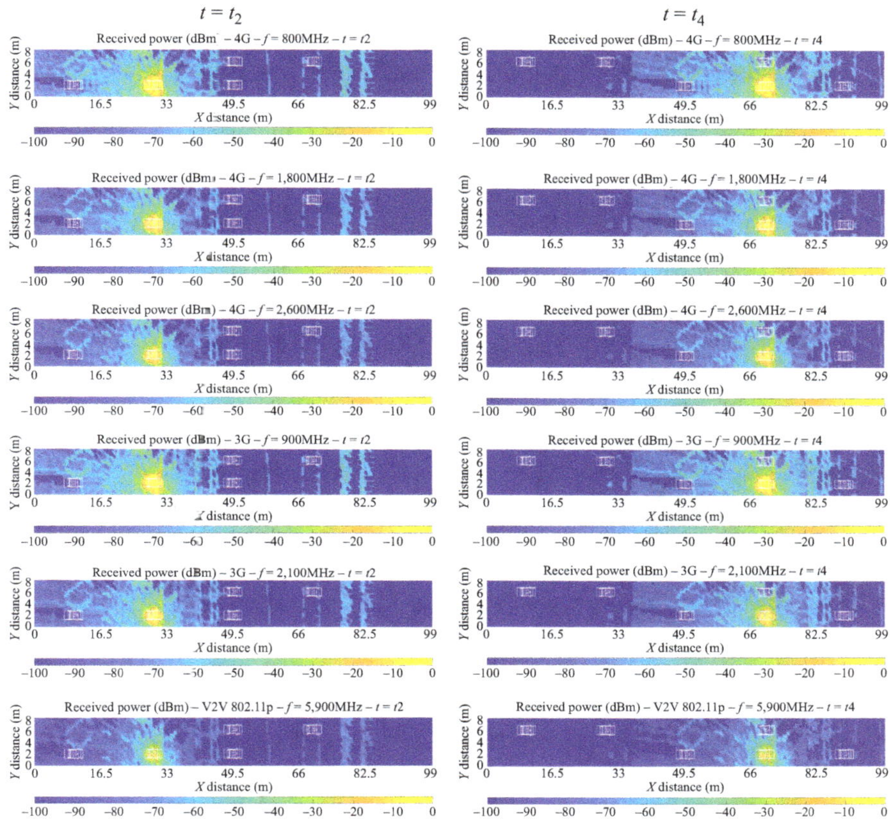

Figure 5.16 Comparison of received power (dBm) for the bi-dimensional planes of 1.5 m height for the different considered systems within the tunnel at the instant of times t_2 and t_4

First, we present the received power for a bi-dimensional plane of 1.5 m for two instants of time for the considered scenario, as is presented in Figure 5.16 for instants of time t_2 and t_4. The influence of the movement of the vehicle in signal fading and changes in the received power in the considered environment is shown. It is observed that the V2V communication link will be feasible only if the vehicles are at certain distance, depending on the frequency under consideration. In the case of the 5.9 GHz frequency band, it can be seen that the distance is shorter, as at this frequency band, absorption is higher.

To get knowledge about the time-varying nature of the channel caused by the different velocities of the vehicles considered, the Doppler shift (f_d) at the receiver can be obtained by

$$f_d = \frac{v}{\lambda}\cos(\psi) \tag{5.1}$$

where ψ represents the angle between the direction of arrival of the specific examined propagation path and the receiver's movement, v is the speed of the vehicle and λ is the wavelength of the signal.

Figure 5.17 presents the Doppler shift for the nearest vehicle for the considered scenario (RX location in Figure 5.15) for two different vehicle velocities: 22.22 m/s (80 km/h) and 27.77 m/s (100 km/h) for the instant of time t_2. It can be seen that for

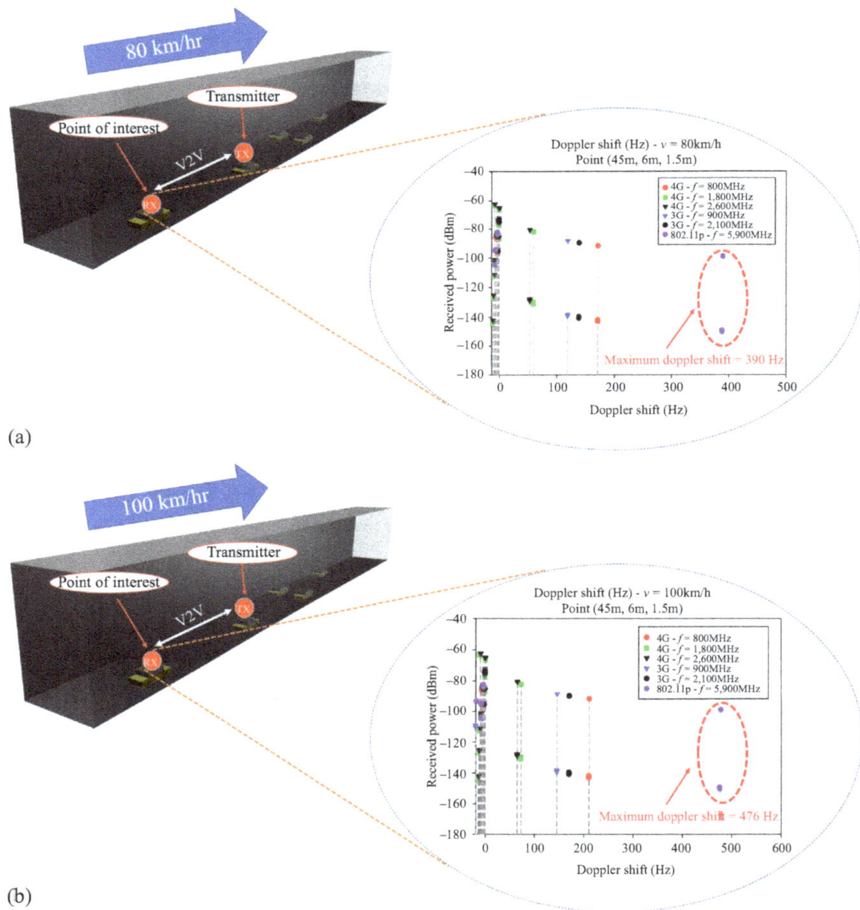

Figure 5.17 Doppler shift (Hz) for fast-moving vehicles within the tunnel at the instant of time t_2 (a) 80 km/h and (b) 100 km/h

22.22 m/s velocity, we can observe that the maximum Doppler shift is around 400 Hz for 802.11p system, which use a higher frequency band than the other considered systems. This means that systems working in higher frequency bands have more impact in the Doppler shift than systems in lower frequency bands (4G 800 MHz and 3G 900 MHz).

We can also observe that if we increase the speed to 27.77 m/s (100 km/h), the Doppler shift also increases, being once again the maximum for 802.11p system in approximately 500 Hz, increasing around 100 Hz when compared with 22.22 m/s.

To get an insight of the impact of Doppler shift in the considered scenario, the Doppler spread has been calculated for all the spatial samples of the scenario and it is shown in Figure 5.18. The Doppler spread has been defined for each spatial sample as the difference of the maximum and the minimum frequency shifts. The entire tunnel has been considered for the analysis, but we have previously seen that the transmitted signal only arrives in the vicinity of the vehicle. From a certain distance, there is no frequency shift since there is no signal level.

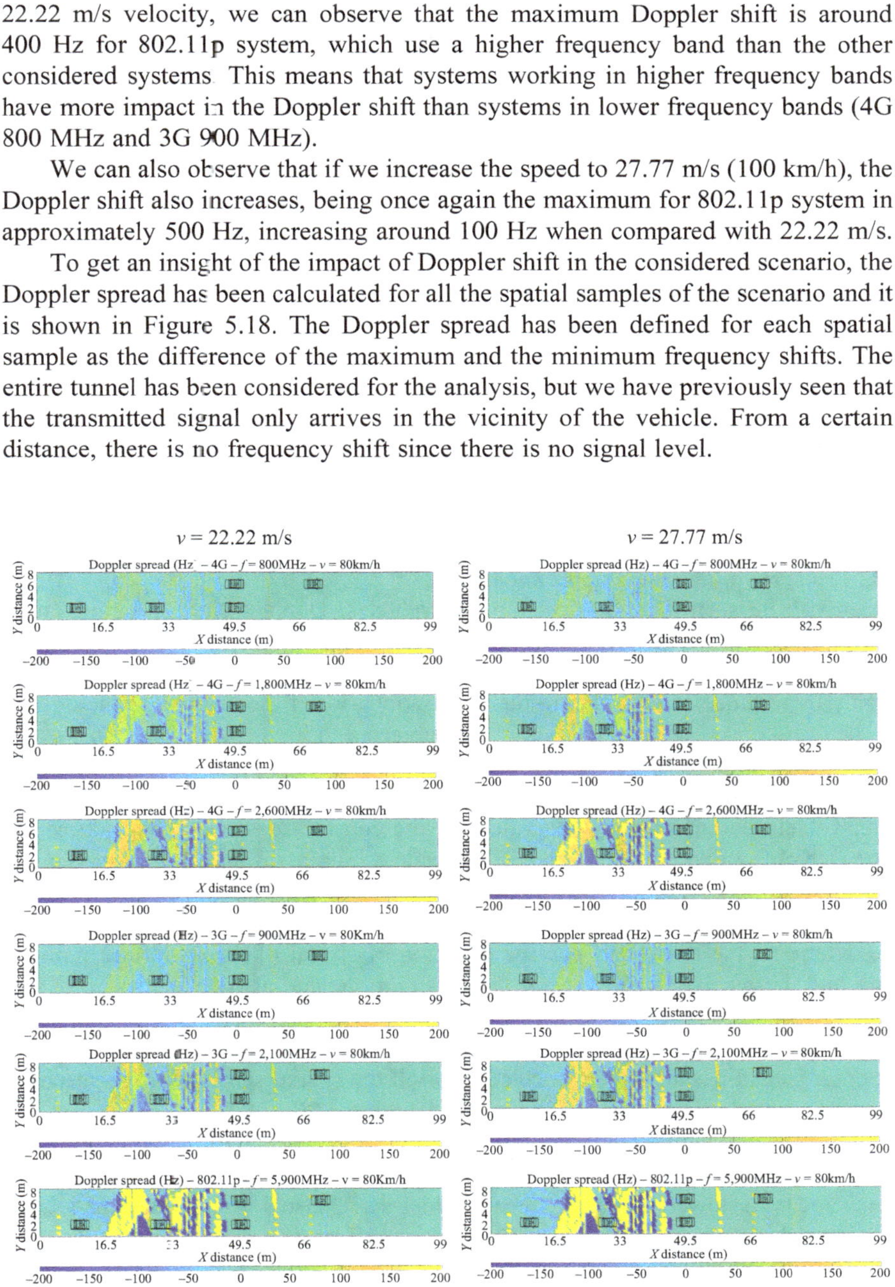

Figure 5.18 Doppler spread (Hz) for fast-moving vehicles within the tunnel at the instant of time t_2

For the case of 27.77 m/s (100 km/hr), we can see that the values of Doppler spread are higher than for the speed of 22.22 m/s (80 km/h), which verify that an increase in the velocity of the vehicle will imply a higher frequency variation.

It can also be observed that for the technologies with higher frequency (5.9 GHz), Doppler spread levels are greater than for the rest of the cases, being the lowest for the 800 and 900 MHz frequency bands. The frequency shift can affect the reception of the signal for underground environments, depending on the speed of the vehicle for V2V communications.

5.2.2 Slow-moving vehicles (in underground tunnels)

In order to compare with different velocities, Doppler shift and Doppler spread has been analysed for the same point (RX location in Figure 5.15) for slow-moving vehicles: 5.55 m/s (20 km/h) and 11.11 m/s (40 km/h). Figure 5.19 presents the Doppler shift for slow-moving vehicles for the instant of time t_2. It can be seen that for 5.55 m/s velocity, we can observe that the maximum Doppler shift is around 100 Hz for the 802.11p system, which use a higher frequency band than the other considered systems. We can also observe that if we increase the speed to 11.11 m/s (40 km/h), the Doppler shift also increases, being the maximum once again for 802.11p system in 200 Hz approximately.

It can be seen that for slow-moving vehicles, Doppler shift is significantly lower than fast-moving vehicles, ranging approximately between 100 Hz for 5.55 m/s and 500 Hz for 27.77 m/s for the 802.11p system. These results state the highly importance of considering the non-stationarity of the channel in the design of the transceivers when it is going to be subjected to high-speed variations.

If we compare these results with the in-vehicle Doppler shift obtained in chapter 4, section 4.2, we can observe that the Doppler shift is higher in the case of inter-vehicle communication regardless of the considered speed. This is because inter-vehicle communication analysis considers bigger distances than in-vehicle communications, thus, Doppler shift has more impact. However, Doppler shift values are going to depend on the position of the transmitter and receiver, in which we determine the angle between the direction of arrival of the specific examined path and the receiver's movement. It is going to depend also on the speed of the vehicle and the wavelength of the signal.

To get an insight of the impact of Doppler shift in the considered scenario, the Doppler spread has also been calculated for all the spatial samples of the scenario for slow moving vehicles and it is shown in Figure 5.20.

As in the previous case of fast-moving vehicles, we can see an increase of the Doppler spread when higher velocities are considered. Nevertheless, in this case, as the velocity is lower, Doppler spread values have also lower values than fast-moving vehicle case; noteworthily, the scale is different than in Figure 5.20.

In order to have a better knowledge about the comparison between the different considered systems and different velocities within the tunnel, Figure 5.21 shows the Doppler shift comparison for the different communication systems considered and different velocities within the tunnel at the instant of time t_2.

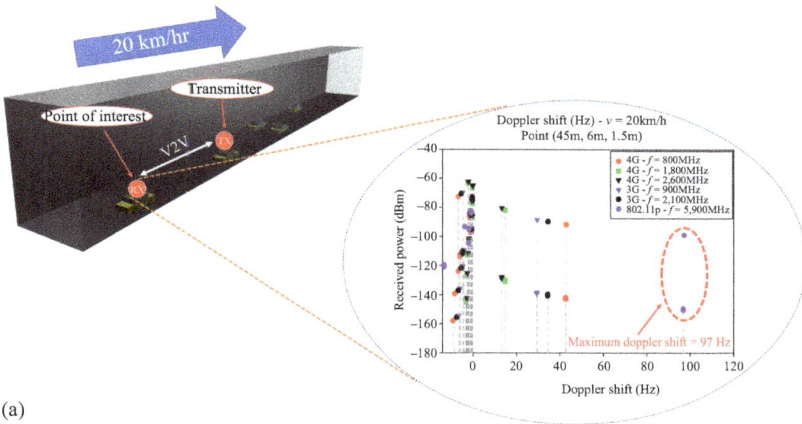

(a)

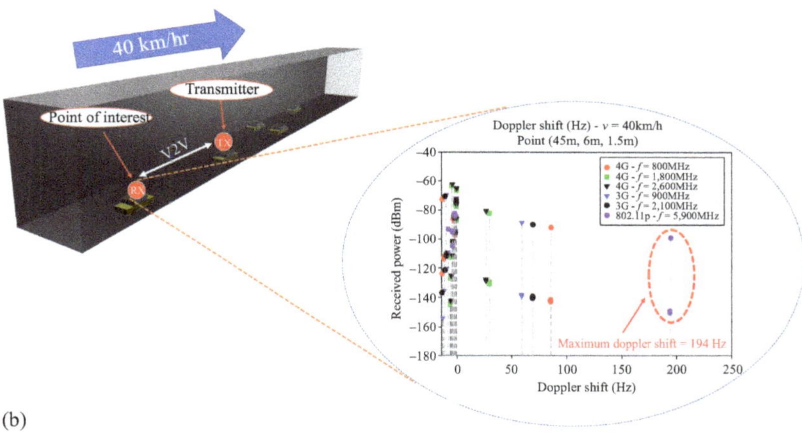

(b)

Figure 5.19 Doppler shift (Hz) for slow moving vehicles within the tunnel at the instant of time t_2 (a) 20 km/h and (b) 40 km/h

It is observed that for all the considered systems, the highest Doppler shift is always when the highest speed is considered: 27.77 m/s (100 km/h), being the highest around 500 Hz approximately for the 802.11p system. The lowest Doppler shift is when we consider the lowest speed: 5.55 m/s (20 km/h). It is important to note that these values have been obtained for vehicles moving within a tunnel with concrete walls. A tunnel with different walls' material may have different behaviour.

5.3 Inter-vehicle radio planning analysis

In this section, radio planning analysis within the urban and rural environments presented before has been performed in terms of potential communication

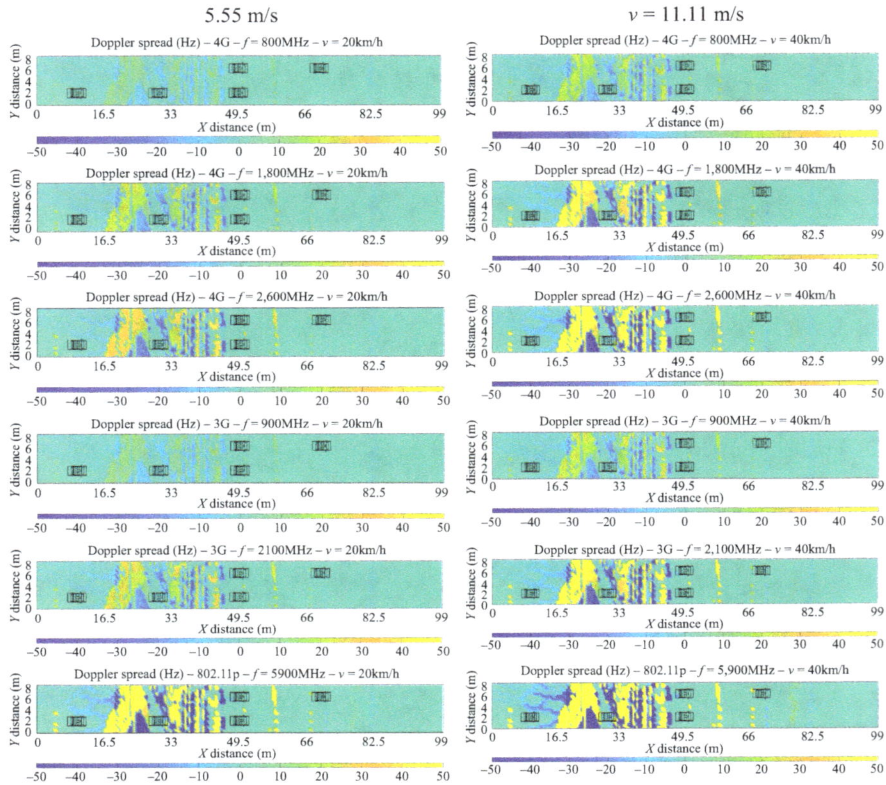

Figure 5.20 Doppler spread (Hz) for slow-moving vehicles within the tunnel at the instant of time t_2

transceiver placement for the deployment of the considered systems. In addition, an empirical modelling based on simulation results is presented in this section for the different analysed scenarios.

5.3.1 Empirical modelling

The path loss (PL) has been analysed and modelled under the different considered cases presented in section 5.1 for the different scenarios (urban and rural). As stated in chapter 4, section 4.3, PL is defined as the ratio of the effective transmitted power to the received power, calibrating out cable losses and antenna gains. The modelling method used for PL, taking into account the complex environments, is given by the empirical model of log-distance [11] expressed as follows:

$$PL_{\mathrm{dB}}(d) = PL_0 + 10n\log_{10}(d) + X_\sigma \tag{5.2}$$

where PL_0 is the intercept, d is the transmitter–receiver separation distance in metres and n is the PL exponent dependent upon the specific propagation

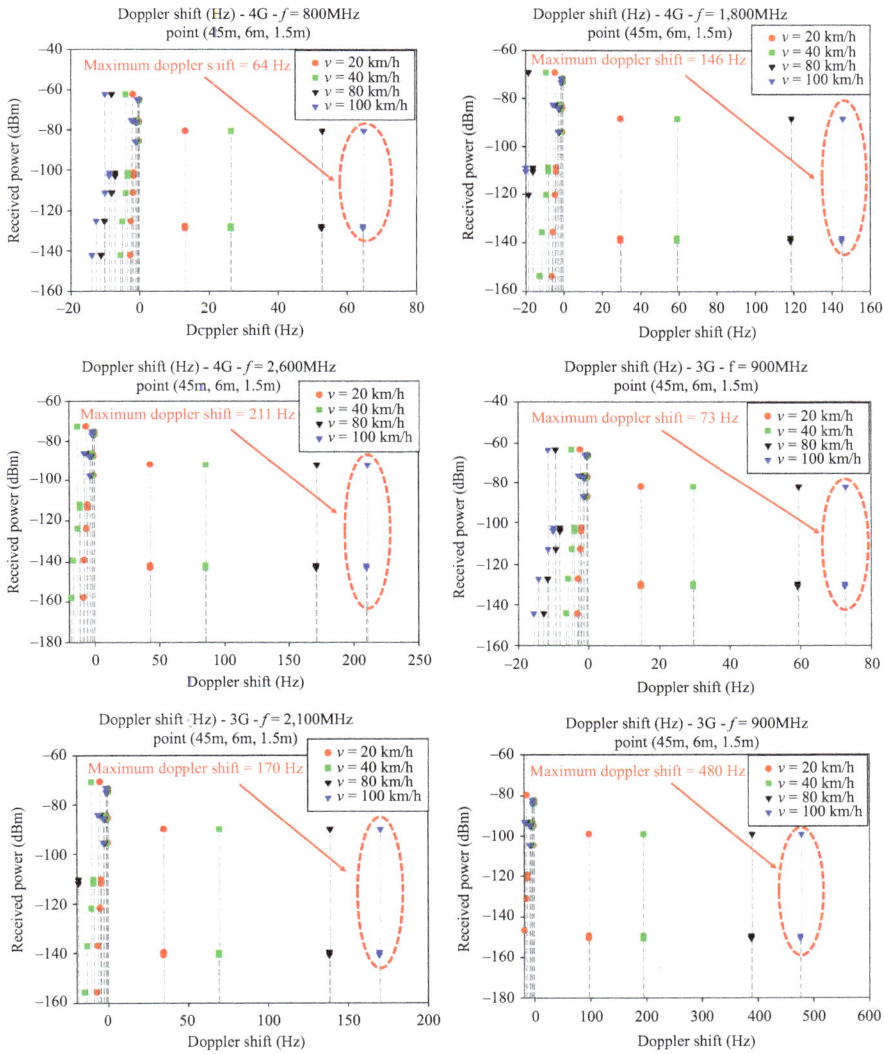

Figure 5.21 Doppler shift (Hz) comparison for different communication systems and different velocities within the tunnel at the instant of time t_2

environment indicating the rate at which PL increases with distance. In free-space propagation, n equals 2. X_σ denotes shadow fading, which is considered to be a zero-mean normally distributed random variable with standard deviation σ (dB). The values of PL_0 and n are estimated by using the linear regression method. Estimated PL values have been calculated within the urban and rural scenarios and compared with the lognormal distribution. From the results of the received power stated in section 5.1, PL can be predicted for each spatial point of the environment applying:

$$PL_{\mathrm{dB}}(d) = P_t(\mathrm{dBm}) - P_r(\mathrm{dBm}) \tag{5.3}$$

With the antenna gains included in P_t(dBm) and P_r(dBm).

5.3.1.1 Urban environment

The modelling results for the I2V communication links along AV-1 and AV-2 (see Figure 5.1 for reference) for the different C-V2X and 802.11p systems are shown in Table 5.4 for a height of 1.5 m. Table 5.4 presents the different values of PL_0, n and σ for the different systems analysed. It is noteworthy that n values range from 1.5 to 3 depending on the system and the analysed area. These values are in accordance with other urban environments extracted from measurements [12,13].

In Figure 5.22, an aerial view of the selected urban scenario is presented with dashed lines along the X-axis for different values of Y, which correspond with AV-1 and AV-2 (see Figure 5.1 for reference). These dashed lines represent the lines which correspond with the linear distribution lines of received power which are represented in Figure 5.23, for 1.5 m height. It can be seen that for all systems,

Table 5.4 Empirical modelling within the urban environment for the I2V link

I2V link	Description	PLo (dB)	n	σ (dB)
4G 800	Along AV-1 (y = 81 m) x: (0 to TX1/TX1 to 220 m)	69.58/51.45	2.42/1.60	9.96/10.21
	Along AV-2 (y = 70 m) x: (0 to TX1/TX1 to 220 m)	70.05/51.07	2.99/1.38	11.42/12.16
4G 1800	Along AV-1 (y = 81 m) x: (0 to TX1/TX1 to 220 m)	76.35/58.07	2.47/1.56	9.38/9.72
	Along AV-2 (y = 70 m) x: (0 to TX1/TX1 to 220 m)	74.68/56.98	2.48/1.53	10.34/11.13
4G 2600	Along AV-1 (y = 81 m) x: (0 to TX1/TX1 to 220 m)	79.30 / 59.65	2.65/2.05	8.88/15.40
	Along AV-2 (y = 70 m) x: (0 to TX1/TX1 to 220 m)	79.96/59.96	2.98/1.55	10.63/11.28
3G 900	Along AV-1 (y = 81 m) x: (0 to TX1/TX1 to 220 m)	68.35/46.11	2.44/2.07	9.73/9.27
	Along AV-2 (y = 70 m) x: (0 to TX1/TX1 to 220 m)	67.26/45.74	2.42/1.83	10.03/9.69
3G 2100	Along AV-1 (y = 81 m) x: (0 to TX1/TX1 to 220 m)	73.94/54.28	2.37/1.89	8.87/8.64
	Along AV-2 (y = 70 m) x: (0 to TX1/TX1 to 220 m)	74.02/53.09	2.29/1.82	9.32/9.12
802.11p	Along AV-1 (y = 81 m) x: (0 to TX1/TX1 to 220 m)	83.66/68.58	2.19/1.60	8.13/9.93
	Along AV-2 (y = 70 m) x: (0 to TX1/TX1 to 220 m)	84.59/64.94	2.64/1.59	10.13/10.94

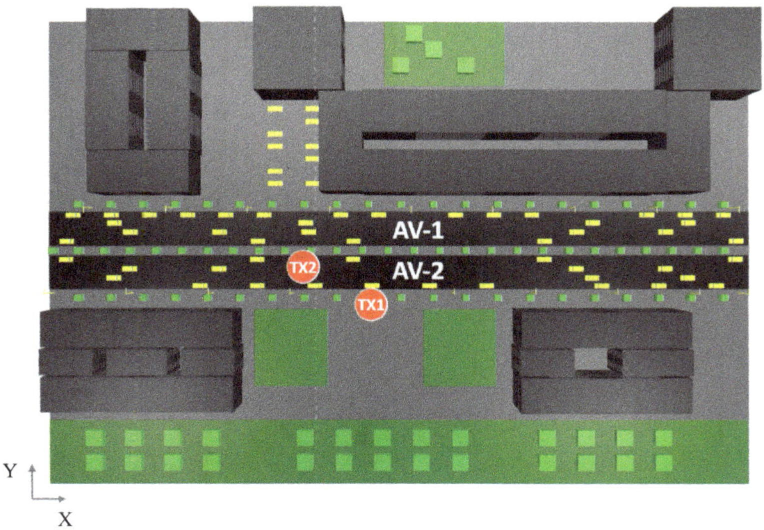

Figure 5.22 Schematic view of Avenues 1 and 2 in the urban environment for the representation of linear distribution of received power levels in Figures 5.23 and 5.24

almost all the spatial locations of both avenues have received power levels higher than sensitivity levels (−100 dBm), so the I2V communication will be feasible. It is noteworthy that in this case, almost all the point along the road are in LoS with the transmitter antenna, which is at a higher height placed at the streetlight (see Table 5.1 for reference).

Table 5.5 presents the different values of PL_0, n and σ for the V2V communication link with the 802.11p system. In this case, it is observed that n values range from 1.97 to 3.31 depending on the analysed area. In this case, n and σ values are slightly higher than in the I2V communication link. This is because in this case, the transmitter and receiver antenna are at the same height (on the rooftop of a vehicle), so there are more scatterers nearby around the communication link that can affect radio propagation, such as other cars, trees and streetlights.

In Figure 5.24, the linear distribution lines of received power along AV-1 and AV-2 for the V2V 802.11p communication link is presented. It can be seen that almost all the road has received power levels above the sensitivity level, but up to certain distance there are some fading's which can cause the link loss.

5.3.1.2 Rural environment

In this subsection, PL modelling for the rural environment for the different considered systems has been performed and it is presented in Tables 5.6 and 5.7 for the I2V and V2V communication links, respectively. These models can be used for

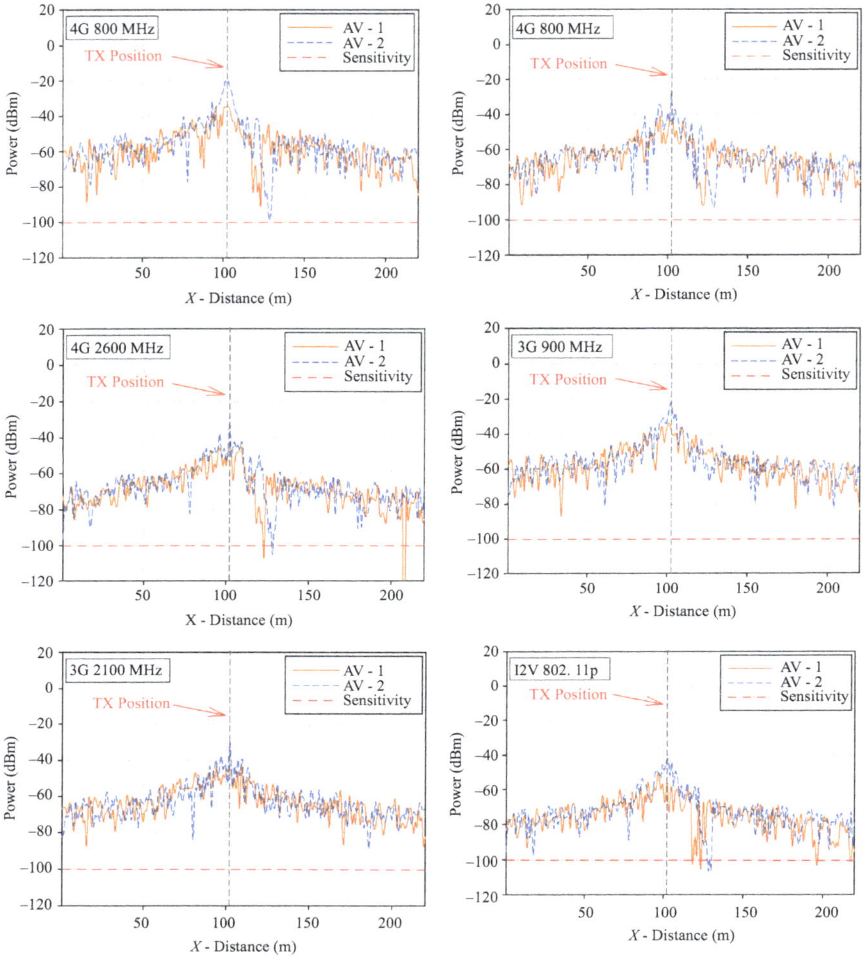

Figure 5.23 Comparison of linear distribution of received power levels for the different considered systems for the I2V link within the urban environment

Table 5.5 Empirical modelling within the urban environment for the V2V link

V2V link	Description	PLo (dB)	*n*	σ (dB)
802.11p	Along AV-1 ($y = 81$ m) x: (0 to TX2/TX2 to 220 m)	81.72/63.85	2.22/1.93	16.26/11.62
	Along AV-2 ($y = 70$ m) x: (0 to TX2/TX2 to 220 m)	82.31/62.11	3.31/1.97	11.10/12.25

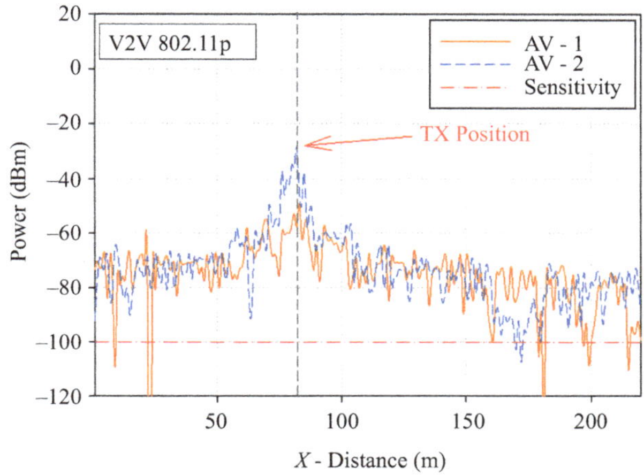

Figure 5.24 Comparison of linear distribution of received power levels for the different considered systems for the V2V link within the urban environment

Table 5.6 Empirical modelling within the rural environment for the I2V link

I2V link	**Description**	**PLo (dB)**	***n***	**σ (dB)**
4G 800	Along AV-1 (y = 81 m) x: (0 to TX1/TX1 to 220 m)	68.75/46.19	2.26/2.87	9.22/24.07
	Along AV-2 (y = 70 m) x: (0 to TX1/TX1 to 220 m)	68.75//46.26	2.65/2.66	10.93/25.66
4G 1800	Along AV-1 (y = 81 m) x: (0 to TX1/TX1 to 220 m)	77.12/53.29	2.56/2.80	9.73/22.79
	Along AV-2 (y = 70 m) x: (0 to TX1/TX1 to 220 m)	73.80/54.08	2.43/2.45	10.03/23.81
4G 2600	Along AV-1 (y = 81 m) x: (0 to TX1/TX1 to 220 m)	79.17/57.16	2.65/2.79	8.91/22.51
	Along AV-2 (y = 70 m) x: (0 to TX1/TX1 to 220 m)	78.59/57.71	2.67/2.44	10.08/23.61
3G 900	Along AV-1 (y = 81 m) x: (0 to TX1/TX1 to 220 m)	67.63/46.79	2.35//2.00	9.58/9.62
	Along AV-2 (y = 70 m) x: (0 to TX1/TX1 to 220 m)	67.36/45.03	2.46/1.85	10.05/10.10
3G 2100	Along AV-1 (y = 81 m) x: (0 to TX1/TX1 to 220 m)	72.96/54.99	2.18/1.76	8.51/9.21
	Along AV-2 (y = 70 m) x: (0 to TX1/TX1 to 220 m)	72.86/53.30	2.11/1.87	8.69/10.35
802.11p	Along AV-1 (y = 81 m) x: (0 to TX1/TX1 to 220 m)	84.05/68.23	2.25/1.58	8.36/9.91
	Along AV-2 (y = 70 m) x: (0 to TX1/TX1 to 220 m)	84.41/63.84	2.57/1.81	10.03/12.07

Table 5.7 Empirical modelling within the rural environment for the V2V link

V2V link	Description	PLo (dB)	n	σ (dB)
802.11p	Along AV-1 (y = 81 m) x: (0 to TX2/TX2 to 220 m)	80.98/59.52	1.97/2.95	16.00/25.05
	Along AV-2 (y = 70 m) x: (0 to TX2/TX2 to 220 m)	81.39/61.40	3.14/1.80	10.84/10.96

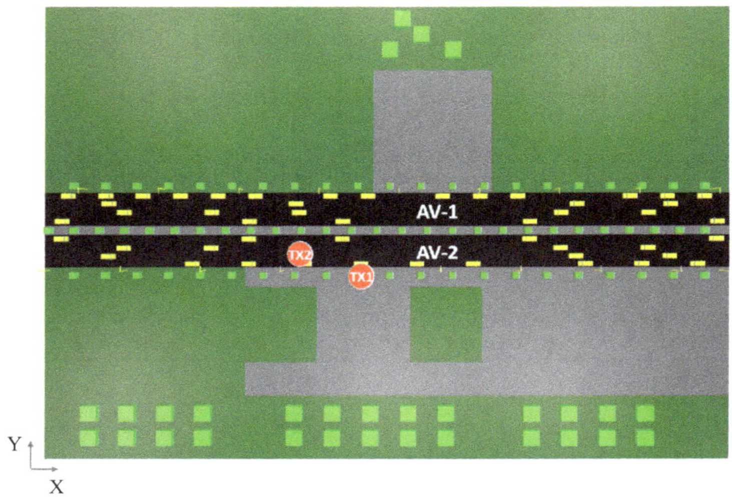

Figure 5.25 Schematic view of Avenues 1 and 2 in the rural environment for the representation of linear distribution of received power levels in Figures 5.26 and 5.27

radio planning purposes before the deployment of a wireless communication system within the scenarios under consideration. It can be seen that n values vary depending on the considered scenario. This is due to the fact that, as stated previously, the morphology and topology of the scenario have a great impact in radio wave propagation. In this case, the absence of buildings makes the values of n to change.

In Figure 5.25, an aerial view of the selected rural scenario is presented with dashed lines along the X-axis for different values of Y, which correspond with AV-1 and AV-2 (see Figure 5.1 for reference). These dashed lines represent the lines which correspond with the linear distribution lines of received power which are represented in Figure 5.26 for the I2V communication link and in Figure 5.27 for the V2V communication link, for 1.5 m height. From Figures 5.26 and 5.27, the differences in received power levels for the different analysed systems can be seen, which can aid designers before the deployment of wireless communication systems in this type of scenarios.

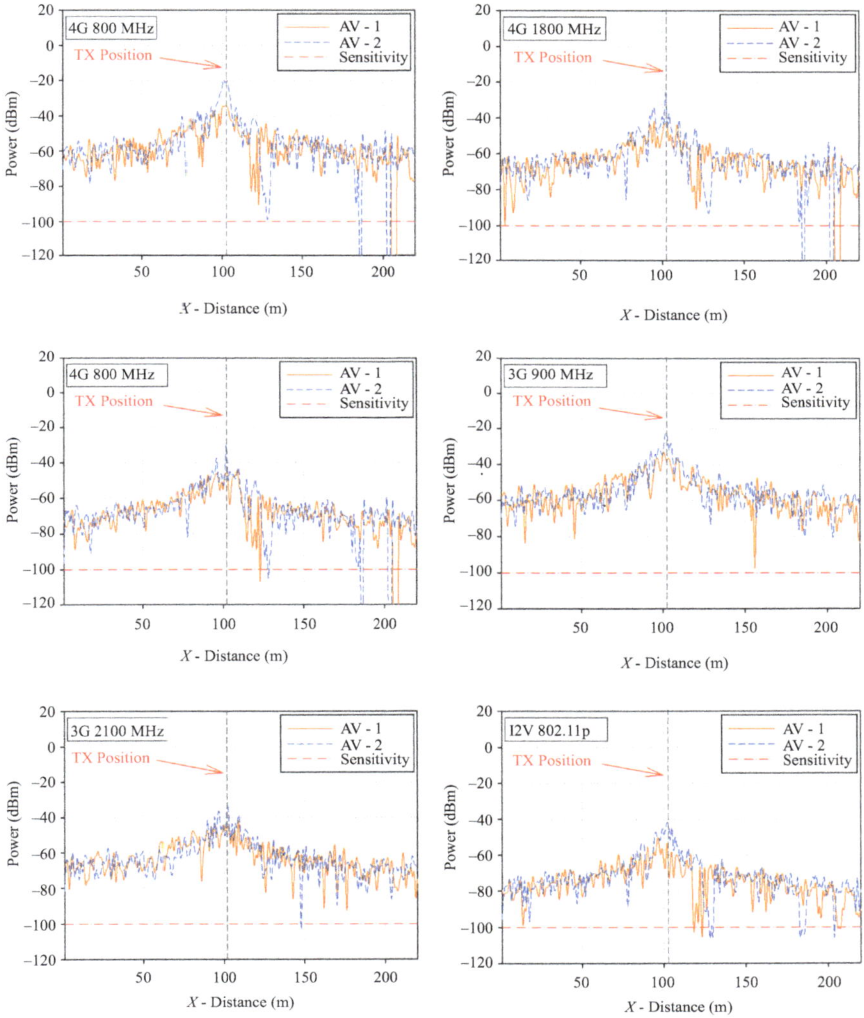

Figure 5.26 Comparison of linear distribution of received power levels for the different considered systems for the I2V link within the rural environment

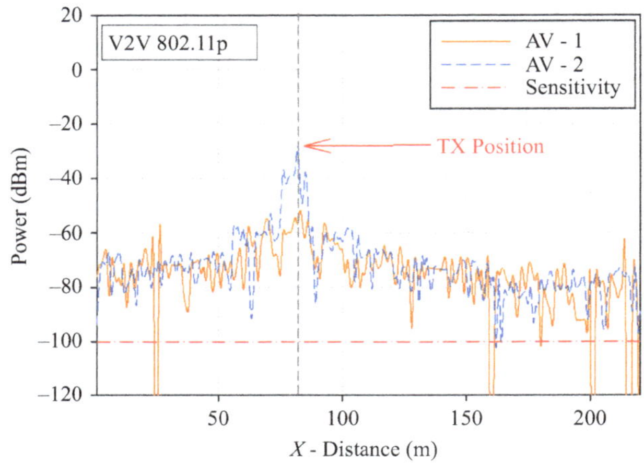

Figure 5.27 Comparison of linear distribution of received power levels for the different considered systems for the V2V link within the rural environment

References

[1] De la Roche, G., Alayón Glazunov, A., and Allen, B., *LTE-Advanced and Next Generation of Wireless Networks: Channel Modelling and Propagation*, Wiley, New York, 2013.

[2] The Institute of Electrical and Electronics Engineers, *IEEE Std 802.11 p-2010. Part 11: Wireless LAN Medium Access Control (MAC) and Physical Layer (PHY). Specifications-Amendment 6: Wireless Access in Vehicular Environments*, The Institute of Electrical and Electronics Engineers, New York, NY, 2010.

[3] Maurer, J., Fugen, T., Schafer, T., and Wiesbeck, W., 'A new inter-vehicle communications (IVC) channel model', *Proceedings of the 6th IEEE Vehicular Technology Conf.*, Los Angeles, CA, USA, 2004, 9–13.

[4] Wiesbeck, W., and Knorzer, S., 'Characteristics of the mobile channel for high velocities', *Proceedings of the 2007 Int. Conf. Electromagnetics in Advanced Application*, Torino, Italy, 2007, 116–20.

[5] Emmelmann, M., Bochow, B., and Kellum C. C., *Vehicular Networking: Automotive Applications and Beyond*, Wiley, New York, NY, 2010.

[6] Balanis, C. A., *Advanced Engineering Electromagnetics*, Wiley, New York, NY, 1989, 205.

[7] Azpilicueta, L., Falcone, F., and Janaswamy, R., 'A hybrid ray launching-diffusion equation approach for propagation prediction in complex indoor environments', *IEEE Antennas and Wireless Propagation Letters*, 2017, 16: 214–17.

[8] RedPine Signals, 'Driving Wireless Convergence. WaveCombo Module. 802.11p V2X', 2018. [Online]. Available: http://www.redpinesignals.com/Products/802.11p_V2X_Connectivity/802.11p_V2X_Module.php. [Accessed: 28-Feb-2018].

[9] Kapsch TrafficCom, 'In-vehicle products (5.9 Ghz). EVK-3300. V2X Evaluation Kit', 2018. [Online]. Available: https://www.kapsch.net/ktc/downloads/. [Accessed: 28-Feb-2018].

[10] Ublox, 'VERA-P1 series. Host-based V2X transceiver modules. Product Information.', 2018. [Online]. Available: https://www.u-blox.com/en/product/vera-p1-series. [Accessed: 28-Feb-2018].

[11] Rappaport, T., Wireless Communications: Principles and Practice, 2nd ed., Communications Engineering and Emerging Technologies Series, Prentice-Hall PTR, Upper Saddle River, NJ, 2002.

[12] Cheng, L., Henty, B., Bai, F., and Stancil, D., 'Highway and rural propagation channel modeling for vehicle-to-vehicle communications at 5.9 GHz', *Proceedings of the Antennas and Propagation Society International Symposium*, 2008, 1–4.

[13] Cheng, L., Henty, B., Stancil, D., Bai, F., and Mudalige, P., 'Mobile vehicle-to-vehicle narrow-band channel measurement and characterization of the 5.9 GHz dedicated short-range communication (DSRC) frequency band', *IEEE Journal on Selected Areas in Communications*, 2007, 25: 1501–16.

Chapter 6
Vehicular communications channel modelling

Vehicular communications technology is promising for supporting multimedia services of high capacities and data rates, inasmuch as it offers wide bandwidths with an unforeseen mobility and ubiquity. However, the combination of large bandwidths and outsized mobility results in severe signal degradation with increased multipath problems and the consequent loss of the quality of service. A precise knowledge of radio channel behaviour is a basic step to define effective models valid to design, build and deploy the vehicular communication systems that support services with guarantee of a correct performance.

Unlike cellular channels, which are stationary in at least one end of the radio link, vehicular radio channels exhibit both non-stationary random spatio-temporal behaviour and frequency dependence. Thus, its characterization by means of widely known mathematical formulation enunciated in [1] is not strictly valid. Channel characterization can either be achieved by radio channel measurements, or when these are not possible, with simulation-based models that describe these channels in detail.

In this chapter, the concepts of channel measurement, modelling and estimation for vehicle-to-everything (V2X) communications are described. The channel models most used for inter-vehicular communications (V2V, vehicle-to-vehicle), such as geometry-based models, are described. Other models for vehicular communications are also described. Critical features of the channel models besides their importance are analysed, such as spatio-temporal characteristics, stationary and non-stationary and three-dimensional (3D) modelling. Channel models for multiple-antenna systems multiple-input multiple-output (MIMO) are also introduced. A discussion is included on specific concerns that need to be considered for channel modelling, such as channel sounders to be used and parameter estimation to be accomplished. The most important techniques of radio channel sounding are presented, both for narrowband and wideband characterization. At the end of the chapter, fundamentals of channel prediction and estimation algorithms of parameters of interest are introduced.

6.1 Channel modelling, measurement and estimation

First and foremost, a definition for 'radio channel' must be stated. In Figure 6.1, the difference between the propagation environment and the radio channel has been

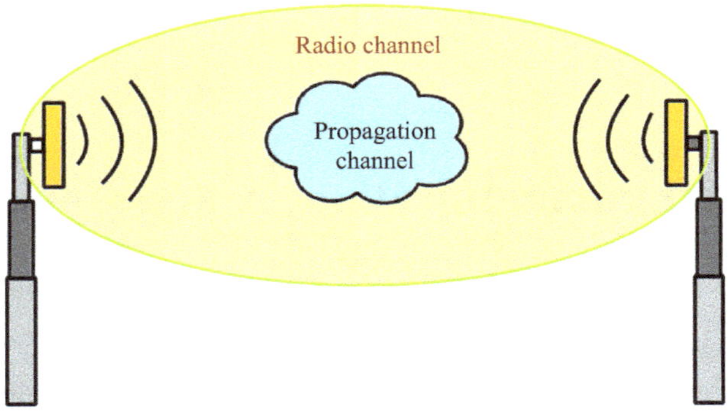

Figure 6.1 Concept of radio channel

illustrated. Wireless radio propagation takes place when an electromagnetic wave travels between two antennas through a specific environment or channel composed of a propagation medium (air, water, troposphere, soil, etc.) and the set of elements located within it. The transmitted signal might undergo diverse propagation phenomena when travelling through the propagation channel or environment, such as reflection, refraction, diffraction, scattering or diffusion. The combination of the wireless propagation environment or channel and the antennas used for transmission and reception of the electromagnetic signal is called 'radio channel'.

The concept of radio channel model denotes a transfer function that involves the electromagnetic phenomena undergone by the signal travelling between transmitter and receiver antennas, and the propagation channel [2].

The term radio channel characterization has been traditionally used in the field to describe the models, theory, and experimental or simulated collected data that summarize overall knowledge of a wireless radio channel in a specific type of environment, typically resulting a function of range, antenna polarization, channel bandwidth and center frequency, among other factors [3]. Thus, the channel modelling implies the quantification of the complete set of parameters that affects the transmitted electromagnetic wave from transmitter to receiver in the frequency band of interest and over the spatial region of the propagation channel [3].

Radio channel characterization has been a regular research activity since the 1920s and 1930s, when researchers recognized the need to characterize wireless channels as a way to quantify its effects, with pioneer studies developed for ionospheric propagation and, for radar systems in the 1950s and 1960s [4–7]. Since then, many research works have appeared providing more results to understand radio propagation, the most relevant of which those due to Cox [8] and Bello [1], which would become the theoretical basis of radio channel characterization to the present day. Finally, the success of the Committee on Science and Technology (COST) actions in the 1990s [9] propitiated and encouraged further collaborative channel modelling activities by standards groups, and so far, it has been assumed as an essential activity in radio engineering.

A description of radio channel behaviour may be done in terms of a system of equations which provide an approximation to the underlying physical mechanism that takes place. The random behaviour makes necessary the development of mathematical models to represent the specific channel characteristics. In general, the channel can be considered linear, and therefore, a representation of the channel as a linear system is a good assumption. Radio channel also undergoes time variations, so the mathematical models used will have to describe the channel as a linear time-variant system. Moreover, the time variation occurs as a random process, so a statistical characterization of the channel is needed [10].

The characterization of these random time-variant systems was enunciated by P.A. Bello in [1]. Various references can be found in literature to this formulation, which establishes relations between the input and the output signal to the channel. These relations are described through eight system functions, but only four of them are usually defined. The other four functions are resulting from the autocorrelations of the four basic system functions.

The four basic system functions enunciated by P.A. Bello, are the following:

- time-variant impulsive response function, $h(t, \tau)$,
- time-variant transfer function, $H(t, f)$,
- Doppler shift-delay function or scattering function, $S(\nu, \tau)$ and
- frequency-shift transfer function, $H(\nu, f)$,

where the variable t means time, τ is the path delay, f is the frequency and ν is Doppler frequency. Figure 6.2 shows a representation of relations between these four basic functions. By assuming wide-sense-stationary and uncorrelated-scattering for the channel, a direct or inverse Fourier transform (IFT) applied to one function in a determined variable or domain drives to the desired function.

In [1], P.A. Bello established that actual channels present, with a reasonable approximation, the properties of being WSS (*wide sense stationary*) in the variable

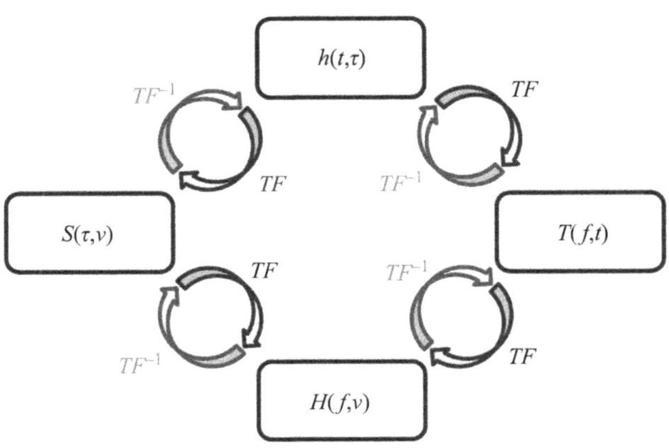

Figure 6.2 Relations among random time-variant system functions

t and US (*uncorrelated scattering*) in the variable τ. When the system can be assumed as WSS, the average of the processes associated to the characteristic functions will be constant in time, and the autocorrelation in two different time instants, t_1 and t_2, will depend only upon the time difference $t = t_2 - t_1$.

The US hypothesis implies that the contributions of elemental scatters with different delays associated are uncorrelated. The time spread is due to the presence of multipath. Along different paths, reflections and diffractions are produced on different bodies, and therefore, the signals belonging to these different paths will be normally uncorrelated. So, uncorrelated time spread can be assumed in the radio channel.

It can also be considered that the channel presents slow time variations superposed to fast time variations, simplifying the analysis. Frequently, fast time variations have a Gaussian behaviour because scatters near to the extremes of the link produce more propagation paths with similar delays. Therefore, these signals will be added coherently within the coherence time. If the fast time variations can be characterized with Gaussian statistics, the correlation function will be enough to describe completely these variations.

Gaussianity condition, together with wide sense stationary uncorrelated scattering (WSSUS) conditions, allows defining an ergodic channel model. A stochastic stationary process is ergodic if the statistical averages or expectations equal the time averages. In an analogue way, a randomly linear time-variant system will be ergodic if the statistical averages pertaining to the system functions equal the time averages.

The characterization of a mobile radio channel by means of [1] functions is purely theoretical. Actual channels present random time variations, and therefore, the system functions described above become in stochastic processes. A rigorous characterization of these systems is complex, and in practice, it is reduced to the autocorrelations corresponding to the system functions.

If ergodicity can be assumed in the channel, the transfer functions of the system can be derived from time averages instead of expectations. Under this assumption, estimations of correlation functions are reliable. According to the properties of all ergodic processes, the channel characteristics can be represented by means of the parameters estimated from any data ensemble.

The general equations of the system functions described by Bello are difficult to obtain by measurements and even simulations. For this reason, only canonical channel parameters, derived from them for null value of some parameters, are derived from measurements or simulation. These channel parameters of time and frequency variation are the following:

- Channel impulse response (CIR) function, $h(t, \tau)$.
- Power delay profile (PDP), $P(\tau) = R_{hh}(\tau,\Delta t = 0)$, that provides the RMS delay spread σ_τ that measures the channel spreading of a input delay component.
- The spaced-frequency correlation function or simply frequency correlation function (FCF), $R_{HH}(\Delta f,\Delta t = 0)$, that defines the channel coherence bandwidth B_C indicating the frequency selectivity of the channel.

- The spaced–time correlation function, $R_{HH}(\Delta f = 0, \Delta t)$, which measures the correlation between channel effects separated in time by Δt also called the channel coherence time T_C that indicates the time selectivity of the channel. During that time frame, the channel response can be considered as invariant.
- Doppler spectra or scattering, $S_S(0, \nu) = S_S(\nu)$, whose width the Doppler spread f_D measures the channel spreading of a frequency input tone.

These sets of parameters do not suffice to characterize fading, multipath and frequency selective effects in Gaussian WSSUS (GWSSUS) models. The complete information is contained in the CIR and FCF functions, and not in the remaining system functions most of them impossible to be measured for specific environments. For instance, the value provided by the delay spread is not representative of the number of multipath components (MPCs) and it cannot indicate the angle of arrival (AoA) of the paths additional to the direct ray. Those parameters are merely valid to obtain an overview of the radio channel at a glance in a quantitative form, and they can be used to easily compare results between researchers and even to validate measurements. The statistical model given by the time-varying CIR, and its related parameters, becomes a useful characterization widely used for analysis, computer simulations and experiment.

The radio channel characterization obtained by P.A. Bello and B.D. Nelli's mathematical formulation is not general and results dependent on the measurement conditions (centre frequency, bandwidth, polarization, etc.) and it is environment specific. For this reason, it is considered a deterministic characterization named stochastic random model.

As mentioned later in Section 2.2, the vehicular radio channel does not meet the WSSUS condition since it is affected by a double non-stationarity (NS) [11]. However, it is admitted being considered WSSUS for short periods of time and distance so that a model can be provided. Besides, vehicular radio channel does not respond to only one general model, and different models have been provided according to specific environments, applications and other criteria.

In any case, it seems clear that radio channel characterization can be achieved by radio channel measurements. The objective is to obtain a model that describes the channel in detail and the parameters of time and frequency variation.

For developing a measurement-based channel model, some decisions must be made a priori to determine which aspects of channel features that produce propagation impairments are critical and must be gathered for its analysis. Among others, a model may reflect model channel parameters such as fading, long- and short-term variability, time and frequency variation, AoA distribution, delay spread and Doppler shift [3,4].

Many times, models are not describing general scenarios and they are reduced to site-specific, losing so generality of application. Applications of the technology or cases of use must contribute to determine which channel parameters need to be analysed: operation centre frequency and bandwidth, data rate, bit error limits, etc. Finally, the transmitter and receiver antenna parameters – e.g. radiation pattern, polarization, beamwidth and side-lobe level – also determine the validity of the channel model estimated. Ideally separate the distortions introduced by the wireless channel (which are captured by the channel model) from the distortions introduced by the antennas (which are captured by the antenna model). A large set of additional aspects will also affect the channel model: mobility of the transmitter and/or receiver ends, heights of the involved antennas, existence of line of sight (LOS), ground morphology, environment (urban, suburban, rural, sea, vegetation, scatterers), distance between transmitter and receiver ends, and many others. Additionally, most of them act as inter-dependent variables.

In order to determine which channel parameters will be considered in the enunciation of the channel model, the main criterion consists of determining the likely contribution to the signal impairment. This decision usually reduces the list of channel parameters which will be treated as random variables to be physically and mathematically characterized.

Then, a new stage of the channel characterization starts. It is required to determine a way to provide values to the channel parameters. Here three main forms of obtaining data can be applied: by means of measurements, by simulation or by formulation-based techniques (finite difference time domain (FDTD), finite element method (FEM), etc.). A fourth option would be possible also and it consists of mathematically deriving a close expression to mathematical theoretical formulation.

If the measurement option is decided as the data set gathering form, it will be needed to configure and build a suitable channel sounder. It consists of specific equipment required to perform radio channel measurements. Several sounding techniques have been developed in order to measure radio channel impulse or frequency responses. A review of these techniques has been presented in this chapter in Section 6.2.2, for narrowband and wideband characterization.

Next, the measurement campaign starts in the selected environments. As described later in Section 6.2.1, there are different scenarios and links for vehicular communications with specific features. Given that a vast amount of measurement data is required to achieve statistically reliable results [4], so the channel sounder will require a digital storage to save raw data for later processing that usually becomes a bottleneck in the measurement process.

Once required measurement data are collected, it must be reduced and validated. Data set reduction and processing will extract the channel parameters of interest [4]. The vast amount of raw data makes necessary the application of some algorithms to facilitate and provide accuracy in the prediction and estimation stage, as described in Section 6.5.

All of the above stages needed to achieve a measurement-based model are depicted in Figure 6.3. Usually transitions between stages are not unidirectional. Measurements under controlled scenarios, such as an anechoic chamber, can be used to evaluate the channel sounder performance, to fine-tune the equipment and to define or improve the data collection protocol before real scenario measurements take place [4]. Controlled measurements are also useful to identify potential models to be fit from the measurement dataset reduction [4].

As described in Figure 6.3, at the end of the process a radio channel model is obtained. It will offer a quantitative and comprehensive description of the

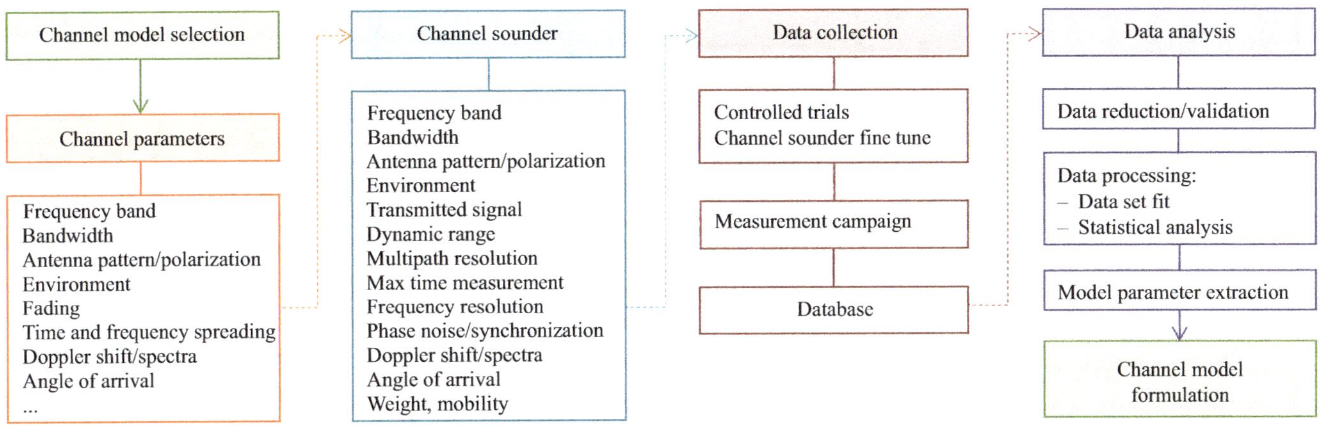

Figure 6.3 Development of a radio channel model based on measurements

Table 6.1 Basic radio channel classification [12]

Parameter	Channel type	
Frequency	Dispersive	Non-dispersive
	Narrowband	Wideband
Time	Time-variant	Time-invariant
Environment	2D	3D
Data origin	Measurement	Simulation
Validity	Deterministic	Stochastic
Applicability	Link-level	System-level

underlying mechanisms occurring to the signals travelling through it. In Section 6.4, the main models developed for vehicular communications are reviewed.

Many channel classifications have been done taking into account different points of views. In Table 6.1, a basic classification illustrates potential ways of classifying models; for example, with regard to the type of channel that is being considered, channel models are classified in wideband versus narrowband [12].

Finally, the usability aspects to assess for selection of a proper vehicular channel model are discussed in Section 6.6. A model is expected not be so complex as to limit its helpfulness. It must be formulated in a convenient form to be used in analysis, computer simulations and experiments [12].

6.2 Radio channel measurements

Despite the NS and the time-dependent variability, the vehicular channel is assumed as linear and therefore a measurement-based model can be derived by achieving the CIR and the FCF functions from data sets. Performing vehicular channel measurements is particularly challenging due to the fast mobility and dynamic environment that alongside the low height of antennas [13–15] make small- and large-scale signal statistics vary rapidly.

The specific features of the vehicular urban environments cause the propagated signal to interact with a large number of surrounding objects from the transmitter to the receiver [13–15], leading to a large number of rays to reach the receiver due to the reflections and diffractions. To all of the above must be added the high density of objects that combined with the high mobility of the communicating vehicles and their surroundings makes of characterizing the vehicular channels far from trivial [13–15].

Vehicular communications differ from conventional wireless and cellular radio systems with semi-static features where only one terminal, usually the user device, is moving and the other one, base station (BS) or access point, stays fixed [13–15]. Therefore, channel models developed for semi-static wireless mobile systems are not well-suited for vehicular systems, due to the unique nature of vehicular channels. For instance, differences in the relative height of the transmitter and receiver antennas cause significantly different signal propagation behaviour [13–15].

Therefore, for the correct design and performance evaluation of vehicular channel models, measurement campaigns must be conducted for the different application scenarios where the communications will be deployed.

Generally, measurements campaigns reported in literature have shown that propagation phenomenon in analogous environments with different frequencies can vary considerably [3,13–18]. Then, with the same level of relevance than the environments, measurements must consider carrier frequencies, frequency selectivity, radiation pattern and polarization of antennas, antenna emplacement, type of communicating vehicles, link distance or range, directions of motion for transmitter and receiver, traffic density, and surrounding objects, always according to the aimed model channel and statistics to derive from the data sets [3,13–18].

Following the IEEE 802.11p standard [19], measurement campaigns should be conducted at carrier frequencies in the 5.9 GHz band, for narrowband and wideband channels. As described in Section 1.3, the Federal Communications Commission has allocated 75 MHz of licensed spectrum in the 5.9 GHz band for dedicated short-range communication (DSRC) design dedicated to safety applications for vehicular communication systems. However, as indicated in [13,14], for improved design of non-safety applications measurement campaigns are still required for V2V systems at other frequency bands outside the standard, such as the IEEE 802.11a frequency band 2.4 GHz, and around the IEEE 802.11a frequency band at 5–5.2 GHz.

Unlike wireless cellular systems, vehicular communications operating at 5.9 GHz establish radio links of short distances in the range of 100 to 500 m, whereas current wireless cellular systems operate between 700 and 2,100 MHz and over longer distance reaching up to tens of kilometres [13,14,20]. Some research has been reported focused on millimetre wave (mmWave) bands [3,4,21–28]. Although it shows a large potential due to the large bandwidth provided in this spectrum, it will hardly be used in the short term.

In the position document [29], the European Automobile Manufacturers Association (ACEA) proposed considering the 3.4–3.8 GHz band. Among the reasons for this proposal, it should be noted that Intelligent Transportation System (ITS)-G5 and Long Term Evolution (LTE)-V2X (see Chapter 1) interfere with each other if operated on the same frequency channel due to the different solutions adopted regarding the access to the communication channel, i.e. medium access control, MAC. The ACEA considers that a frequency band with a lower carrier frequency would be beneficial for V2X also in terms of antenna placement. ACEA also indicates that communication in the 3.4–3.8 GHz provides the advantage of not producing harmonics interfering in the 5.9 GHz band. In summary, the 3.4–3.8 GHz band is considered by ACEA as an optimal compromise between high (5.9 GHz band) and low (3.4–3.8 GHz band) carrier frequencies in terms of radio propagation characteristics as well as antenna size and allocation [29].

Commonly, measurements are taken along linear paths or radials to explore the dependence on the distance, with spatial resolutions along those paths between a quarter and eight of wavelength in order to detect the small-scale fading rapid fluctuations of the amplitude, phases, or multipath delays of a radio signal over a short period of time or travel distance fast-fading component [30]. Additionally, pan and tilt scanning is combined to the linear displacement in order to achieve angular information and then obtaining a 2D or 3D channel model. In occasions, the amount of measurement points to be performed may result overwhelming and impossible to achieve in terms of time.

Due to high operating frequency, vehicular channels exhibit higher signal attenuation, particularly for the additional paths due to mechanisms of reflection, diffraction and scattering [3,18]. This turns the multipath propagation mechanism dominant. This fact combined with the high vehicle mobility and high density of surrounding objects leads to a time-varying number of MPCs that must be captured in order to ensure correct channel modelling.

The channel bandwidth proposed by the IEEE 802.11p standard is of 10 MHz that corresponds to a delay bin width of 100 ns equal to the reciprocal of the signal bandwidth [3,18]. This delay bin width value implies that only MPCs separated in delay by an amount larger than 100 ns are individually distinguished and therefore resolvable. Those MPCs can then be compensated, otherwise they would result combined in the receiver via complex addition likely resulting in signal distortion [3,18].

By far, Doppler shift is the major concern in vehicular communication systems. If the carrier frequency f_c is 5.9 GHz, for pedestrian speed, a relative velocity of 20 m/s, or equivalently 72 mm/h, produces a maximum Doppler shift of $f_{D,max}$ = 393.3 Hz smaller than the f_c value [3,18]. For a car speed of 200 km/h, maximum Doppler shift would be $f_{D,max}$ = 1,092.6 Hz, also smaller than f_c.

The licensed spectrum of 75 MHz allocated by the Federal Communications Commission in the 5.9 GHz for DSRC includes seven channels of approximately 10 MHz bandwidth. Those channels will exhibit frequency-selective features (or wideband) channels [13,14]. Then channel characterization based on narrowband measurement results is not sufficient. Wideband measurement campaigns are therefore essential for understanding the frequency-selectivity features of vehicular channels and designing high-performance systems [13,14].

Measurement campaigns must not only be focused on single-input single-output (SISO) systems consisting of single antenna at both link ends. Measurement campaigns should be conducted with MIMO systems dotted of multiple antennas at both link ends given that this technology has been identified as a firm candidate for future communication systems in IEEE 802.11p and WAVE standards since multiple-antenna elements can be easily employed in the large surfaces of the communicating vehicles [4]. Moreover, MIMO channel measurements should be carried out in wideband configuration [4].

6.2.1 Vehicular environments

Similarly, to the cellular case, the vehicular channel must be represented by a set of models for different environments, or channel classes: urban, suburban and rural were established by the COST in 2007 [3,18,31]. Initial vehicular models employed these classes, with subclasses for other settings, such as traffic circles, urban plazas, and walled roadways, tunnels, street intersections, overpasses/underpasses parking garages, ramps and slopes.

Thus far, a comprehensive classification of all the vehicular channel classes does not exist. The most complete classification can be found in [32]. The scheme in [32] makes use of the approach employed in the COST 259 effort for cellular [33,34], in which a hierarchy of multiple cell types contains multiple radio environments, within which are multiple propagation scenarios [14].

It has been demonstrated that the radio propagation is strongly influenced by the type of environment where the communication is taking place, and more particularly by the objects that make up that environment, as well as their number and characteristics (e.g. size, density, static and mobile) [14].

It results easier to identify object types than undertaking the classification of vehicular environments that those objects create [14]. Considering the main type of objects of the environments where vehicular communication occurs are most often qualitatively classified as highways, suburban areas and rural areas [14]. These become the most often analysed propagation environments

Any classification of environments should be considered under a sceptical perspective. As stated in [14], deviations in the presence, positions, density and variety of objects, along with variations in the velocity and density of vehicular traffic will significantly modify the signal propagation whatever the environment class.

Different measurement research studies have reported extremely variable and even inconsistent results in channel parameters as path loss (PL) exponents or spread delays, for the same environment class [14]. Environments reported in many studies may be rare, uncommon and not repeatable: urban area with open spaces similar to highways, or neighbourhoods with low-rise buildings arranged similarly to a typical suburban setting [14]. It can be thought that most of existing channel models are more site-specific, and for this reason, they do not properly work when applied to another environment of the same class.

Therefore, rather than being environment-specific, a radio channel model should be defined considering the specific objects present in the environment, along with their accurate features, dimensions and locations [14].

Whereas it is widespread the realization of propagation measurements in the most common environments, wherein vehicular communications occur, other environments remain underexplored. Among others, measurements in multilevel highways, tunnels, parking garages, bridges and roundabouts are scarce. Further research is necessary for those scenarios of innovative applications [14].

6.2.1.1 Vehicular radio link types

As described in Chapter 1, vehicular communications can be classified according to the link type, also called operation mode, defined by the type of vehicle of one of the communication ends. Among others, the following links could be listed [35–37]:

- V2V: The connection between vehicles that allows V2X enabled self-driving cars to communicate with one another for speed control, emergencies, entertainment, etc.
- Vehicle-to-infrastructure (V2I): The link that allows self-driving cars to interchange information from smart units located on buildings, bridges, roads

and traffic lights, and provides access to information such as traffic conditions, Internet access information, download content, communications and emergencies.

- Vehicle-to-roadside (V2R): Vehicle connections with elements located at the roadside (roadside units, RSU) such as road signs, emergency signalling equipment and payment system.
- Vehicle-to-pedestrian (V2P): The link established between the car and the pedestrian, planned to support Vulnerable Road User safety applications like pedestrian detection systems that can work with a car doted of advanced driver-assistance systems.
- Vehicle-to-home (V2H): The link in which smart homes can send and receive information directly from the car.
- Vehicle-to-network (V2N): The mobile connection between the car and a cellular network provider.
- Vehicle-to-cloud (V2C): The link that provides direct access to cloud networks using secure TCP/IP connections.

In Figure 6.4, some of the above listed vehicular link types are illustrated. Each link type exhibits very specific signal propagation properties. Then, the radio channel model must combine the vehicular link type and the environment. Generally, each link shows differences in mobility, fading, shadowing, and height of the transmitter and receiver antennas. All of these issues introduce conditions leading to significant variations in reflections, diffractions and scattering patterns of the propagating signal [13,14,38,39].

- In V2V channels, the transmitter and receiver antennas use to be located on the vehicle rooftop, and both vehicles are mobile. Links are non-stationary in time or distance.

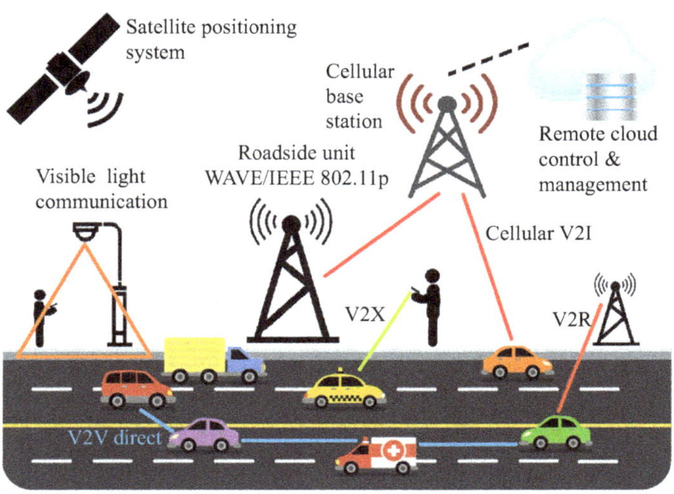

Figure 6.4 Vehicular communications: different types of links [36]

- In V2I channels, the transmitter (BS or access point) is stationary whereas the vehicle is mobile, and the antenna is elevated with respect to the receiver similarly to a picocell.
- In V2P channels, established links are very short with both communications ends in movement. Links are non-stationary in time or distance.

6.2.1.2 Radio link range

Vehicular links can be classified according to the distance that the link established between transmitter and receiver ends can reach [13]. This concept is similar to the classification used for conventional cellular systems. Three main scales are determined according to the radio link range:

- Large spatial scale (LSS): The link range larger than 1 km, used for broadcasting or geographic broadcasting (geocasting).
- Moderate spatial scale (MSS): The link range from 300 m to 1 km.
- Small spatial scale (SSS): The link range smaller than 300 m, used for broadcasting, geocasting or unicasting.

Most of use cases of vehicular communications pertain to MSS or SSS categories, mainly V2V applications. For these reasons, measurement campaigns usually are conducted in these scenarios [13,40–44]. Channel modelling should further consider measurements to explore LSS scenarios.

6.2.1.3 Vehicles types

When planning measurement campaigns, it must be considered the different types of vehicles that usually are present in the environment of interest. Radio channel models defined for environments with preponderance of one vehicle type might not be valid if the vehicle type changes. In [14,45,46], the following types of vehicles are listed as potential users of vehicular communications: personal vehicles, commercial vans, trucks, scooters and public transportation vehicles. Railways, bicycles, motorbikes and UAVs should be also included. The list will always depend on the use case under modelling.

Each vehicle type presents particular features like dimensions and mobility dynamics that impact the channel modeling regardless of the vehicle role – transmitter, receiver - or surrounding object [14,45,46].

6.2.1.4 Environment surrounding objects

Regardless of the communication link types, vehicular environments mainly consist of a number of different types of objects that impact the signal propagation. Along with mobile objects, accounting for static objects is critical for modelling vehicular channels, since both types of objects are sources of shadowing, reflections and diffractions [13,14,47].

For each environment, mobile objects can be predominant as in highways; whereas in roadside environments, according to the scenarios categorized as urban canyon, suburban street and expressway, the predominant objects located on the roadside are static, like buildings, bridges, trees, parked cars, etc., which affects the channel statistics [13,14,48–51].

In case of vehicular communications, the three most important type of objects that influence the propagation are buildings, vehicles and vegetation [13,14]. A combination of different object types, as well as their number, size and density, has a profound impact on the radio propagation.

6.2.1.5 Vehicles direction of motion

Directions of motion of both vehicles immersed in a vehicular communication also affect channel statistics due to the Doppler effect occurring between the transmitter and the receiver ends [13,14]. Generally, channel characteristics are not analysed when communicating vehicles are moving in opposite or transversal directions.

The direction of motion depends and varies on the scenario topology. For vehicular channels in highway environments, the communication between the transmitter and the receiver vehicle takes place over the road surface, i.e. within 1D topology. In urban environments, the communicating vehicles are likely to be on different streets and the scenario shows 2D topology [14].

6.2.1.6 Vehicular traffic density

In addition to considering the locations and density of surrounding objects, static or mobile, the changes in the vehicular traffic density (VTD) will introduce noteworthy variations in the propagation conditions that affect the channel modelling of the environment under study [13,14]. Very especially the VTD strongly affects the channel statistics for MSS and SSS scenarios [13,14]. In general, it has been determined that the shorter link range, the deeper impact of the VTD will be [13,14]. On the contrary, vehicular channels typically show non-isotropic scattering except in cases of high VTD.

Then measurement campaigns for MSS and SSS scenarios should consider different VTDs, at least when the transmitter and the receiver vehicles move in opposite directions [13,14].

Variations of the velocity of the mobile surrounding objects also impact the propagation signal and, if applicable, should be considered in the development of a vehicular channel model.

6.2.1.7 Antenna placement

One of the main concerns of the ACEA is the antenna placement on the vehicle. In the position document [29], the ACEA transmits the importance of performing more research to optimize the antenna installation for any size of vehicle. This concern is due to the interference than the antennas actually located on the rooftop element known as shark fin can produce interference with the antennas dedicated to V2X systems [29].

The search for another antenna placement installation to achieve maximum performance of the V2X system at 5.9 GHz is not considered a trivial matter by ACEA. For large vehicles as trucks, the antenna cannot be placed on the roof of the cabin given that the trailer or bodywork will shadow the radio coverage. Putting the antenna on the cabin will cause that the LOS component will always be blocked backwards, although coverage in front of the truck will be good [29]. In small cars, the ideal placement would be in the middle of rooftop; however, it is not possible due to the curvature of the vehicle rooftop.

6.2.2 Channel sounding

Throughout the previous sections, it has become clear the need of obtaining accurate measurements for a correct modelling of the vehicular channel. The key element of the measurement process is the channel sounder to mimic V2X channels.

A channel sounder is the ad-hoc built equipment that mimics the radio channel features under analysis. For the vehicular case, the sounder should meet the following requirements: 10 MHz channels, data rates from approximately 3 to 27 Mb/s [3,52], maximum transmit power level of up to 29 dBm [3,53] and link range less than 1 km like planned in DSRC [3].

Generally, channel sounders can be classified in two main basic groups: narrowband and wideband. This classification depends not only on the transmitted signal bandwidth, but also on the radio channel behaviour within the occupied frequency band.

Narrowband sounders are intended to measure a radio channel that will be used by a narrowband system. Communication systems in which the signal distortion affects equally the occupied bandwidth are considered narrowband. Once time variation of signal level is measured and the channel parameters estimated from the measurement data, it can be decided if frequency dispersion affects the radio channel so that it can be considered narrowband or not.

Wideband sounders are intended to characterize channels for wideband systems that must provide information both about the strength level field, and about the impulsive response or the frequency response. If the channel response varies within the transmitted signal bandwidth, the channel is considered as wideband. Equivalently in the time domain, a wideband system exhibits time dispersion, whereas a narrowband system does not. Most multimedia radio systems fall into the wideband category because high data rates are required and hence large bandwidths are occupied, in order to accommodate enhanced multimedia services. In this group of sounders, two subcategories are distinguished, according to the domain chosen to perform channel measurements:

- Measurements in frequency domain: Radio characterization is obtained from the channel frequency response $H(t,f)$.
- Measurements in time domain: Radio characterization is obtained from the channel impulsive response $h(t, \tau)$.

Several classifications of channel sounders can be found in literature. Many of them are based not only in the type of channel to be measured, narrowband or wideband. According to the probe signal transmitted and the method used to process the response measured, it is possible to make other classifications. In Table 6.2, the most extended techniques used in radio channel sounding are listed, resulting an ample general classification of sounding techniques based upon a criterion different to the channel behaviour. In the next sections, some of them will be described in greater detail.

Table 6.2 Classification of radio channel sounding techniques

	Transmitter	Receiver
Pulse Short Pulse	Short Pulse	Envelope detector
Long Pulse	Phase codes (Golay, Barker, PRBS, Gold)	Heterodyne detector and correlation technique
	Frequency sweep (Chirp)	Heterodyne detector or matched filter SAW
Continuous waveform	Single tone	Heterodyne detector and correlation technique
	Multitone	Heterodyne detector and correlation technique
	Frequency sweep	Heterodyne detector or matched filter SAW and DFT
	Direct Sequence (PRBS)	Heterodyne detector and correlation technique

Figure 6.5 Narrowband sounder scheme

6.2.2.1 Narrowband sounding

In a narrowband system, the bandwidth is small enough to consider that the channel features are constant over the occupied band. In these cases, channel characterization in a single frequency within the band is sufficient because the channel behaviour will be identical for other values of frequency in the same bandwidth.

Characterization is then reduced to measure both the time variation of the channel, in the time domain, and to obtain the Doppler spectrum. Frequency response is not meaningful because it will be constant over the bandwidth, and so will occur with the impulsive response that would equal a Dirac's delta.

Narrowband characterization can be achieved by means of two stations working independently. One station acts as transmitter providing a single carrier pertaining to the frequency band to characterize. The second station will act as receiver, registering the level variation of signal received at each time instant. One of the two stations may be mobile, travelling within the coverage area to measure. If this occurs, the position and speed of the mobile station (MS) must be recorded to obtain the relation between power level and position, which is termed as spatial characterization. Spatial variation provides information about the Doppler spectrum.

A typical scheme for implementation of a narrowband sounder is shown in Figure 6.5. In this case, the receiver side is based on an envelope detector. An alternative implementation can be achieved with a super heterodyne receiver.

This type of sounders can be used in several environments, adjusting the transmission amplifier to the size of the desired coverage area. The amplitude of received signal with respect to the noise level determines the dynamic range of the system. A single carrier is transmitted enabling the use of narrow filters in reception, which helps to reduce the noise power.

Practical implementations of narrowband sounding can be achieved in actual systems. In working BSs, it may be sufficient the installation of a field strength indicator, with digital data output connected to a computer, to register or process this information. A mobile version of this system may consist of placing the two elements, receiver and computer, in a vehicle alongside an odometer.

Spacing, time or spatial, between samples cannot be too large because information about channel may be lost. Yet it cannot be also too small since the samples would be correlated, providing redundant information. Sample spacing must be slightly smaller than the named correlation time T_c. Several authors [30,54,55] recommend sample spacing between $\lambda/4$ and $\lambda/8$, for macrocell environments.

Once registered the signal in the time domain, it can be extracted information of channel in the form of empirical laws of mean power variation with distance ($\sim d^n$), probability distributions of amplitude envelope for small and large scale and even, if MS speed is known, the Doppler spectrum after applying a Fourier transform.

An example of signal registered by a narrowband sounder is shown in Figure 6.6 (a) [56,57]. It is also shown the adjustment of mean power variation with the distance to the transmitter station as kd^{-n}. It is noted that a limitation in the measurement exists due to the noise level of the system, in this case with a value of -74 dBm, or -17.55 dB relative to the median value -56.45 dBm. The Doppler spectrum has been also estimated and is shown in Figure 6.6(b) [56,57]. The statistics of fast fading can be obtained from the previously shown measurements [56,57].

6.2.2.2 Wideband sounding

A channel is wideband if the relative delays of the received components are larger than the basic unit of information. In other words, the distortion that the transmitted signal undergoes varies over the signal bandwidth. It is also possible to interpret as wideband channel if the occupied bandwidth of the transmitted signal exceeds at least 20% the value of the centre carrier frequency used.

The characterization of radio channels for wideband systems requires obtaining the time-variant impulse response of the channel, $h(t, \tau)$, or, alternatively, the time-dependent frequency response, $H(f,t)$.

According to the theory of linear systems [1,58], the transfer function representation of a system states that, for a given input $x(t)$ to a channel, the output $y(t)$ is the convolution of $x(t)$ with the time-variant CIR $h(t, \tau)$, as shown in Figure 6.7 and in the following equation:

$$y(t) = x(t) * h(t, \tau) \tag{6.1}$$

The transfer function of a system is ideally obtained if a signal $x(t) = \delta(t)$, with $\delta(t)$ being the Dirac's delta function, is applied to the system input considered

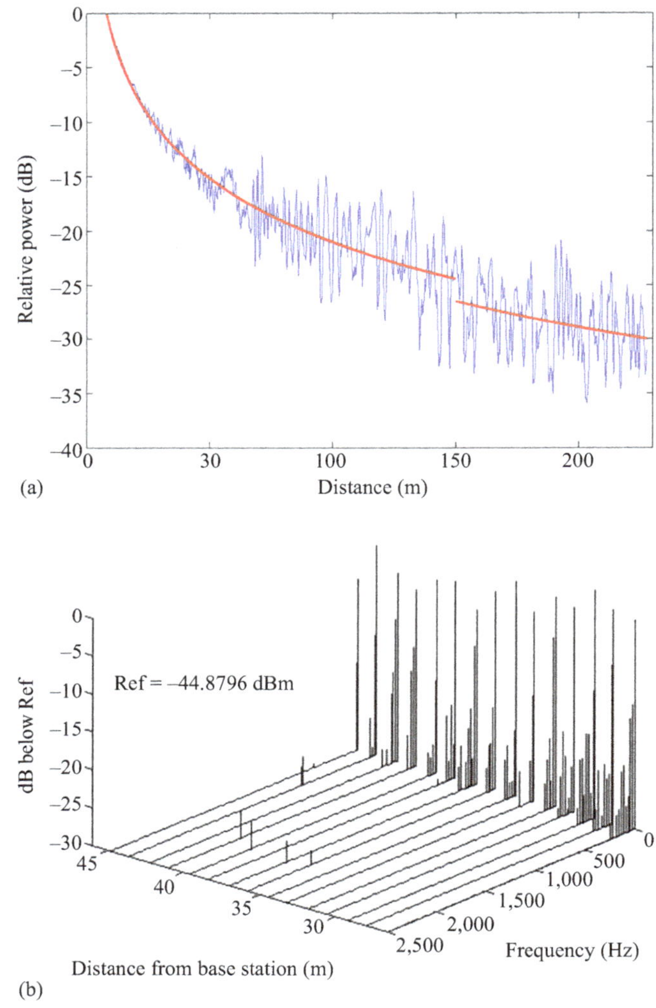

Figure 6.6 (a) Example of signal registered by means of a narrowband sounder; (b) Doppler spectrum obtained from measurement of (a) [56,57]

linear. The system output $y(t)$ equals the transfer function $h(t, \tau)$ due to the convolution property of the delta functions. However, in practice, it is not possible to shape a strict delta function, since it would require generating a signal of infinite amplitude, infinitesimal duration and correspondingly infinite bandwidth [59].

Therefore, several methods have been developed to measure $h(t, \tau)$, or equivalently $H(f,t)$, approximating an impulse function. Among them, the following wideband sounding techniques are the most widely used: periodic pulse (also known as Direct Pulse), frequency swept, frequency hopping, pulse compression and swept time delay cross-correlation [30].

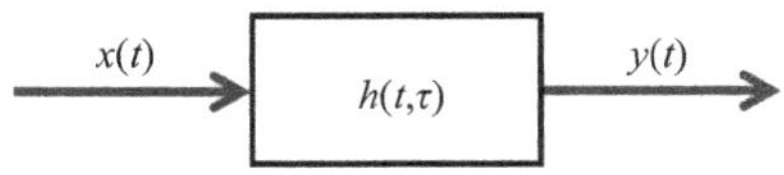

Figure 6.7 Graphical representation of channel transfer function

Table 6.3 Classification of wideband radio channel sounding techniques

Domain	Transmitted signal
	Short pulse
Time	Pulse compression
	Chirp signal
Frequency	Multitone
	Frequency sweep
	Frequency hopping

It should be noted that, technically, the output of the measurement system is not exactly $h(t, \tau)$, but rather $h(t, \tau)$ convolved with the shape of the pulse that is used to sound the channel [60]. Nonetheless, if the bandwidth of the channel sounder is large enough compared to that of the channel under evaluation, then the output at the channel sounder may be approximated as $h(t, \tau)$ for all practical purposes [60].

A classification of different types of wideband channel sounders is presented in Table 6.3. Two types of wideband sounders are found depending on if the channel is to be characterized in the time (impulse response) or in the frequency (frequency response) domain. Transformation from one domain to other is direct if WSSUS conditions stated in [1,58] are met.

Time domain wideband radio channel sounding techniques

CIR can also be measured with time-based techniques. The input to the channel is a single carrier chosen from within the frequency bandwidth of interest, and the output is the CIR. As shown in Table 6.3, according to the transmitted signal, three main time-based wideband sounders can be considered.

Short pulse sounders

As mentioned above, the impulse response of a linear system is obtained at its output by applying a Dirac's delta signal at its input. Short pulse sounders are based in the transmission of a like-delta impulse by means of a very narrow pulse $x(t)$ of duration T_c, through the radio channel.

As in (6.1), the convolution of the transmitted pulse $x(t)$ with the CIR $h(\tau)$ results the received signal $y(t)$. Pulse transmission is repeated with period T_p to capture the temporal variations of the channel frequency response. The delay resolution obtained with this sounder is T_c, and the maximum non-ambiguous delay is T_p. A scheme of signals involved in a short pulse sounder is presented in Figure 6.8.

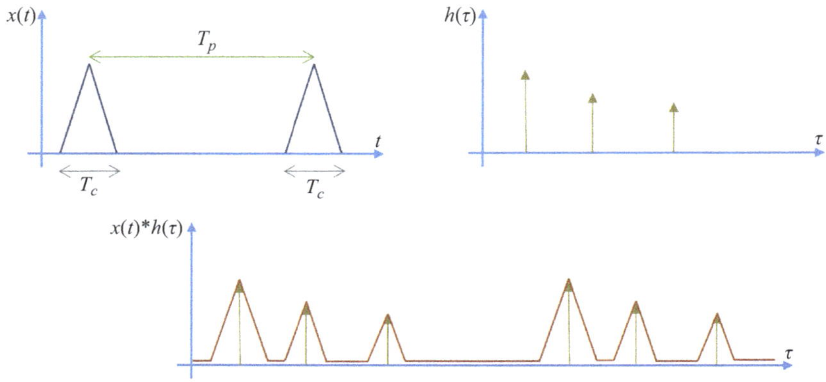

Figure 6.8 Signals involved in periodic pulse sounding

A disadvantage of this type of sounder is the inefficient use of transmitted power, since an important part of the signal power will exceed the operating bandwidth. Another disadvantage is the high noise level at the receiver because the receiver filter bandwidth is as wide as that of the system bandwidth. Finally, as the peak-to-average power ratio (crest factor) of the transmitted signal is large, linear power amplifiers will require operating with a large power backup.

Pulse compression sounders

Some of these problems associated with short pulse techniques are avoided by using pulse compression sounders. These sounders are based on the correlation properties of linear systems: the cross-correlation of the output signal and the input signal of the linear system equals the convolution of the impulse response of the channel and the autocorrelation of the input signal. If the autocorrelation of the input signal is a delta function (case of white noise), the cross-correlation will equal the CIR [30,61,62].

This type of sounders has some advantages: the crest factor is reduced, there is a processing gain factor and the delay resolution is proportional to the chirp duration. They do not require a cable link between transmitter and receiver, but highly stable clocks are needed at both sounder ends to keep the phase reference.

Cross-correlation-based wideband sounders have become very popular among the pulse compression sounders family, even for 5G channel modelling. They are also called sliding correlation sounders. These sounders usually transmit a carrier modulated by a pseudo-noise (PN) sequence like pseudorandom binary sequences (PRBSs), which can be viewed as a continuous periodic waveform with a narrow autocorrelation. At the receiver, the signal is correlated with a replica of the transmitted signal. An example of signals involved is shown in Figure 6.9.

Chirp sounders

An alternative pulse-compression technique consists of the transmission of a linearly frequency modulated (LFM) carrier, or chirp signal, with an almost constant

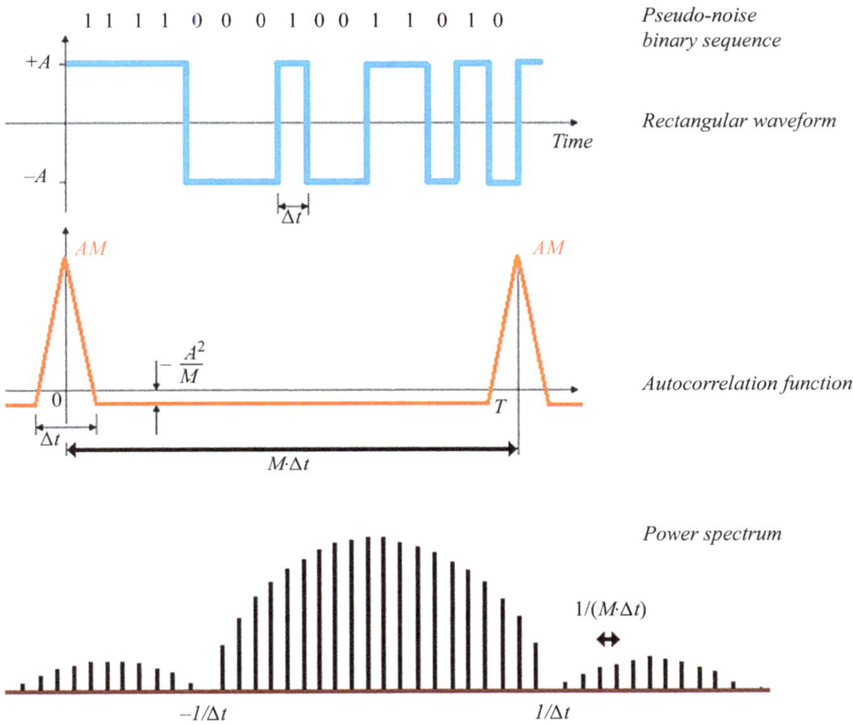

Figure 6.9 Signals involved in pulse-compression sounding

frequency spectrum. A matched filter or a heterodyne detector at the receiver is used to compress the chirp pulse. The advantages of using chirp waveforms have long been recognized in high frequency (HF) radar and sounders [63].

An improved Doppler resolution and the capability of post-processing bandwidth variation are some of its advantages, whereas the compressed pulse high-side-lobe levels at reception and the complex hardware required are its main disadvantages. An example of signal employed in chirp sounders is presented in Figure 6.10.

Frequency domain wideband radio channel sounding techniques
Instead of the CIR, the channel frequency response can be measured. Feeding the channel at its input with a number of carriers with equally spaced frequencies within the bandwidth of interest, the channel frequency transfer function will be obtained at the output. The frequency response will exhibit magnitude and phase variations corresponding to multipath propagation.

The techniques based on frequency sweeping requires a finite amount of time to complete the snapshot, particularly for large bands or small frequency step sizes,

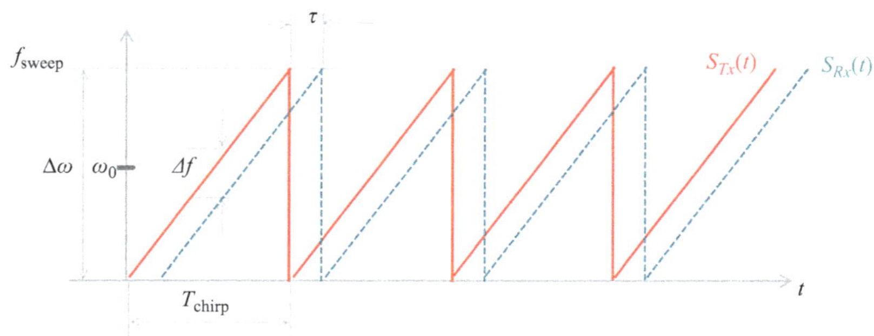

Figure 6.10 Example of chirp signal [64]

meaning that the system carries non-real-time measurements, which has two important implications:

1. The IFT is performed only at the end of the frequency sweep, making the swept frequency system useful only for either time-invariant channels, or channels which vary slowly with respect to the step time. A dynamic channel would change too quickly to be accurately characterized by this type of measurement system.
2. No Doppler shift or power spectra can be obtained from the system due to the frequency-stepping technique [65,66].

Despite these limitations, the frequency domain techniques remain popular due to the simplicity of implementation, and its use is still prevalent, even for 5G characterization. As shown in Table 6.3, frequency-based wideband sounders apply three main techniques generate the transmitted signal, as described in the next.

Frequency swept sounders

A vector network analyser (VNA) is frequently used to build up frequency domain sounders. A frequency swept is conducted in order to get the channel frequency response over the required frequency band. The delay resolution and the maximum resolvable delay are inversely proportional to the swept bandwidth and the frequency spacing, respectively. Compared to other channel sounders, the received noise is reduced because the receiver filter is much narrower than the bandwidth to be measured.

However, it presents two main disadvantages. First, if the bandwidth of interest is large, the time consumed for stepping a tone with a low-frequency separation can be significant, and the radio channel should be kept stationary during frequency swept. Second, the VNA connects permanently the transmitter and receiver by means of a cable to keep the phase reference, which usually limits its use to indoor environments. The dynamic range, sensitivity and noise threshold are also an important limitation of a VNA-based sounder. Additionally, memory size and sampling speed of the VNA equipment are also important issues.

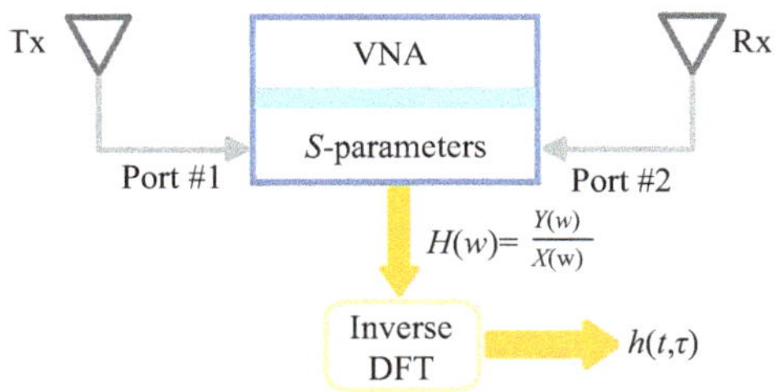

Figure 6.11 Frequency swept sounder built with a VNA

A scheme of a frequency swept sounder based upon a VNA is presented in Figure 6.11.

Frequency hopping sounders

Linear frequency swept sounders transmit sequentially a sinusoidal wave whose frequency is increased in constant steps over time. Yet, other frequency time patterns are possible. The frequency-hopping spread spectrum transmission is an example, in which a PRBS controls a synthesizer to produce a pseudorandom sequence of discrete carrier frequencies (hops).

The cost and complexity of frequency-hopping equipment is larger than linear frequency sweep sounders. Hopping systems cannot cover a frequency band as fast as sweeping frequency, because of the settling time that must be allowed between hops. On the other hand, frequency-hopping sounders usually use narrow-band transmission and receiving techniques, thus covering long distance links with excellent noise immunity and interference rejection capabilities.

Multi-carrier sounders

To reduce the time required for frequency sweep, the channel can be excited with a multi-carrier signal, allowing the simultaneous measurement of the frequency response over all the bandwidth. The problem is that multi-carrier signals have a large crest factor that has to be reduced to a predefined value by optimizing the phases of the spectral lines.

This probing method to achieve wideband channel sounding involves transmitting a signal composed of multiple carriers with specific amplitude and relative phase that forms the complex frequency spectrum [4]. Once received, the IFT of the complex envelope is applied to derive the CIR. For this technique, the resolution of the channel sounder and the maximum resolvable delay are fixed by the number of carriers and the frequency interval between them [4].

6.2.2.3 MIMO sounders

Including multiple antennas in the transmitter and receiver sounding system ends enables the characterization of channel features dependent upon the distribution of angles at which signals travels from the transmitter to the receiver. As an example, the correlation between the fading signals observed at each antenna element in spatial diversity and MIMO transmission schemes [4,67].

MIMO channel characterization involves estimating the AoA distribution observed at the receiver for rays launched at each of the possible angles of departure (AoD) from the transmitter antenna. The resultant characterization is denoted as a double-directional channel model [4,67].

In order to achieve the estimation of the channel bidirectional parameters for all of the possible angular propagation paths, a MIMO channel sounder requires being equipped with one of the following antenna schemes [4,12,67]:

- Antenna arrays in both the transmitter and the receiver ends [68–70].
- A highly directional antenna in both the transmitter and in the receiver ends, mounted on mechanical multi-axis positioners and controlling their movements across a grid or sequence of closely spaced points in space as a way to generate a virtual array. This scheme allows characterizing the channel spatial correlation, temporal correlation and AoA distributions [4,71].

The scheme with an antenna array has been widely used to construct measurement-based models, for example [12],

- WINNER spatial channel models (SCMs) [72],
- IMT-advanced models [73] and
- COST2100 MIMO models [74].

The solution with a single antenna has been also commonly used for modelling 5G wireless channels [12,75–82].

In the scheme of multi-element spatial arrays, the underlying antenna array can be formed by with one of the following arrangements [12]:

- a virtual array obtained as above-mentioned by placing a single antenna at specific spots in space [69,83];
- physical multiple-element antenna arrays in the transmitter and receiver ends, with two usual measurement configurations:
  - *Switched channel sounding* [84]: This scheme includes a high-speed radio frequency (RF) switch at both the transmitter and receiver ends. These switches connect the RF front end to individual elements in the antenna arrays. Figure 6.12(a) sketches a switched MIMO channel-sounding system. The advantage of this configuration is the low cost and the real-time recording capability.

    A plausible disadvantage of a switched sounder is the trade-off between the delay estimation range and the highest absolute Doppler frequency [12,85,86]. A correct switching strategy must be set so that the absolute Doppler frequency estimation can be up to half the inverse of the sensing period [12,87–89].
  - *Parallel channel sounding* [12,85,86]: The multiple transmitting antennas send signals simultaneously and the multiple receiving antennas receive signals simultaneously; then, the signals dispatched by different transmitting antennas should exhibit suitable autocorrelation and cross-correlation properties to be separated correctly at the receiver end. Figure 6.12(b) shows a parallel MIMO channel sounder scheme.

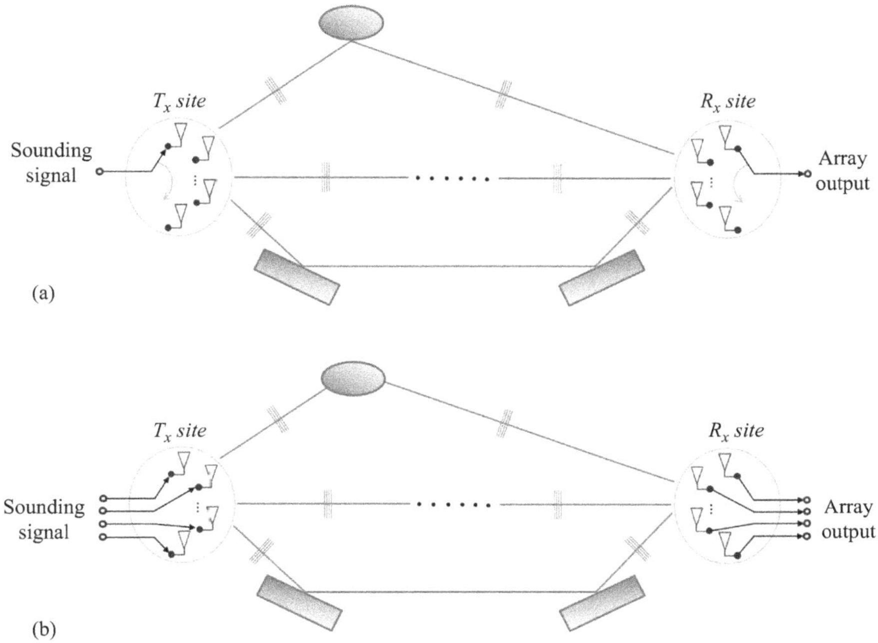

Figure 6.12 MIMO channel sounder schemes: (a) switch and (b) parallel [12]

The main features of some known wideband channel sounder are summarized in Table 6.4. An example of a wideband MIMO sounder for V2V channel modelling is that of the HHI channel sounder, developed by the Fraunhofer Heinrich Hertz Institute (HHI). It consists of a wideband measurement system with 1 GHz bandwidth operating at a carrier frequency of 5.7 GHz [90,91]. The measurement bandwidth enables a delay time resolution of 1 ns valid to resolve MPCs with high resolution. The channel sounder consists of a transmitter unit and a receiver unit that can be installed in conventional passenger vehicles and tested in actual traffic environments. The transmitter vehicle is equipped with two roof-mounted antennas omnidirectional and vertically polarized on the left and right edges of the vehicle. One measurement record of the HHI channel sounder contains 10.000 snapshots [90,91].

The MIMO sounder built by WIDES [92] is based on software-defined radio components. It counts on antenna switching capability to enable the measurement of 8 × 8 MIMO channels at 800 MHz and 5.9 GHz, with a 160 MHz bandwidth. The schematic of this wideband V2V sounder is shown in Figure 6.13.

6.2.2.4 Features of an ideal channel sounder

The channel sounder should ensure that the measurements performed accomplish a set of expected features. Regardless of the selected measurement technique, the

Table 6.4 Wideband channel sounders

Reference	Features
Cox [8]	Sliding correlator sounder Outdoor suburban environments at 910 MHz
Devasirvatham [93]	Sliding correlator sounder Office buildings at 850 MHz, 1.7 GHz and 4 GHz
Newhall [94]	Sliding correlator sounder Large train yard at 915 MHz
Kivinen [95]	Wideband sounder at 2.4 GHz
Nielsen [96]	Wideband MIMO sounder at 5.8 GHz
Salous [97]	Parallel receiver channel sounder Spatial and MIMO characterization
Ranvier [98]	VNA-based MIMO Indoor measurements at 60 GHz
Alejos [30]	Sliding correlator sounder Outdoor, indoor, LOS and NLOS at 41.5 GHz Phase-shift keying signal modulated by a PN sequence
Felhauer [99]	Commercial by Siemens, SIMOCS 2000
Heddergott [100]	24 GHz, ECHO 24 by the Swiss Federal Institute of Technology Zurich
MEDAV [101]	RUSK sounders by MEDAV GmbH
Trautwein [102]	Multiple-frequency periodic
Hentila [103]	2.55 GHz, MIMO channel sounder by Elektrobit Propsound CS™
Kyosti [72]	WINNER sounder, spatial
Liu [74]	COST2100, MIMO sounder
ITU [73]	IMT-Advanced by ITU
HHI [91]	5.7 GHz, MIMO, Fraunhofer Heinrich Hertz Institute

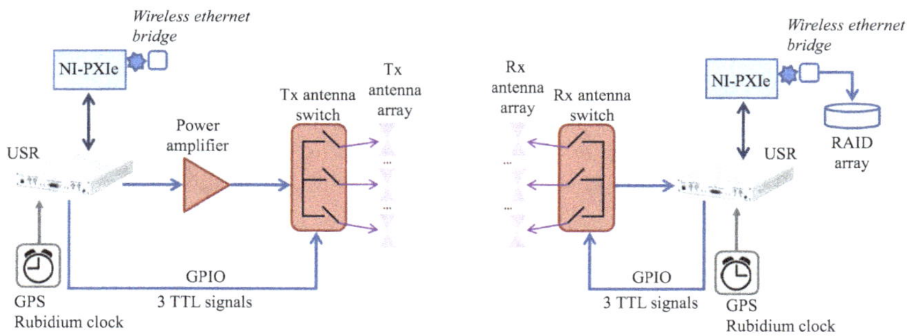

Figure 6.13 8 × 8 MIMO V2V sounder [92]

following main requirements enunciated in [104] would be fulfilled by an ideal channel sounder:

1. *Synchronization and timing*: There should be an accurate timing synchronization between transmitter and receiver and a reasonably constant vehicle speed during acquisition.

2. *Complex sounding*: In order to extract Doppler information or the Bello's system functions [1,58], the phase information of the received signal must be measured. This will require that both In-phase and Quadrature (IQ) components must be extracted, or that received signals can be acquired in a not baseband frequency.
3. *Amplitude, phase and dynamic range requirements*: The amplitude and phase balance of the quadrature components should be appropriate to the dynamic range.
4. *Doppler and RF bandwidth requirements*. The sounder, including antennas, should exhibit flat amplitude response and linear phase over the band of interest. The impulse response snapshot must be measured in a time short enough to remain the channel response essentially constant during acquisition and sampling. The number of impulse response snapshots per second should be fixed according to the expected maximum Doppler shift.
5. *Centre frequency and interference*: The system should be capable of tuning to any desired frequency band in case of detecting interference to other users of the same radio spectrum; this would allow the use of continuous wave pulse compression techniques.
6. *Power efficiency*: Signal transmission should require minimum peak power and constant envelope to enable the use of C-class amplifiers, in order to minimize interference to other users.
7. *Mobility*: The sounder receiver or transmitter end must be portable, and the hardware construction should be compact and robust. Measurement devices should therefore be lightweight, and the controlling and analysis tools should provide user friendly graphical interfaces.
8. *Digital realization*: Analogue signal processing should be reduced to an absolute minimum, since the use of digital techniques allows greater flexibility in processing, easier calibration and increased accuracy.
9. *Data formats and scalability*: Data acquisition and data format should allow minimizing post-processing, so that, ideally, results could be available for inspection at the time of measurements.

Additional aspects summarized in [105,106] should be considered for the correct design or optimization of a wideband channel sounder:

1. *Large bandwidth*: The minimum resolvable delay between MPCs is equal to the pulse width of the probing signal. Then the probing signal composed of a single pulse, or not, must exhibit short time duration to ensure that its frequency bandwidth is larger than the channel coherence bandwidth. Thus, the sounder resolution allows detecting and capturing MPCs. Then, it is necessary to adequate the probing pulse width to the measurement resolution requirement. However, short-time pulses are not easy to generate and require of large transmission power.
2. *Large time-bandwidth product* (TBP): The product of the signal duration and its occupied bandwidth is constant; however, a compromise is required between the time period and the bandwidth due to the involved power required. For example, a longer *m*-sequence means a longer period and bigger TBP, for a

constant bandwidth given by the inverse of the pulse duration. A higher transmission power is required when the signal duration is larger, which results in a higher signal to noise ratio (SNR) and smaller error rate at the receiver. This will also improve the correlation properties of the sounder.
3. *Signal duration*: To maintain the channel stationarity, a larger TBP is obtained by increasing the signal duration; however, the signal duration should not be longer than the coherence time. It is considered that the ideal signal duration is larger than the maximum excess delay of the channel, due to the required TBP, and shorter than the coherence time.
4. *Power spectral density* (PSD): As mentioned above, the PSD of the sounding signal should exhibit the same absolute value for all frequency components over the considered bandwidth to ensure no shaping of the channel response estimation [30,60].
5. *Low peak-to-average ratio* (PAR): Efficient transmit power amplifier implies a lower PAR. By adjusting the phases of the frequency components, a low PAR is obtained.
6. *Good correlation properties*: For correlation-based channel sounder development, the autocorrelation function (ACF) should be as close to ideal as possible. This will require, among other settings, a zero-mean biasing in the receiver, choosing the correct m-sequence type and length, alongside noise reduction.

Finally, in [4] a checklist is provided about the features to be met by a channel sounder:

1. It should work whether the channel is static or time varying.
2. The nature of the antennas must be carefully analysed, as well as the frequency dependence on the antenna radiation pattern. The degree of mutual coupling between co-located antennas should be determined, especially in multi-antenna, multisite or MIMO sounders.
3. The non-linearities of the transfer functions due to active devices used in the equipment must be carefully characterized to disassemble its effect from the captured data, especially if multi-carrier or other complex wideband signals are used as transmitted waveforms.
4. The phase noise due to the reference signals generated by local oscillators included in the sounder must be known in order to ensure that the registered signal is phase stable during the measurement interval. Otherwise, phase noise and instability should affect the Doppler analysis.
5. The sensitivity of the equipment to the environment conditions must be well known, especially temperature and humidity. As an example, power amplifier exhibits linearities due to heat increase.
6. Calibration of the channel sounder is required to avoid distortions inherent to the hardware devices composing the equipment that can influence the accurate estimation of the channel parameters. The most extended calibration procedure is the back-to-back connection which extracts the system transfer function, or a replica of the PN sequence, later used to correct the recorded raw data sets or

performing the cross-correlation in the processing stage. Among the faults that a calibration can solve it is found phase noise or clock drifting in synchronization. More information can be found in [30,60].
7. Other technical difficulties might be the requirements of high-quality anti-aliasing filters, the minimization of local oscillator phase noise and other intrinsic noise sources.

Generally, all of the above requirements must be fulfilled to achieve that the channel sounder perform effective and accurate measurements.

6.2.3 Limitations of measurement-based methods

Despite being widely used, and the undeniable several of advantages that the direct measurement of wireless channel offers, the methodology has several limitations [4]:

1. *Large time and effort consuming*: To collect and reduce a statistically significant amount of data requires a significant amount of time and effort, even if redoing measurements is not required.
2. *Adequate portability*: the size, weight and power consumption of the measurement equipment must be set according to the environment to be sounded. Actually, ultraportable VNAs have come to help simplify this task, opening new possibilities.
3. *Antenna distortion*: distortions introduced by the channel can only be separated from distortions introduced by the antenna if the width of the AoA distribution is larger the antenna beamwidth, otherwise the antenna will mask arrival directions.
4. *Precise phase reference*: In order to accurately measure the phase difference between the transmitted and received signals frequency drift and phase noise must be substantially reduced. This can be accomplished through a direct connection between the transmitter and receiver or by using high-quality frequency standards, like Rubidium or Cesium clocks.
5. *Cost*: Required budget increases according to the complexity of the measurement system that responds to the measurement aims: 3D, MIMO, angular scanning, VTD, environments and so on.
6. *Adequate environment selection*: The measurement environment must be as similar as possible to the use case environment; otherwise the resulting model and statistics will not be of valid application.

In spite of all the above, the measurement-based methodology has demonstrated to provide valuable results and so it continues to be the most widespread method for wireless channels characterization. Among other reasons [4], this methodology provides results of direct use and application. The measurement data sets are also useful to validate the results obtained from simulation-based methods. Measurement- and simulation-based channel modelling methods have been gradually more used as complementary approaches to ensure the validity of the results provided by software tools [4].

6.3 Vehicular channel characterization

As seen so far, channel models for vehicular environments outcome from a combination of four main features specific of the environment and related to the potential use case [4]: (1) the link range (short for V2V V2P, long for V2I), (2) the speed of the communicating vehicles, (3) the type of surrounding objects and (4) the traffic density and speed.

A large number of combinations are possible, and usually, only a subset is chosen for characterization given the time and effort necessary for this process. As already mentioned, the radio links behaviour may be determined either by ray-tracing simulation including realistic models of surrounding objects, [107] or by direct measurement using a channel sounder deployed in representative environments [4].

Generally, vehicular channel is time-varying and dispersive, and it also varies spatially, although still linear. Just as in the cellular channel, obstructions from buildings, terrain, other human-made structures, vehicles and foliage, present in the environment, destroy the homogeneity of the channel and essentially render it anisotropic [3,108]. The combination of all the factors introduce attenuation, reflections, diffraction and scattering on the signal that yield multipath propagation, which causes dispersion, alongside mobility of the communicating vehicles and of other surrounding objects yields time variation.

Since the settings are distinct, channel models obtained from conventional cellular systems cannot be directly used for V2X systems. As an example, the WINNER model [72] is not directly applicable to the V2V case. Worth noting is the V2I case.

The differences between mobile cellular and vehicular channels can be summarized as follows [14]:

1. V2X communications use a setup peer-to-peer with the transmitter and receiver antennas placed at the same or similar height, and in the identical or alike environment. Conversely, in mobile cellular links the BS is located above the mobile user station that is located at street level. Therefore, the dominant propagation mechanisms producing the MPCs strongly differ.
2. In V2X environments, scattering mainly occurs around link ends, whereas for mobile cellular channels, the scatterers located around the BS area are negligible.
3. The distance between link ends is considerably smaller in V2X channels than in typical mobile cellular scenarios (~1 km).
4. In cellular channels, only one of the link ends is mobile, and moving scatterers have less impact; whereas in a V2V channel, both link ends, alongside many relevant scatterers in the environment, are moving with a severe impact. This produces that the signal fluctuations is faster in the V2X channels than in the cellular case.
5. V2V systems operate at 5.9 GHz frequency band, whereas cellular systems occur regularly at 700–2,400 MHz. Due to the higher operating frequency,

V2V channels exhibit larger signal attenuation, and specific propagation processes like diffraction are less efficient than in cellular radio.

6. The propagation characteristics of vehicular channels are highly dependent on the existence of the LOS path, as indicated by empirical measurements performed by [109]. Regardless of the propagation environment (e.g. highway or urban scenarios), non-LOS (NLOS) channels show noticeably larger RMS delay spreads than the LOS channels [14]. This is due to the stronger signal attenuation and multipath effects caused by an increasing number of reflections and the reduced contribution of the diffracted components at large frequencies.

Distinctive features of vehicular channels are due to the impairments undergone by the signal propagating through these scenarios, and are time- and space-variant, potentially high Doppler shift, and frequency-selective fading due to surrounding objects, mobile or static [14]. Each of these problems must be analysed individually to be incorporated into the channel model.

6.4 V2X channel models

Radio channel classification supports many options. Channel models can be classified as wideband or narrowband, time-variant or time-invariant, 2D versus 3D, stationary or non-stationary, measurement data based or purely theoretical, completely deterministic or stochastic [12]. Channel models can also be distinguished according to the environment, the link type (V2V, V2I, V2P, V2C) and range, the vehicle speed or the VTD [4,12,14].

The most extended classification [12,14] considers the implementation approach and the availability of geographical information. By this criterion, models can be categorized as geometry-based deterministic models [110] and stochastic models (SMs). The second category can be divided into non-geometry-based stochastic models (NGSMs) [17,43] and geometry-based stochastic models (GBSMs) [39,42,111–115].

The categories of the implementation-based classification exhibit specific features described in the following [12,14]:

1. *Non-geometry-based stochastic models*:

   An NGSM determines the physical parameters of a V2V channel in a completely stochastic manner without presuming any underlying geometry. The SISO NGSM proposed by [17] was the origin of the V2V channel model standardized by IEEE 802.11p [19].

   NGSM models are based on measurement campaigns performed in specific environments (urban, suburban, rural, highway, etc.). The impact of VTD and vehicle speeds on channel statistics is also investigated. The channel characteristics for a specific environment are derived from measurements and then the parameters of the PL, shadowing and the small-scale fading are fitted [14]. Among the most widely-used NG models is the tapped delay line (TDL) model that was included in the document by European Telecommunications

Standards Institute (ETSI) [116] with each tap representing signals received from several propagation paths with specific delay and Doppler spectrum type (flat, round, classic-3-dB or classic-6-dB shape).

The NS of the channel is incorporated by modelling MPC persistence via Markov chains. The Markov chain technique is used to model the multipath. The persistence process accounts for the finite duration of the resolvable propagation paths. In [117], it is proposed a TDL-based channel model with birth/death processes to account for a sudden appearance of the LOS component [12,14].

The NGSM can easily capture the effect of an abrupt disappearance of strong MPCs, caused by express blockage or obstruction due to another vehicle or surrounding obstacles. However, because of being based on the WSSUS assumption, the TDL model does not account for the drift of scatterers into different delay bins and then the probabilities of transition in the Markov model used for the persistence process is no longer accurate. The NGSM cannot accurately represent the NS of actual V2V channels [12,14].

2. *Geometry-based deterministic models* (*GBDM*):

GBDMs characterize V2V physical channel parameters in a completely deterministic manner [12–14]. The methodology consists of replicating the actual physical radio propagation mechanisms within a specific environment. For these models, a real environment is composed of two main parts to be modelled: the dynamic road traffic formed by the set moving vehicles (cars, vans, trucks, etc.) and the roadside surroundings (buildings, parked cars, road signs, trees, etc.).

Figure 6.14 illustrates the modelling concept of GBDMs [12–14]. Once the environment is defined, an accurate signal propagation modelling is achieved by generating all possible signal paths or rays between the transmitter and the receiver under the geometrical optics rules.

Generally, ray-tracing models [14,108] make up the majority group within the GBDMs for the vehicular channel modelling; 2D or 3D ray-tracing methods require a detailed description of the propagation environment to replicate the actual physical propagation mechanisms in order to accurately calculate the channel statistics [12–14].

The simulated complex impulse response contains full channel information: the statistical NS of the channel. It also allows evaluating the impact on channel statistics due to VTD or the elevation angle for 3D cases [12–14]. Simulations should be later validated with measurements. An example of GBDM based on the ray-tracing method was proposed in [110] for machine-to-machine (M2M) channels that included a 3D ray-tracing approach.

Among the disadvantages of GBDMs are the large amount of computation time required and the need of a detailed description of the propagation environments. Due to the site-specific feature, the results cannot be generalized to a different class of scenarios or environments.

Another subcategory of GBDMs is formed by the simplified geometry-based models that consider the geometric properties or outlines of the surroundings (vehicles, buildings, vegetation, etc.) to simplify calculations and extract channel statistics either from the measurements or from the simulations,

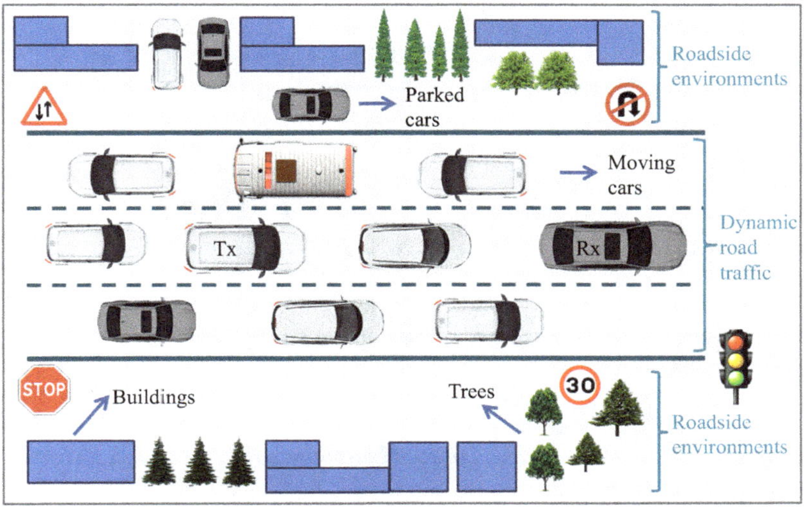

Figure 6.14 Geometrical description of the GBDM for a typical V2V environment [13,110]

including large- and small-scale fading. An example is the GEMV^2 model [45,118] valid for V2V and V2I.

3. *Geometry-based stochastic models*:

   A GBSM is obtained assuming a stochastic distribution of effective scatterers in addition to the theoretical principles of electromagnetic wave propagation [12–14]. These models can be easily modified to fit different environments just by choosing a different shape of the scattering region and/or modifying the probability distribution function or probability density function (PDF) assumed for the location of the scatterers within that region [12–14]. The individual contribution to the complex impulse response of each region of effective scatterers is determined using a simplified ray-tracing method.

   According to the scattering region shape, the GBSMs can be divided into two categories [12–14,39]:

   - *Regular-shaped GBSMs (RS-GBSMs)*: This category assumes that all effective scatterers are placed within areas with a regular shape (one/two-ring, ellipse, etc.). The RS-GBSMs are generally used for the theoretical analysis of channel statistics, as well as for theoretical design and comparison of applications and systems [12–14]. An illustration is shown in Figure 6.15.
   - *Irregular-shaped GBSMs (IS-GBSMs)*: Effective scatterers are randomly placed within regions with irregular shapes meeting specific statistical distributions. This category is valid to replicate physical reality so that it can be evaluated the need to adjust the location and/or features of the effective scatterers of RS-GBSMs [12–14]. An illustration is shown in Figure 6.16.

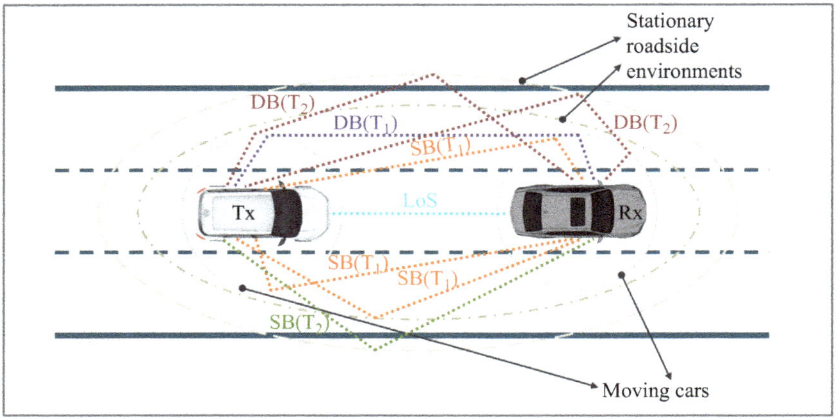

Figure 6.15 SB: single-bounced; DB: double-bounced; T1: Tap 1; T2: Tap 2 [13]

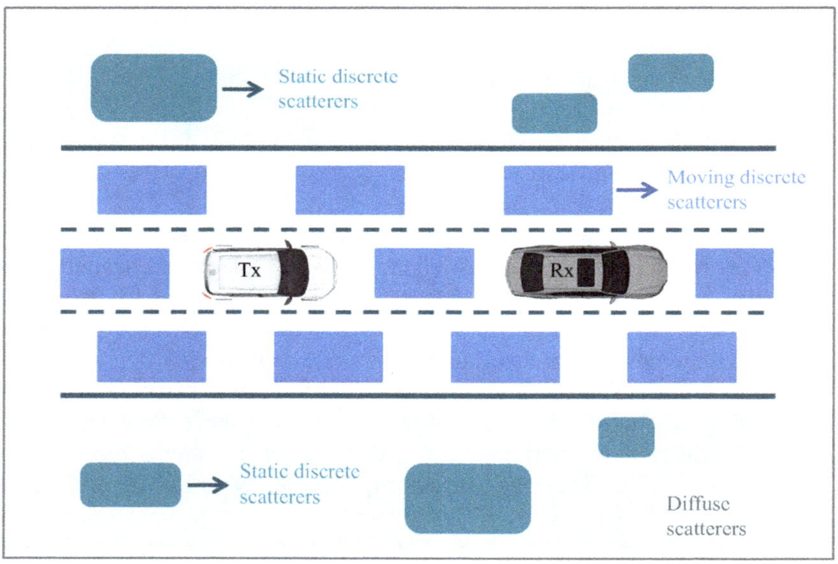

Figure 6.16 Geometrical description of the IS-GBSM in [39] for t V2V environment in Figure 6.15. [13]

6.4.1 General use models

In wireless communications, deterministic PL models are usually considered as the simplest option. These are the free space path model, the two-way ground reflection model and the PL model [116].

The free space PL model estimates the signal power loss when the transmitted wave propagates through free space, from the transmitter to the receiver without nearby obstacles [116]. The free space PL is given by

$$FSPL(d,f) = 20\log_{10}(d) + 20\log_{10}(f) + +20\log_{10}\left(\frac{4\pi}{c_0}\right) - G_{Tx} - G_{Rx} \tag{6.2}$$

where d is the distance between transmitter and receiver, f is the carrier frequency, G_{TX} and G_{RX} are the transmitting and receiving antenna gains, respectively. This theoretical model fails to fully characterize the PL experienced by V2X channels due to its obvious key propagation impairments like the diffractions and reflections produced on the road or on the surrounding vehicles and objects [116].

The free space propagation model only considers the direct or LOS component. However, due to the inherent features of the V2X environments – especially the low antenna heights and the short range – in an LOS propagation case, at least two dominant signals components should be considered: a direct or LOS component and a ground-reflected ray [116].

Then, the two-way ground reflection model with the corresponding reflection coefficient is considered a more appropriate option when modelling the PL model for LOS V2X channels [116,119,120]. It is based on a single street with one transmitter and one receiver, without obstacles in between and surrounding. In this scenario, the LOS path interferes with the ground-reflected path. The rays corresponding to both paths arrive at the receiver with phase and amplitude difference [116].

The different phase produces in the resulting receiver signal constructive and destructive alternating interference depending on the distance at which the reflection on the ground occurs, d_{LOS}, somewhere between the receiver and the transmitter, and that varies on the relative antenna heights and operating frequency. The interference pattern fades out by increasing the range link and totally ends at the break point, d_b. Beyond this distance, the length difference between the two signal paths is under one half of the wavelength and the AoA of the wave reflected on the ground is negligible with a corresponding phase shift of 180° or close that finally leads to destructive interference in the receiver [116].

Log-distance PL model is an extension of the free space PL model, where the PL decay occurs according to a linear fitting or exponent does not necessarily equal two as should be for the free space propagation case. The PL model is dependent on the environment and transmitter and receiver surroundings [116]. This model is explained in more detail in the following.

In order to respond to the different effects occurring to the signal propagation in the radio channel in terrestrial wireless communications, like V2V systems, it has become of more interest to model the effects of the channel as composed of three generally distinct mechanisms [3,4]: PL, shadowing and small-scale fading.

The three-block set forms a general model with a wide range of use. In line with the description given in [3,4], each of these blocks is presented as follows:

- PL figures out the distance-dependent decay of received signal strength. In free space environment, power of propagating signal decays according to the factor $1/distance^2$. However, in urbanized terrestrial environments PL exhibits average attenuation, due to obstruction of the first and consecutive Fresnel zones, and thus power decays as $1/distance^n$. The PL exponent n can vary between 2 and 6 and it must be obtained from site-specific measurement data by fitting of the power versus distance curves.
- Shadowing, also referred as obstruction or blocking, is a large-scale fading effect due to the attenuation caused by surrounding obstacles of large size compared to the carrier wavelength, like terrain or buildings. In V2X settings, for lower frequencies, these obstacles would also include buses, trucks and any other large vehicles. However, at very high frequency (VHF) and higher frequencies like the 5.9 GHz band, the size of those obstacles is plenty larger than one wavelength, so producing that shadowing effects are different from that due to the small-scale effects of multipath fading. For this case, it is quite common combining shadowing and PL effects in the same approach [3].

The most extensively used model for characterizing together the PL and shadowing is the log-distance PL model [3]. According to this model, the relationship between PL and the distance d between the transmitter and the receiver fits a power-law relation and is described by the following:

$$PL(d) = PL_0 + 10n \log_{10}\left(\frac{d}{d_0}\right) + X_\sigma \tag{6.3}$$

where $PL(d)$ expresses the PL value in decibel at distance d, PL_0 is the PL value (in dB) produced at reference distance $d = d_0$, n is the PL exponent or distance exponent, and shadowing is modelled by means of the random variable X_σ, usually considered as a zero mean Gaussian random variable with standard deviation σ_X dB. A typically used reference distance value is $d_0 = 1$ m for V2V settings, although any convenient value can be used.

The variable PL_0 uses to be around the value of the attenuation produced at d_0 under free space conditions. The PL exponent n is usually obtained by a linear fit of the curve defined by the first two terms in (6.3).

If the log-distance PL model is applied to LOS environments which lack shadowing fading effect, X_σ represents the deviation of measurements from the linear fit of the first two terms [3]. The location variability due to the random variable X_σ is generally attributed to differences in the degree to which the path is obstructed at different points within the coverage area [4].

PL can be assumed as the most fundamental measure of channel quality [4]. In decibels, PL, $PL(d)$ is defined as

$$PL(d) = P_t + G_t + G_r - P_r \tag{6.4}$$

where P_t and P_r are the time-averaged power levels (in dBm) at the output of the transmitter and the input of the receiver, respectively, and G_t and G_r indicate the transmitting and receiving antennas gains (in dBi) [4].

For those narrowband systems with fractional bandwidths $\Delta f/f_0$ less than 20%, where Δf is the occupied bandwidth of the signal and f_0 is the carrier frequency, PL can generally be anticipated as constant over the band. For systems with large fractional bandwidths, there exists a frequency dependence of PL. In such cases, the frequency and distance dependence of PL are separable, as indicated in (6.5) [4]:

$$PL(f,d) = PL(f) \cdot PL(d) \tag{6.5}$$

The relationship between PL and frequency meets a power-law relation modelled by (6.6):

$$\sqrt{PL(f)} \propto f^{-k} \tag{6.6}$$

with k the frequency exponent and $k = 1$ in free space [4].

Log-distance PL with appropriate PL exponent and shadowing deviation was experimentally shown to model well the PL for V2V links in NLOS cases but only for a limited set of scenarios [45,120]. For V2V links at 5.9 GHz, the following values can be used [45,116]:

- $d_0 = 1$ m,
- $PL(d_0) = 47.8649$ dB,
- $n = 2.5$ (slight obstruction by building),
- $n = 3$ (strong obstruction by building).

Other features of the log-distance model are [3,4] the following:

- A log-distance model can be derived from measurements or ray tracing should exhibit a good agreement with measurements
- In presence of a break point, a log-distance model will exhibit two adjacent segments as in (6.3) with a different PL exponent for each slope. A multi-slope model can be also found.
- Log-distance PL is a simple model that requires a minimum number of input parameters
- More complex models can be derived by taking into account specific additional information about the environment and geometric relationships of the surrounding objects (distances from the cars to the intersections, street and intersection widths, distances from cars to the buildings nearby the street, relative antenna heights, and frequency). For these multivariable cases, PL exponents are fit as dependent on the environment parameters. One example of these models is found in [121] developed for NLOS PL in the 5.9 GHz band [116].
- Exponent PL n generally exhibits values close to those of the two-ray model: 2 and 4.

- Small-scale fading is due to the destructive addition of MPCs also in V2X channels [3]. The number of MPCs varies upon the transmitted signal bandwidth. The small-scale fading considers the amplitudes of the MPCs (αs) as a random variable distribution. MPC phases are considered uniformly distributed in the interval $[0,2\pi)$, except for the non-fading dominant MPC, like the LOS component [3]. The rate of change of the MPC amplitudes and phases directly depend upon the velocities of the involved vehicles and scatterers. The maximum rate of change, for both amplitude and phase variables, is given by maximum Doppler shift [3]. The MPC amplitude and phase random processes can be generated by using the filtered Gaussian noise method [3,122] or the sum-of-sinusoid method [3,123].

Figure 6.17 shows the structure of the general use model for V2V derived from combining the effects of PL, shadowing and small-scale fading. The order of the blocks is not relevant due to the linearity of the operations. Each block requires inputs parameters specific to the V2V conditions whether from measurements such as simulation [3].

The PL and shadowing blocks can be implemented following the expressions indicated in (6.2)–(6.6). The dispersion block represents the small-scale fading and requires specification of the relative average powers of all MPCs and the fading statistics of each, as well as any inter-MPC correlations [3].

An example of an implementation of the general use model is shown in Figure 6.18. PL is symbolized by $l[d(t)]$, shadowing as $\beta(t)$, and the dispersion is represented by a two-ray model [3]. Generally, shadowing is spatially correlated that can be obtained by low-pass filtering an uncorrelated random shadowing process [3].

The model in Figure 6.18, in particular the portion that accounts for dispersion, is termed a TDL. This is the traditional model type for dispersive channels [116]. TDL model can be classified into the group of non-geometry-based stochastic channel models given that it only provides statistical parameters to describe the paths between transmitter and receiver with no reference to the path geometry [116].

In TDL models individual MPCs are characterized individually. Each MPC termed as tap exhibits specific fading statistics – Rician, Rayleigh or other – and

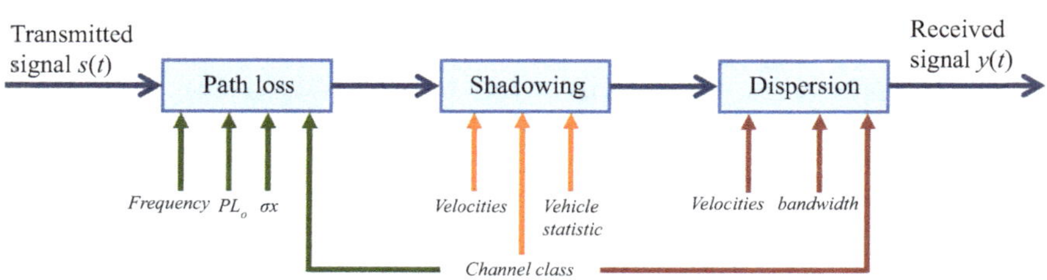

Figure 6.17 Structure of general use model for V2V channel [3]

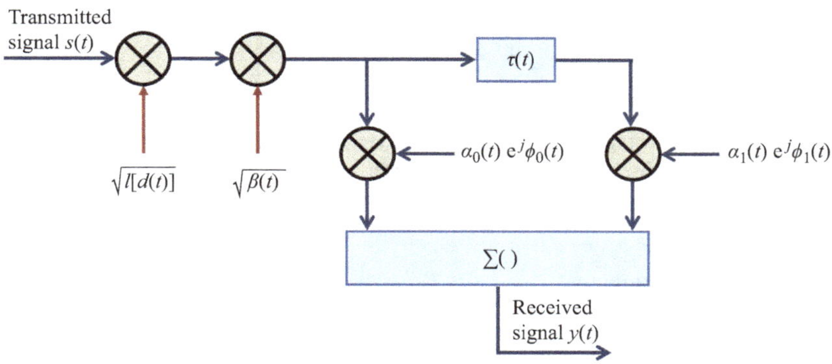

Figure 6.18 Example of an implementation of the general use model shown in Figure 6.17

Table 6.5 TDL models for a set of scenarios [116]

		Power (dB)	Delay (ns)	Doppler (Hz)	Profile
Urban approaching LOS	Tap 1	0	0	0	Static
	Tap 2	−8	117	236	HalfBT
	Tap 3	−10	183	−157	HalfBT
	Tap 4	−15	333	492	HalfBT
Urban crossing NLOS	Tap 1	0	0	0	Static
	Tap 2	−3	267	295	HalfBT
	Tap 3	−4	400	−98	HalfBT
	Tap 4	−10	533	591	HalfBT
Rural LOS	Tap 1	0	0	0	Static
	Tap 2	−14	83	492	HalfBT
	Tap 3	−17	183	−295	HalfBT
Highway LOS	Tap 1	0	0	0	Static
	Tap 2	−10	100	689	HalfBT
	Tap 3	−15	167	−492	HalfBT
	Tap 4	−20	500	886	HalfBT
Highway NLOS	Tap 1	0	0	0	Static
	Tap 2	−2	200	689	HalfBT
	Tap 3	−5	433	−492	HalfBT
	Tap 4	−7	700	886	HalfBT

individual phase shift. Each tap even features a distinct Doppler spectrum. The individual analysis of the MPCs provided by the TDL increases the accuracy of the channel model compared to the traditional approach that consider, for example, a Rician fading model assuming that only one signal reaches the receiver. However, TDL model fails to characterize the specific objects surrounding the vehicle environment. Yet, a TDL model could be tailored to a specific scenario [116].

In Table 6.5, extracted from the ETSI Technical Report [116], TDL models are provided for several scenarios [109,124,125].

A large variety of models have been developed to parameterize MPCs statistics. Following the most important are reviewed.

6.4.1.1 Spread function

Due to multipath propagation, a radio channel may result dispersive in multiple dimensions. Those dimensions of the dispersion are the following [12]:

- geometrical dimensions:
  - delay,
  - Doppler frequency,
  - direction of arrival (DoA) regarding the azimuth and elevation of arrival,
  - direction of departure (DoD) regarding the azimuth and elevation of departure;
- complex attenuation and fading;
- polarimetric properties due to the polarization of a path;
- propagation mechanisms: reflection, diffraction and scattering.

The set of dispersion dimensions define the spreading of the propagating wave, as illustrated in Figure 6.19 [12]. If the measurement equipment or the simulation software has limited capacity to resolve the channel parameter estimation of some dispersion dimensions, such dimension should not be considered into the generic channel model [12].

Multiple specular path components produced within a propagation channel are due to waves departing from the location of the transmitter and impacting in a region surrounding the location of receiver. The spread function is one of the system functions described by [1,58] that describe the behaviour of diffuse scattering, created by a large amount of impinging waves arising from the diffuse scattering on objects with a large geometrical volume [12].

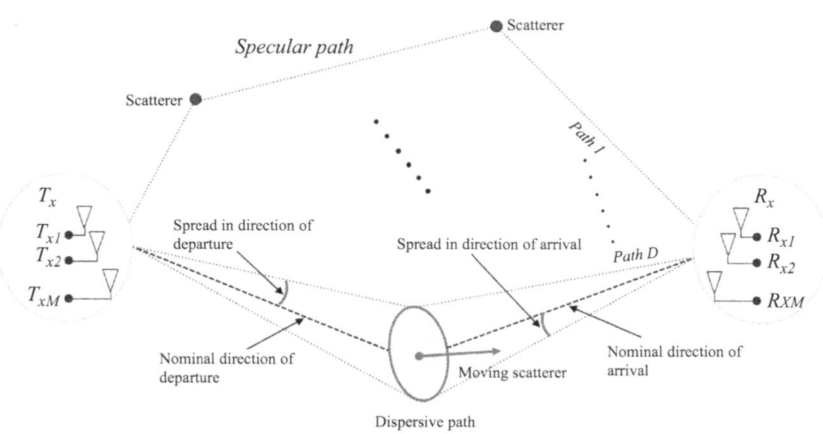

Figure 6.19 Multiple dimensions of radio channel dispersion [12]

The output Doppler-spread function $D(f,\nu)$ is obtained by applying the Fourier transform to the time-variant transfer function $H(f,t)$ with respect to the time variable t as shown in (6.7). It describes the channel frequency response to the frequency $f+\nu$ [108,126].

$$D(f,\nu) = TF_t\{H(f,t)\} = \int_{-\infty}^{+\infty} H(f,t) \cdot e^{j-2\pi\nu t} dt \tag{6.7}$$

The delay-Doppler-spread function is obtained by the Fourier transform of the time-variant impulse response $h(t,\tau)$ with respect to the time variable t as shown in (6.3), or by taking the inverse Fourier-transform of the output Doppler-spread function $D(f,\nu)$ with respect to frequency. The delay-Doppler-spread function describes the complex gain of the channel in the delay interval $[\tau + d\tau]$ and the Doppler-shift interval $[\nu + d\nu]$ [108,126].

$$S(\tau,\nu) = TF_t\{h(\tau,t)\} = \int_{-\infty}^{+\infty} h(\tau,t) \cdot e^{j-2\pi\nu t} dt \tag{6.8}$$

The function $S(\tau,\nu)$ is termed the delay-Doppler spreading function of a linear time-variant channel that describes the spreading of the transmit signal in the time and frequency variables. The value of the spreading function $S(\tau,\nu)$ at a given delay-Doppler point (τ,ν) characterizes the overall complex attenuation and scatterer reflectivity associated with all paths of delay τ and Doppler ν, and it describes how the delayed and Doppler-shifted version $s(t-\tau)\cdot e^{j2\pi\nu t}$ of a transmitted signal $s(t)$ contributes to the received signal [127].

The delay-Doppler spread function was extended to include DoA in [128], and in [12] even included the channel bidirectionality considering both DoA and DoD.

For linear system, the received signal can be considered to be the superposition of infinitesimal elementary plane waves originating from the elementary surfaces of the scattering objects. Then, when the channel can be decomposed into summable multiple specular path components, the bidirectional-delay-Doppler spread channel function $h(\Omega_{Tx},\Omega_{Rx},\tau,\nu)$ is expressed as in (6.9) [12]:

$$h(\Omega_{Tx},\Omega_{Rx},\tau,\nu) = \sum_{l=1}^{L} \alpha_l \delta(\Omega_{Tx} - \Omega_{Tx,l}) \delta(\Omega_{Rx} - \Omega_{Rx,l}) \delta(\tau - \tau_l) \delta(\nu - \nu_l) \tag{6.9}$$

where the parameters are the complex amplitude α_l, the delay τ_l, the DoA $\Omega_{Rx,l}$, the DoD $\Omega_{Tx,l}$ and the Doppler frequency ν_l of the lth impacting wave [12]. The spread function in (6.9) includes the transmitting and receiving antennas.

The bidirectional-delay-Doppler power spectrum is obtained as in (6.10) [12]:

$$P(\Omega_{Tx},\Omega_{Rx},\tau,\nu) = E\left[\left|h(\Omega_{Tx},\Omega_{Rx},\tau,\nu)\right|^2\right] \tag{6.10}$$

given that the expectation of the spread function is null [12]. This function describes the way in which the average power is distributed in the dimensions of the bidirectional (departure or arrival), delay and Doppler frequency [12].

Figure 6.20 shows the DoD power spectrum experimentally estimated in an indoor environment. For the delay of the LOS component $\tau = 115$ ns, maximum of the PDP, the power spectrum is concentrated around a single DoD; however, for the delay instant $\tau = 145$ ns, the power spectrum shows four dominant DoDs indicating a larger dispersive channel [12].

6.4.1.2 Specular path model

The specular-path model is based on the assumption that electromagnetic waves propagate in an environment along multiple specular paths [12]. A bidirectional-delay-Doppler-dual-polarization specular path-model has been derived [12], in which the parameters describing a specular path include its delay, DoA, DoD, Doppler frequency and polarization matrix [12]. Other parameters may be also included, as the time-variability of the path for time-variant channels as in V2X scenarios.

In the specular path model, the received signal is considered as the superposition of multiple signals propagating along uncorrelated paths so that the related signals give raise to individual MPCs. In order to be considered uncorrelated paths, those paths should exhibit parameters distinguishable by the resolution of the measurement channel sounder [12].

The estimation accuracy of the path parameters is higher if dual polarization is considered. The estimation algorithms extract the polarization features of MPCs due to the radio channel [12]. The use of dual polarized antenna arrays has been considered to improve the capacity and the performance of a MIMO system [129–133]. This interest has required the channel modelling in the dual-polarization domain [134–137].

For this type of channel characterization, the sounding technique required is called time-division-multiplexing (TDM) [12]. Communication systems provided with transmitting/receiving antennas equipped with dual-polarized configuration transmits and/or receives simultaneously in both polarizations [12]. Channel sounding incorporating multiple dual-polarized antennas can be conducted with a radio-frequency switch scheme in both link ends. Examples of TDM channel sounder are PROPsound [84], RUSK [101,138,139] and rBECS [12,140–144].

6.4.1.3 Dispersive path model

The specular path model considers the individual waves propagation through uncorrelated paths. Under real conditions, propagation paths are likely correlated: in the path, parameters are close, or the resolution of the channel sounder is not enough. A set of correlated paths is termed a dispersive path, and its contribution cannot be accurately described by the specular path model [12].

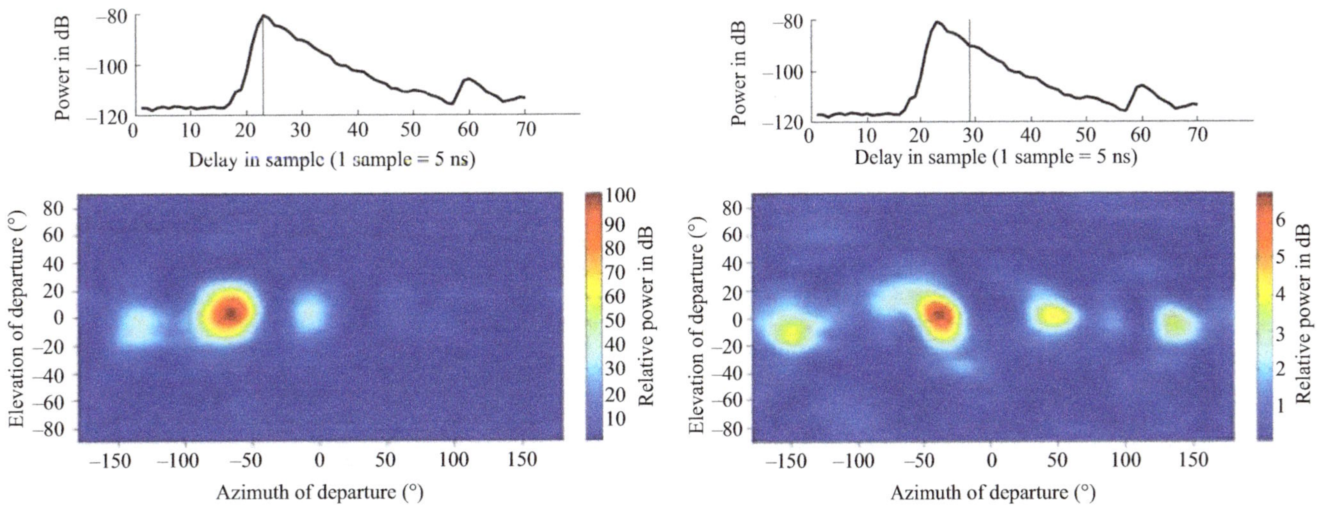

Figure 6.20 Estimated power DoD samples [12]

Dispersive path is also called slightly spatially distributed source estimation, and for its estimation, there exist two groups of approaches [12]:

- Approximate models: The first-order Taylor expansion approximation model of [145] and the two-ray model described in [146].
- High-resolution estimators: distributed signal parameter estimator (DSPE) [147], dispersed signal parametric estimation (DISPARE) [148,149], and spread root-multiple signal classification (MUSIC), estimation of signal parameters via rotational invariance techniques (ESPRIT) and method of direction estimation (MODE) [146].

6.4.1.4 Time evolution model

The time variability of propagation paths [150,151] directly leads to the time-variant characteristics of path parameters. This fact has been considered an additional degree of freedom for the path clustering characterization as proposed in [12,151]. The time evolution of the path clusters has been documented in [12,150] using measurement-based data.

The path-evolution characteristics have been usually obtained indirectly from the path parameter estimated from individual observation snapshots [12]. Among the estimation methods used, it can be mentioned the space-alternating generalized expectation-maximization (SAGE) algorithm [150] and the Unitary ESPRIT [151]. As explained in [12], these methods are configured assuming that the path parameters in different observation snapshots are independent. However, this is not a realistic assumption and it results in an erroneous estimation of the path temporal evolution. It even causes that a time-variant path can be erroneously considered to be several paths. These effects influence the performance of clustering algorithms and the effectiveness of the channel models based on them.

Therefore, it is required to design algorithms capable of directly estimating the temporal characteristics of paths, and further research is needed in this line. As described in [12], some methods have been proposed to track time-variant paths for MIMO channel sounding [152–154]:

- Tracking the azimuths of arrival (AoAs) of time-variant paths by means of recursive expectation-maximization (EM) and recursive SAGE-inspired algorithms [152]. This method is only applicable to a linear observation model.
- Tracking of the delays, DoAs, DoDs and complex amplitudes of time-variant paths using the extended Kalman filter (EKF) in [153,154]. This method is only applicable to a linear observation model.
- Tracking any of the parameters of time-variant paths by a sequential Monte–Carlo method so-called article filter [142,143]. This method can be applied even if the observation model is non-linear.

In the above-listed methods, the state-space modelling of the time-variant path parameters has been adopted [12]. The state-space model describes the time-evolving path parameters, and the observation model for the received signal in the receiver of the sounding channel [12].

6.4.1.5 Power spectral density

A dispersed component is modelled using a cluster of MPCs obtained from measurement data, especially in parametric models of MIMO wideband channels [12,150,155,156]. A cluster of MPCs is composed of a set of specular components conveniently identified. The parameters characterizing a cluster of MPCs, called the dispersion parameters, are the nominal directions, the delay spread and the direction spread. However, this characterization results not accurate due to the use of generic models of specular components that do not correctly identify the dispersive components [12,157].

Many algorithms used for the estimation of the dispersion parameters of individual components conforming a cluster of MPCs [149,158–160] calculate the PSD of each component of the cluster. The PSD related to an individual component can exhibit irregular shape in real environments wherein the channel measurements are performed. Then other parameters are used to describe the dispersive characteristics, such as the gravity centre and PSD spread. These parameters are estimated by comparing the shape of the normalized component PSD to standard PDFs in AoA [149,158,160] and AoD [159]. The performance of these estimation algorithms has not been corroborated by means of measurement data [12].

6.4.1.6 Keyhole channel

The keyhole phenomenon first described in [161,162] confirmed the existence of the worst propagation case for MIMO transmissions that is called rank-1 channel [12]. The keyhole effect is observable when the propagation paths between the transmitter and the receiver share a 1D diffraction component [12,163].

The existence of the keyhole was experimentally demonstrated by [164] that presented the keyhole phenomenon as a generalization of the pinhole scenario. As a result of the keyhole effect, the radio channel is considered degraded to a low-rank channel type [12]. The pinhole effect can be observed in many scenarios and propagation situations typical of different wireless communication systems, such as diffraction by building edges and corners, wave guiding in street canyons and corridors, or above-roof propagation [12].

In general, both the keyhole and pinhole effects usually occur when a propagation channel can be divided into two independent sub-channels and the propagation through the interface of both sub-channels has a small number of degrees of freedom [12]. This occurs in outdoor-to-indoor propagation through windows, and propagation through two environments connected by a hallway or a tunnel. The keyhole effect can also be used to describe the propagation in a relay network where a repeater or a relay station (RS) receives signals from an environment and transmits the signals to another isolated area. In Figure 6.21, there are two examples of propagation where a keyhole channel can be formed.

The conventional keyhole model focuses on the impact of the keyhole on the statistics of the channel matrix [12]:

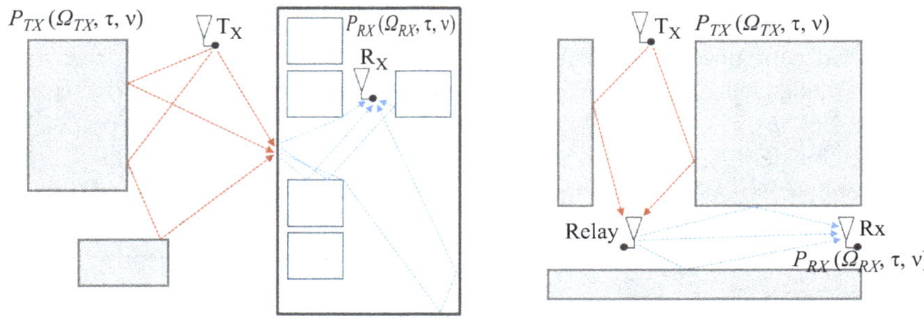

Figure 6.21 Two favourable scenarios to the keyhole or pinhole effect [12]

- In [165], it's described the keyhole channel using the MIMO correlated fading channel. The entries of a keyhole channel transfer matrix are modelled as an i.i.d. double-Rayleigh distribution.
- In [166], it's described the distributions of the eigenvalues of the keyhole and multi-keyhole channel matrix.
- In [167], it's described the distribution of the outage capacity of correlated keyhole MIMO channels.

These models allow estimating the impact of a keyhole channel in an analytical form. However, it is still required a complete description of the space–time–frequency characteristics of wideband keyhole channels [12].

In [12], a generic model is outlined for characterization of the dispersion of the keyhole propagation channel in delay, DoD, DoA and Doppler frequency. This model introduces the marginal dispersive function of the keyhole channel and the analytical expressions for its transfer matrix and correlation matrix. The more valuable result is the understanding of the underlying propagation mechanism and the complete characterization of the keyhole channel. The model in [12] verified the existence of keyhole effect using measurement data recorded with a wideband MIMO channel sounder [12]. For example, since relay networks are expected to be widely extended in future wireless communication systems as 5G and V2X networks, further research is needed to accomplish ray-tracing or measurement-based characterization of keyhole phenomenon for such scenarios [12].

The scenario considered in this model is depicted in Figure 6.22. Table 6.6 summarizes the generic models for channel spread functions in keyhole scenarios. In [168], these generic models were evaluated by using measurement data.

6.4.2 Geometry-based models

As reviewed in Section 6.4, GBMs can be categorized as GBDMs and GBSMs. The second category can be divided into NGBSMs and GBSMs.

The GBSM approach belongs to the so-called scattering modelling category, which also includes geometry-based deterministic modelling. Compared with its

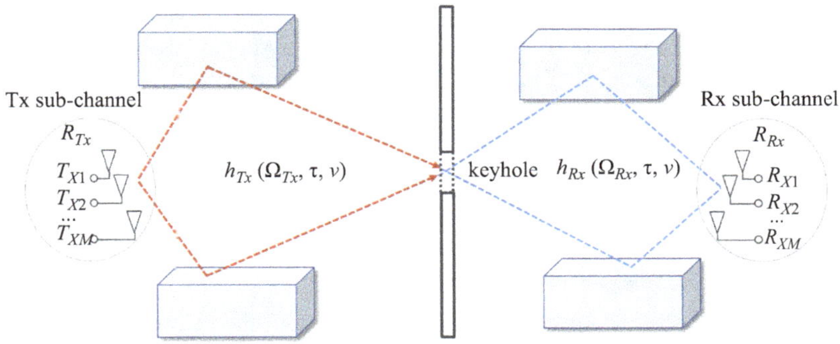

Figure 6.22 Keyhole channel splitting into Tx and Rx sub-channels [12]

Table 6.6 Generic models of keyhole scenarios [12]

Function	Expression
Spread	$h(\Omega_{Tx}, \Omega_{Rx}, \tau, \nu) = \alpha_k \cdot h_{T_x}(\Omega_{Tx}, \tau, \nu) * h_{R_x}(\Omega_{Rx}, \tau, \nu)$
Delay spread	$h(\tau) = \alpha_k \cdot h_{RT_x}(\tau) * h_{R_x}(\tau)$
	$h_{T_x}(\tau) = \int\int h_{T_x}(\Omega_{Tx}, \tau, \nu) d\Omega_{Tx} d\nu$
	$h_{R_x}(\tau) = \int\int h_{R_x}(\Omega_{Rx}, \tau, \nu) d\Omega_{Rx} d\nu$
Doppler frequency function	$h(\nu) = \alpha_k \cdot h_{RT_x}(\nu) * h_{R_x}(\nu)$
Spread function	$h_{T_x}(\nu) = \int\int h_{T_x}(\Omega_{Tx}, \tau, \nu) d\Omega_{Tx} d\tau$
	$h_{R_x}(\nu) = \int\int h_{R_x}(\Omega_{Rx}, \tau, \nu) d\Omega_{Rx} d\tau$
Bidirectional Spread	$h(\Omega_{Tx}, \Omega_{Rx}) = h_{T_x}(\Omega_{Tx}) \cdot \alpha_k \cdot h_{R_x}(\Omega_{Rx})$
DoD spread	$h(\Omega_{Tx}) = \alpha_k \cdot \alpha_{R_x} \cdot h_{T_x}(\Omega_{Tx})$
DoA spread	$h(\Omega_{Rx}) = \alpha_k \cdot \alpha_{T_x} \cdot h_{R_x}(\Omega_{Rx})$
Power spectrum	$P(\Omega_{Tx}, \Omega_{Rx}, \tau, \nu) = P_{T_x}(\Omega_{Tx}, \tau, \nu) * P_{R_x}(\Omega_{Rx}, \tau, \nu)$

deterministic equivalent, which needs a detailed description of the real communication environment, GBSM approach is simpler and more general and thus has been widely used.

The GBSMs constitutes a simpler alternative to the complex ray-tracing-based models [3]. In a similar way, in GBSMs the environment geometry must be first configured by means of environment model parameters. The probability distributions for the different obstacles must be also provided as an input to the geometry-based stochastic channel model (GBSCM). Once completed the scenario setup, the ray tracer provides the CIR. Usually GBSMs only consider single reflections, even that additional reflections can be incorporated. It was found in [39] that increasing the complexity of the ray tracer did not provide an improved agreement with measurements [3]. As described in [3], for a roadway case, obstacles – buildings, road signs, possibly foliage, and other cars– may be placed uniformly over some section of roadside, being the dimensions of those obstacles taken from other specified distributions [3]. Generally diffuse scattering cannot be modelled by conventional ray tracing since this latter fundamentally models specular reflections. Then diffuse scattering contribution is incorporated by considering its parameters derived upon measurements. These diffuse contributions model multiply-bounced components [3].

A generic structure for GBSMs is shown in Figure 6.23. The simulation engine requires the values for the input parameters and for the tabulated geometric parameters in order to setup the environment and the object distribution [3].

The general modelling procedure involved in GBSCM is summarized in Figure 6.24, and consists of the following steps [12]:

1. *Basic setting of vehicular environment*: This includes the position and/or moving direction and velocity of the Tx/Rx, as well as the classification of effective scatterers, into, for example, moving scatterers and static scatterers [12].
2. *Scatterer placement*: Scatterers are allocated within the predefined scattering region according to a PDF. The resulting shape depends on the scatterers' distribution within the region: regular-shaped (one/two-ring, ellipse, etc.) or irregular-shaped (random) [12].
3. *Parameterization*: There exist two approaches to parameterize scatterers [12]:
   - Considering a finite number of scatterers: fading characteristics are assigned to each based on measurement data.
   - Considering an infinite number of scatterers: It is not possible to assign fading properties to the scatterers. Thus, the channel characteristics are only using

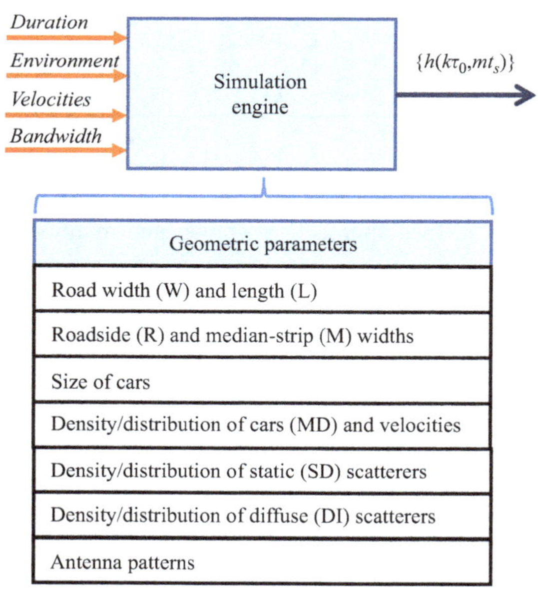

Figure 6.23 GBSCM generic structure [3]

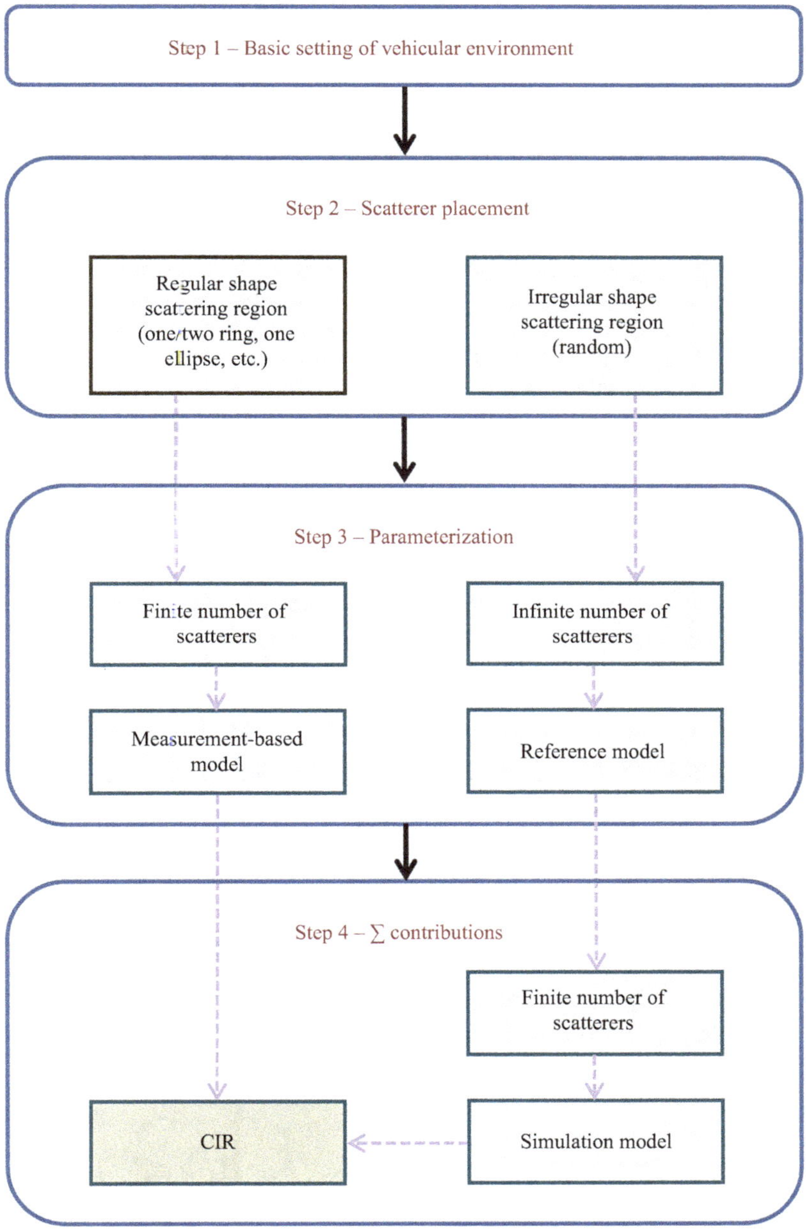

Figure 6.24 A general modelling procedure for the GBSM approach [12]

PDFs. A channel reference model is derived not implementable in practice and only valid for theoretically evaluating the channel characteristics.

4. *Addition of all scatterers' contributions*: All the contributions produced by the above set scatterers are summed up at the receiver end to achieve the model CIR. Due to the infinite number of scatterers of the reference model, the corresponding simulation model must be obtained first to become realizable [12].

The GBSM gives a mathematical function of the CIR representing the electromagnetic properties of environment, and then utilizes the proposed stochastic distribution of the scatterers along with geometrical knowledge to get the statistical properties of the parameters, such as the distributions of AoA, AoD, time of arrival (ToA), angle spread and so on, in the given channel response [12].

GBSMs are also called geometrical-based scatterer models that determine the distribution of scatterers and scatterer density within the propagation environment [169,170]. The GBSMs incorporate the scattering according to different SMs illustrated in Figure 6.25: one-ring, two-ring, elliptical and/or cylindrical scattering models. These models consider the scatterers located within the area delimited by a ring or other geometrical shape. The one-ring and two-ring scatterer models only consider single bounce scattering [171]. A MIMO narrowband channel based on the geometrical one-ring scattering model was first proposed by [172] and was further developed by [173].

The geometry of the one-ring model is illustrated in Figure 6.26 for a 2 × 2 MIMO case. The main properties of the one-ring model are the following [174,175]:

- GBSM valid for the macrocellular wireless network [175–178]. The transmitter is elevated above all surrounding scatterers and the receiver is surrounded by effective or local scatterers.
- Signals reach the receiver in different directions according to predetermined statistical distributions. An efficient scattering distribution depends on the environment surrounding the transmitter and receiver [169,178] and allows determining the AoA and AoD.
- The scatterer locations or the distance of the scatterers are known in order to determine the delay for a MPC propagating through local scatterers.
- The antenna spacing is denoted by d_{tx} for the transmitter side and d_{rx} for the receiver end. Dimensions of the model geometry are defined in [176,178].

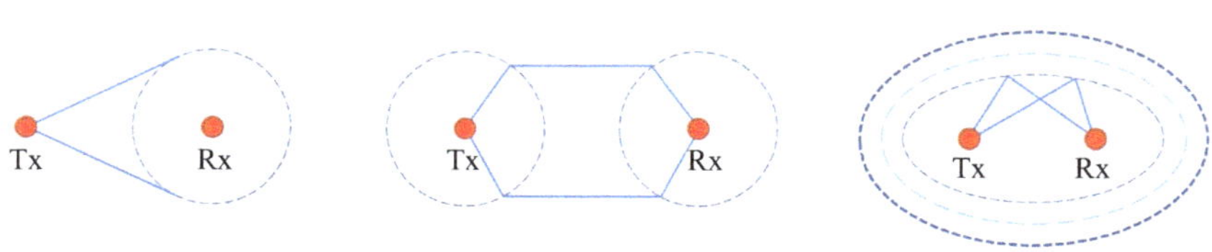

Figure 6.25 Scattering models: (a) one-ring model; (b) two-ring model; (c) elliptical model [12]

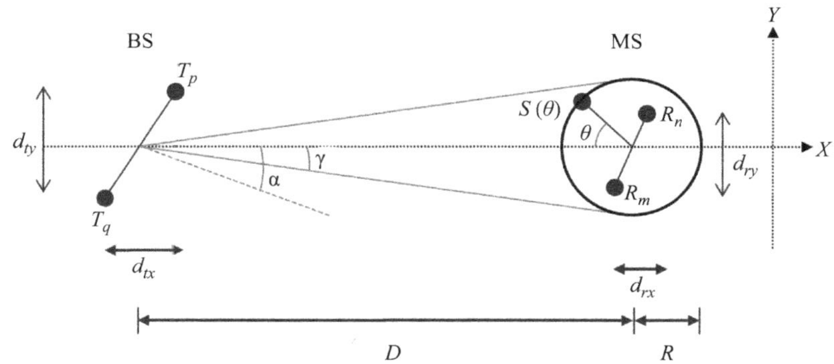

Figure 6.26 One-ring scattering model [174] Patzold2004

- The model assumes only single scattering with all scatterers having equal power.
- The transmitter or BS is not obstructed by local scattering while the receiver or MS is surrounded by scatterers.
- There is no LOS between the BS and MS.
- T_p is the pth antenna element at the BS, and R_n is the nth antenna element at the MS.
- D is the distance between the BS and MS, and R is the radius of the ring of scatterers.
- The effective scatterer on the ring is denoted as $S(\theta)$ with θ the angle between the scatterer and the array at the MS.
- In the model, it is assumed that $S(\theta)$ is uniformly distributed in θ.
- The phase shift $\varphi(\theta)$ associated with each scatterer $S(\theta)$ is distributed uniformly over $[-\pi, \pi)$ and i.i.d in θ.
- Each ray is reflected only once.
- All rays reach the receive array with the same power.
- There are K effective scatterers $S(\theta_k)$, $k = 1,2, ..., K$ distributed on the ring.
- The complex channel coefficient between the pth elements at the BS and nth element at the MS can be expressed as in (6.11) [174]:

$$H_{p,n} = \frac{1}{\sqrt{K}} \sum_{k=1}^{K} \exp\left\{ -j\frac{2\pi}{\lambda} \left(D_{T_p \to S(\theta_k)} + D_{S(\theta_k) \to R_p} \right) + j\phi(\theta_k) \right\} \quad (6.11)$$

with D_{XY} the distance between X and Y and λ is the wavelength.

The two-ring model is characterized by the following properties [174,175]:

- Both the BS and MS are surrounded by scatterers.
- The model consists of two rings with centres at BS and MS locations as shown in Figure 6.27.
- The ring radio represents the average distance between its communication link end and its corresponding scatterers [176,179,180].

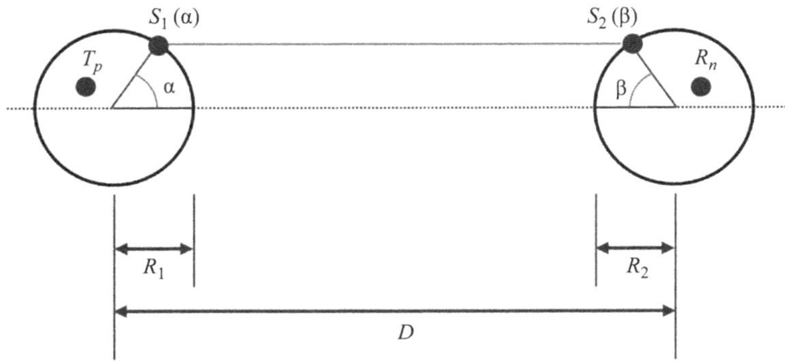

Figure 6.27. Two-ring scattering model [174]

- Scatterers are placed randomly on these rings.
- It is assumed that scatterers surrounding the BS and MS are in control of the AoD and AoA, respectively.
- Each ray/path is reflected twice.
- The channel coefficient for the two-ring model is as in (6.12) [174]:

$$H_{p,n} = \frac{1}{\sqrt{K_1 K_2}} \sum_{k=1}^{K_1} \sum_{l=1}^{K_2} \exp\left\{ -j\frac{2\pi}{\lambda} \left(D_{T_p \to S_1(\alpha_k)} + D_{S_1(\alpha_k) \to S_2(\theta\beta_l)} + \right.\right.$$
$$\left.\left. + D_{S_2(\theta\beta_l) \to R_p} \right) + j\phi_1(\alpha_k) + j\phi_2(\alpha_k) \right\} \tag{6.12}$$

- Drawback of the model: The signals reflected by the scatterers at the receive end are probably not independent.
- The channel coefficient is not zero-mean complex Gaussian, even that the numbers of scatterers K_1 and K_2 arise to infinity.
- The two-ring single-bounce scattering model results easily controllable based on scatterers distribution and clustering [176].

Finally, the characteristics of the elliptical model are the following:

- It defines the distribution of effective scatterers on the eccentricity of an ellipse.
- It is valid for scenarios using the same antenna heights for both transmitter and receiver.
- It is also applicable to channels with a dominant LOS component, reflective surfaces and mobile ends at both communication sides [169,181,182].

The elliptical model for a 2 × 2 MIMO channel is depicted Figure 6.28. In this figure, Tx, Rx and S represent the transmitter, receiver and effective scatterers, respectively. The number of effective scatterers is given as n, where $n = \{1, 2, 3, \ldots\}$.

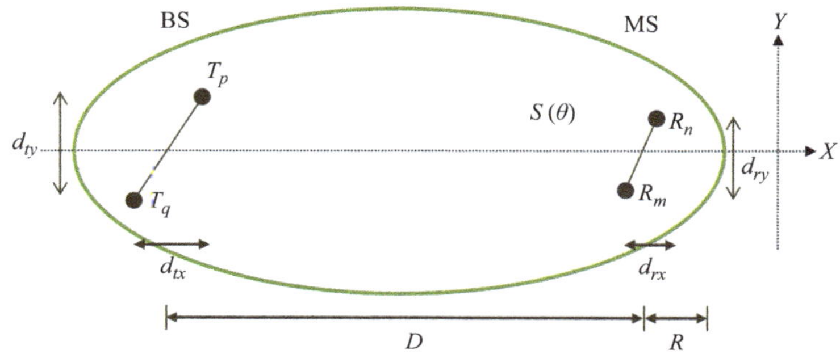

Figure 6.28. Elliptical scattering model

Despite the model considered, scattering is not usually diffuse and then it is difficult to assign realistic scattering coefficients [175,183] and then the single-bounce model does not represent all observed channel scattering effects. Some GBSMs incorporated both single- and double-bounced paths in order to solve the problem [175,182]. A combination of the elliptical model with the two-ring model has been considered to characterize both mobile and stationary scatterers [182]. A non-stationary one-ring model is described in [184].

6.4.2.1 GEMV^2

GEMV^2 is the acronym for geometry-based efficient propagation model for V2V communication. In 2014, Boban *et al.* [45] developed the GEMV^2 model as a geometry-based, efficient propagation approach for V2V and V2I communications.

The main features of this model are listed next as extracted from [45,118]:

- Uses outlines of vehicles, buildings and foliage to distinguish the following three types of links: LOS, NLOS due to vehicles (NLOSv) and NLOS due to static objects (buildings/foliage) (NLOSb).
- Deterministically calculates large-scale signal variations (PL and shadowing) for each link type.
- Calculates the small-scale signal variations stochastically via a GBSM using the number and size of objects around the communicating vehicles.
- Models both V2V and V2I links.
- Validated against V2V and V2I measurements in urban, suburban, highway and open-space environments.
- Imports vehicular mobility from simulation of urban mobility (SUMO) [185,186], and uses floating car data format to generate vehicle outlines.
- Imports outlines of buildings and foliage from OpenStreetMap [187,188].
- Easily extensible to import vehicular mobility and object outlines from other data sources. Real-world traces via GPS can be used.
- Exports the visualization of vehicular communication to keyhole markup language (KML) format for use with Google Earth or NASA World Wind.

- Requires minimum geographic information (the outlines of modelled objects), so it is suitable for implementation in vehicular ad-hoc networks (VANET) simulators.
- Available implementation in MATLAB®, free of charge and openly distributed.
- Available implementation in C++ for OMNet++/Artery.

SUMO is an open-source, highly portable, microscopic and continuous traffic simulation package designed to handle large road networks. It is mainly developed by employees of the Institute of Transportation Systems at the German Aerospace Centre. SUMO is open-source, licensed under the EPLv2 [185,186].

NASA World Wind [189] is an open-source virtual globe API that allows developers to quickly and easily create interactive visualizations of 3D globe, map and geographical information. It can be applied to monitor weather patterns, visualize cities and terrain, track vehicle movement, analyse geospatial data and educate humanity about the Earth [189].

The GEMV^2 package implemented in MATLAB [118] showed its ability to scale and simulate radio propagation for citywide networks with tens of thousands of communicating vehicle pairs in different environments: highway, suburban and urban [45]. GEMV^2 was able to simulate propagation for the entire city of Porto an area of 41 km^2 of surface containing 10.566 vehicles and 17.346 static objects. In Figures 6.29 and 6.30 is shown a Google Earth visualization of the received power, in dBm, in Porto generated by GEMV^2 [45].

As described in [45], GEMV^2 uses computational geometry concepts suitable for geographic data representation in order to ensure efficient processing of the object outlines. More specifically R-trees [190] is the indexing structure used to record information about the outlines of vehicles, buildings and foliage.

As previous works demonstrated [45,191], in V2V propagation the channel characteristics for LOS and NLOS links are deeply different. In [191], surrounding

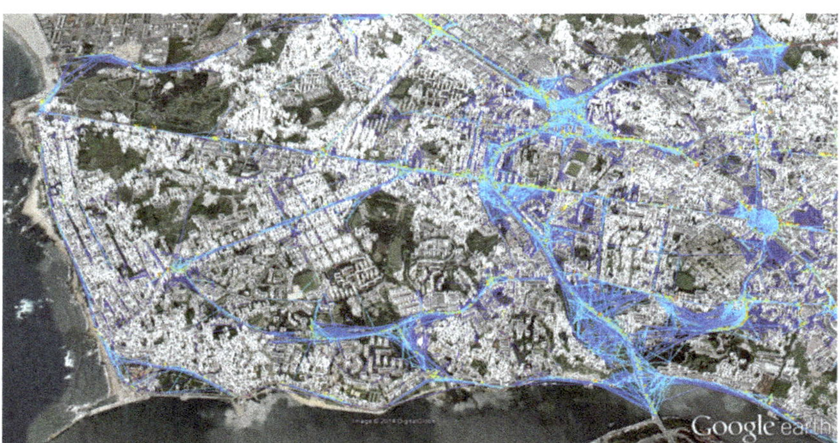

Figure 6.29 Snapshot showing a part of transmit–receive pairs in the city of Porto generated by GEMV^2 (colour bar in dBm) [45,118]

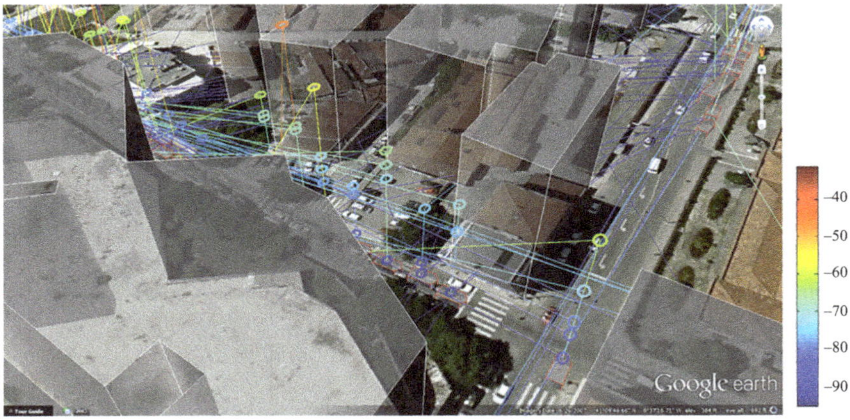

Figure 6.30 Street-level view generated by GEMV² (colour bar in dBm) [45,118]

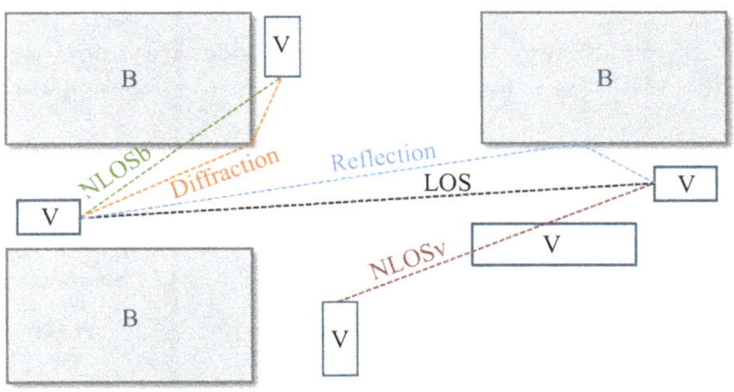

Figure 6.31 Link types and propagation effects considered in GEMV² (B: buildings, V: vehicles) [45,118]

vehicles were identified as a key factor to determine the V2V channel characteristics. Thus, the GEMV² approach considered simple geographical descriptors of the simulated environment – outlines of buildings, foliage, and vehicles on the road – resulting into a classification of V2V links based of three groups [45]:

- LOS: Links with an unobstructed optical path between the transmitting and receiving antennas.
- NLOSv: Links with LOS component blockage due to other vehicles.
- NLOSb: Links with LOS component blockage due to buildings or foliage.

Figure 6.31 depicts the link types and propagation effects considered in GEMV². An extensive measurement campaign was performed to validate GEMV² in urban, suburban, highway and open-space environments [45]. Results reported in

[45] show that the model accurately predicts both small- and large-scale propagation effects for LOS, NLOSv and NLOSb links in all environments.

6.4.2.2 ETSI model

The model completed by ETSI uses the GBSCM developed by the third-generation partnership program (3GPP) [24], which is based on the models by GEMV^2, WINNER [192], WINNER-II [193] and WINNER+ [194]. The ETSI model is valid to simulate any type of vehicular scenario: V2V, V2I, V2P, P2P, V2R and R2R. It has also the ability for modelling the communication with BSs, i.e. V2B, P2B and B2R links.

The channel realizations are obtained by a stepwise procedure composed of three high-level steps, as illustrated in Figure 6.32 [116]:

- *General parameters setting*: The determination of general parameters consists of four steps outlined in Figure 6.33.
- *Small-scale parameters setting*: The determination of small-scale parameters consists of five steps outlined in Figure 6.33.
- *Coefficient generation setting*: The coefficient generation consists of three steps outlined in Figure 6.33.

Explicit V2X parameters for each of the procedure steps are specified in greater detail in [116], even that parameters extracted from [195] are taken as

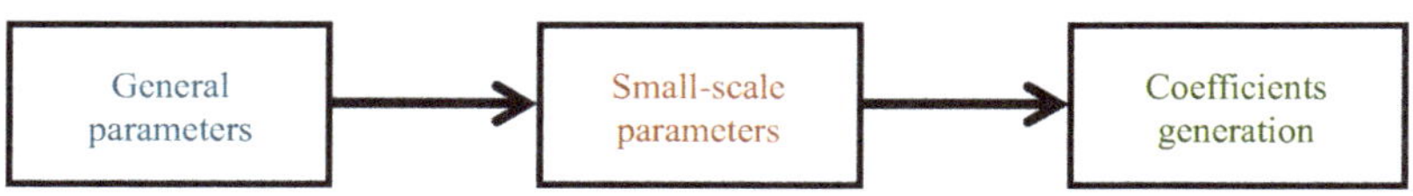

Figure 6.32 Levels of parameters generation in the ETSI GBSM model [116]

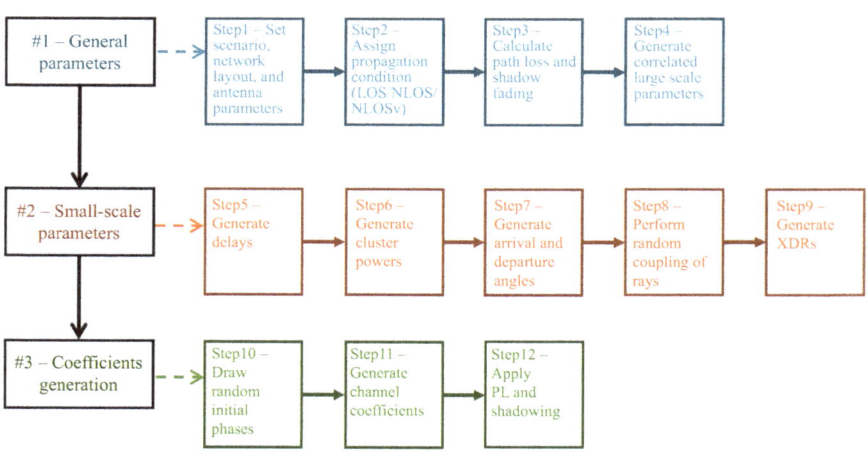

Figure 6.33 Detailed stepped procedure [116]

baseline and they can be modified if needed. As a way to ensure spatial consistency, the model by ETSI includes an approach both to generate correlated large-scale parameters (LSPs) for transmissions between closely located vehicles, and to achieve the channel evolution as vehicles move along an objective route [116]. The ETSI model is projected for system-level simulations (SLS).

The steps required for the three-level procedure are briefly described as follows:

- **Step 1: set scenario.**
  Possible scenarios are urban, highway and rural. Full parameterization is available for 'Urban' and 'Highway', whereas for 'Rural' only TDL models exist. Therefore, for all remaining parameters, either 'Urban' or 'Highway' parameters need to be reused [24].
- **Step 2: assign propagation condition.**
  The V2V channel is modelled according to the following three states: LOS, NLOS and NLOSv. Vehicle location is updated every 100 ms. The probability of LOS and NLOSv states depend on the link range. At each state, each link uses the PL, shadowing and fast-fading parameters corresponding to the current state [116].
- **Step 3: PL and shadowing.**
  PL model for V2V links is provided for each state and these can be reused for the following links as well: V2P, P2P, V2R and R2R. For modelling the PL for communication with BSs (i.e. V2B, P2B and B2R links), the models described in [194] can be used.
  For V2V link under NLOSv state that occurs when vehicles are blocking the LOS link, it is induced an excess blockage loss that can be calculated by means of one of the following two options as described in [116]:
  - *Option 1* is preferred when accurate per-link loss is needed, such as in case of emergency message exchange or platooning in vehicle convoys. It is a deterministic approach for estimating the vehicle blockage loss. A model for vehicles-as-obstacles is presented in [191], where vehicles are modelled using the ITU-R multiple knife-edge diffraction method [196], later described in Section 6.4.2.3. The model estimates attenuation excess caused by each of the surrounding vehicles blocking the LOS link. The estimation of the link blockage state is accomplished geometrically.
  - *Option 2* is used when simulation speed becomes critical. This is a stochastic approach for estimating the vehicle blockage loss. The blocker height is the vehicle height that is randomly selected from three vehicle types according to the ratio of the vehicle types in the simulated scenario.

The shadow-fading process is assumed to follow a log-normal distribution [24] with zero mean and standard deviation σ. The value of σ is fitted to the environment and link type. For V2V links, a measurement-based analysis of σ was completed by GEMV^2 model [45] for both highway and urban environments [116].

- **Step 4: large-scale correlated scatterers.**
  This block generates LSPs: Delay spread, angular spread (AS), Ricean *K* factor and shadow fading [116].
- **Step 5: generate delays.**
  This block randomly draws delays out of the delay distribution defined. The exponential delay distribution is also calculated. For the LOS case, additional scaling of delays is required to compensate for the effect of LOS peak addition to the delay spread [116].
- **Step 6: generate cluster powers.**
  This block calculates the cluster powers assuming a single-slope exponential PDP. Power assignment depends on the delay distribution defined in [116]. For an exponential delay distribution, the cluster powers are calculated. For the LOS case, an additional specular component is added to the first cluster. Clusters with less than −25 dB power compared to the maximum cluster power are removed. The number of rays per cluster is M [116].
- **Step 7: generate AoA and AoD.**
  The composite power angular spread (PAS) in azimuth of all clusters is modelled as wrapped Gaussian. The AoAs are determined by applying the inverse Gaussian function. For the LOS case, additional scaling of the angles is required to compensate for the effect of LOS peak addition to the angle spread. The generation of AoD follows a procedure similar to AoA [116].
- **Step 8: perform random coupling of rays.**
  Couple randomly AoD angles $\varphi_{n,m,AoD}$ to AoA angles $\varphi_{n,m,AaA}$ within a cluster n, or within a sub-cluster in case of two strongest clusters [116].
- **Step 9: generate XPRs.**
  Generate the cross-polarization power ratio (XPR) for each ray of each cluster. XPR is log-normal distributed [116].
- **Step 10: draw initial random phases.**
  Draw random initial phase for each ray m of each cluster n and for four different polarization combinations ($\theta\theta$, $\theta\phi$, $\phi\theta$, $\phi\phi$). The distribution for initial phases is uniform within $(-\pi, \pi)$ [116].
- **Step 11: generate channel coefficients.**
  The method uses at least for drop-based evaluations irrespective of vehicle speeds. The aim is obtaining the CIR. For side link, dual mobility is modelled by Doppler for the LOS path and Doppler for the delayed paths. Inter-cluster delay spread is also modelled. The CIR is given by adding the LOS channel coefficient to the NLOS CIR and scaling both terms according to the desired K-factor [116].
- **Step 12: apply PL and shadowing.**
  In this last step, PL and shadowing are applied [116].

6.4.2.3 Regular-shaped GBSM

As discussed in Section 6.3, RS-GBSMs are mainly used for theoretical analysis of channel statistics and theoretical performance evaluation of V2V communication

Table 6.7 Comparison of main RS-GBSMs [13]

Model	Description
Akki and Haber [111,112]	2D two-ring, only DBR, narrowband isotropic scattering, SISO, V2V, Rayleigh fading channels, LSS scenarios. No analysis of impact of the VTD on channel statistics
Pätzold *et al.* [113]	Two-ring, only DBR, non-isotropic scattering MIMO, V2V, Rayleigh fading, macrocell scenarios
Zajic and Stüber [197]	General 2D two-ring, both SBR and DBR, narrowband non-isotropic scattering, MIMO, V2V, Ricean channels in both LSS (macrocell) and MSS (microcell) scenarios. No analysis of impact of the VTD on channel statistics. No ability to explore per-tap channel statistics.
Zajic and Stuber [115]	Extension of model in [114] to 3D wideband two-concentric-cylinder. No analysis of impact of the VTD on channel statistics
Cheng and Wang [38]	2D non-isotropic scattering wideband MIMO, V2V, extension of narrowband model in [182] with respect to the frequency selectivity. Two-ring model and a multiple confocal ellipses model consisting of LOS, SBR and DBR based on the TDL structure, analyse the per-tap channel statistics. Evaluate the impact of the VTD on channel statistics for every tap

systems. RS-GBSMs assume that all the effective scatterers are located on regular shapes (one/two-ring, ellipse, etc.) [13]. In Table 6.7, the main RS-GBSMs are summarized with a brief description.

The model described in [13,14] enables evaluating the impact of the VTD on channel statistics for every tap in our wideband model. To this aim, the model distinguishes between the moving cars around the transmitter and receiver vehicles described by a two-ring model, and the stationary roadside environments, described as a multiple confocal ellipses model [13].

For a typical V2V environment, Figure 6.34 shows a geometrical description of the 3D two-concentric-cylinder wideband model, which has LOS, single-bounced rays (SBR) and double-bounced rays (DBD). The impulse response for the first tap of the TDL model can be divided into three parts or components [12,13,114,115]:

- The LOS component.
- The SBR generated from the effective scatterers located on either of the two-rings or the first ellipse.
- The DBR produced from the effective scatterers located on both rings.

Whereas for other taps, the impulse response contains two parts [13]:

- The SBR generated from the effective scatterers located on the corresponding ellipse.
- The DBR caused by the effective scatterers from the combined one ring (either of the two rings) and the corresponding ellipse.

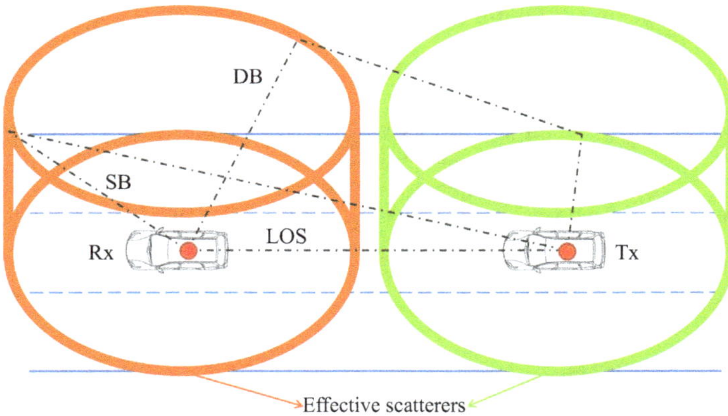

Figure 6.34 Geometry of an RS-GBSM for a V2V environment: SB, single-bounced rays; DB, double-bounced rays [12,14]

According to the above, the complex CIR between the pth elements at the BS and qth element at the MS can be expressed mathematically as follows:

$$h_{pq}(t,\tau) = h_{pq}^{LOS}(t,\tau) + h_{pq}^{SBR}(t,\tau) + h_{pq}^{DB}(t,\tau) \tag{6.13}$$

However, the model in [12–14] cannot explore the NS of the channel due to the static nature of the geometry inherent to RS-GBSMs.

6.4.2.4 Irregular-shaped GBSM

IS-GBSMs aim to replicate the scenario with a realistic location and characteristics of the effective scatterers. IS-GBSMs allocate the effective scatterers with detailed properties at random locations with proper statistical distributions [12]. The geometry of an IS-GBSM is depicted in Figure 6.16 shown in Section 6.4.

The total signal value is obtained by summing up the contributions due to the effective scatterers that are determined by a simplified ray-tracing method [12]. The result is the CIR calculated only for a null elevation angle. As described in [12,39], the CIR is further divided into four parts:

- The LOS component.
- Discrete components due to reflections from mobile scatterers (moving cars).
- Discrete components due to reflections from strong static scatterers (building and road signs on the roadside).
- Diffuse components due to reflections from weak static scatterers present in the roadside.

Therefore, the complex CIR can be expressed mathematically as follows [39]:

$$h_{pq}(t,\tau) = h_{LoS}(t,\tau) + \sum_{p=1}^{P} h_{MD}(t,\tau_p) + \sum_{q=1}^{Q} h_{SD}(t,\tau_q) + \sum_{r=1}^{R} h_{DI}(t,\tau_r) \tag{6.14}$$

with $h_{LoS}(t,\tau)$ the LOS component; $\sum_{p=1}^{P} h_{MD}(t,\tau_p)$ the discrete components due to reflections from P mobile scatterers (MD); $\sum_{q=1}^{Q} h_{SD}(t,\tau_q)$ the discrete components due to reflections from Q static scatterers (SD); and $\sum_{r=1}^{R} h_{DI}(t,\tau_r)$ the diffuse components (DI) due to R diffuse scatterers.

Therefore, IS-GBSMs can be considered GBSMs applicable to an extensive set of V2V scenarios by fine-tuning the statistical distributions of the effective scatterers. The advantage of the IS-GBSMs is that the embedded ray-tracing approach, as in [12,39], can easily handle the NS of V2V channels by prescribing the motion of the mobile communicating vehicles as well as mobile surrounding scatterers. For low VTD, only SBR are needed to be considered as shown per the measurements in [44]. For a high VTD case, DBR should be also included. Unlike NGSMs, the IS-GBSMs can explore the drift of scatterers into different delay bins although at the cost of higher complexity [12,39].

6.4.3 Stationary and non-stationary channel models

The time variability of V2X channels is larger than that in other wireless communication technologies due to the combination of the high velocity of the communicating link ends and the existence of moving scatterers. In most cases, V2X channels are statistically non-stationary and the stationarity condition is kept over a short time period [12]. Until now incorporating the NS into V2X channel models has not been achieved properly and in a simple way [12]. Practical models for NS channel performance do not yet exist [3].

Traditionally, the conditions of Bello for considering a GWSSUS model are assumed without verifying them [60]. The validity of the relationship described must be subjected to the verification of the next three conditions:

1. $h(t,\tau)$ must verify uncorrelated scattering in τ.
2. $H(t,f)$ must be constant mean in f.
3. $h(t,\tau)$ must verify ergodicity in t.

Under the above conditions, it will be sufficient to apply the Fourier transform to a time average of impulse response estimates to obtain correct estimation of the channel functions and parameters. However, the validity of the WSSUS assumption was rarely challenged. The first work to consider NS in V2V channels may have been [198]. In [199], it was demonstrated the importance of considering NS in V2V channel modelling, given that system performance results indeed vary significantly, e.g. bit error rates change by up to an order of magnitude [3].

In order to model NS, the time-frequency analysis by [200] was applied in [201] to describe the NS of CIR or complex transfer function (CTF) and derive the local scattering function (LSF) to represent stationarity region within channel correlation functions for both time and frequency domains [3]. In [202], stationarity intervals were quantified by a correlation matrix distance for different V2V environments. The stationarity intervals fitted a log-normal distribution with mean around 3 m that means about 50 wavelengths at 5.3 GHz [3].

Measurement-based results of the LSF were obtained in [203], demonstrating stationarity intervals of 23 ms found on highways for transmitter and receiver moving in opposite directions and about 1.5 s for same-direction moving case [3]. In [11], the dimensions of the minimum stationarity region were found on the order of 40 ms in time and 40 MHz in frequency. In [204], it is described an adaptive reduced-rank estimation of non-stationary time-variant channels by applying subspace selection.

NS can also be due to the vehicle velocity variation. In [205], it is proposed a non-stationary multipath fading channel model that incorporates the effect due to velocity variations of the communicating vehicles.

The concept of MPC lifetime is related to NS, in which MPCs in the CIR can arise (birth) and disappear (death) over time/space [3,206]. The birth/death process of MPCs may be caused by the mobility of vehicles close to transmitter and receiver vehicles. These MPC lifetimes has been modelled as a birth–death random process by using discrete Markov chains [43,207].

In [208], it was measured and developed models for MPC lifetimes at urban intersections founding two types of log-normally distributed MPCs: a large number of short-duration MPCs and a small number of long-duration MPCs, for a lifetime threshold of 50 ms [3].

NS is also described by delay drift which implies that the delay of an MPC varies in time [3]. This variation is due to the varying link distance as the transmitter and receiver vehicles move. The delay drift is found relatively slow for even highway speeds: no faster than a 100-ns delay variation occurs over a 1 s time period [209,210].

As described in [3,52], in V2V channels NS becomes critical for long transmission durations. Only applications with data rate transfers on the order of 3 Mb would be affected by the NS channel effect [3].

Delay drift can be incorporated in GBSCMs, but not so in TDLs. Controlling the MPC persistence parameter, TDLs could be improved by including this effect [3]. The NS has been considered in NGSMs [13,43] by means of birth–death random process that accounts for the appearance and disappearance of MPCs or equivalently taps in TDL model. In IS-GBSMs [13,39], the NS prescribed the motion of the both link ends and mobile scatterers.

Incorporating NS will trade-off between the accuracy and complexity in NGSMs and IS-GBSMs. Then it is required a comprehensive research in some questions. First, it would be to determine whether the effect of the drift of scatterers into different delay bins on the NS of V2V channels is significant. It should be also determined the advantaged of eliminating for capturing this effect by using IS-GBSMs and the related complexity to achieve its incorporation [13].

6.4.4 MIMO and massive MIMO channel models

In Section 6.2.2.3, it has been discussed the main aspects of the MIMO channel characterization. In Table 6.3, the main works reported for MIMO channel sounders are summarized.

According to [3], the MIMO channel can be analysed as a spatial generalization of the single-antenna channel. The CIR for the MIMO channel is given by a matrix of CIRs with each matrix entry a CIR specified for each transmitter–receiver antenna pair [3].

The MIMO channel characterization aims the relationship between all the antenna-pair channels [3] characterized by the correlation of the MPC amplitudes on these single channels. It will be also required the spatial distributions of MPCs which is quantified via SMs for AoDs and AoAs. V2V measurements in [211] quantified these distributions [3].

The MIMO V2V channel requires radio modelling more complex that for the single-antenna-pair case [212]. The technique that provides an easier form to generate MIMO CIRs is GBSCM [213]. The use of the TDL model would require multiple parallel TDLs, each of them incorporating to the multiple-antenna-pair channels the correlation among MPC amplitudes [3]. MIMO V2V channel measurements and models have been described by [214–216].

MIMO V2V models could include multiple polarizations due to the implementation of polarization diversity, as described by [3,169]. Even though MIMO V2V channel measurements have been reported using dual-polarized antennas, as in [202], the measured values resulting of cross-polarization discriminations (XPD) were not provided [3]. Although including polarization in GBSCMs can be easily achieved, it is not addressed in their analysis [3]. Most of GBSCM reports validated with measurements only consider a single polarization that is vertical [3].

In [217] is provided a review of important research works on MIMO V2V. In [218,219], an urban NS MIMO V2V channel model was derived analysing the impact of fixed and moving scatterers on the statistics [217]. In [220], it is proposed a dynamic geometry-based scattering model for urban wideband MIMO V2V fading channel. This model assumed static and mobile scatterers uniformly distributed on time-varying ellipses and time-varying segment of the road, respectively [217,220]. In [221], a geometry-based stochastic scattering model is described for wideband MIMO V2V channels.

In [12], MIMO channel models are first classified into two main categories: deterministic models and SMs. Within the SMs, MIMO channel models are divided into physical (based on bidirectional radio propagation) and analytical models (based on MIMO channel matrix). The classification of MIMO channel models as in [12] is the following:

- Deterministic models (DM):
  - Ray-tracing and stored measurements
- Stochastic models:
  - Physical models (PM)
    - Geometry-based models (GBM)
      - Geometry-based stochastic model (One-ring, Two-ring Ellipse)
    - Non-geometrical models (NGM)
      - Parametric stochastic model (PSM) (Extension of the Saleh–Valenzuela and Zwick model)
  - Analytical models (AM)
    - Correlated-based models (CBM)
      - i.i.d model, Kronecker model and Weichsellberger model
    - Propagation-motivated models (PMM)
      - Finite scatterer model, maximum entropy model and virtual channel representation.

Additionally, the MIMO channel models can be divided in two categories given in terms of accuracy and methodology: accurate versus simple or physical versus analytical [12]. This classification is illustrated in Figure 6.35 that describes the trade-off between the required complexity of implementation and the type of performed analysis. It is noted that DMs provide the largest accuracy at the expense of requiring setting up the environment at physical level [12].

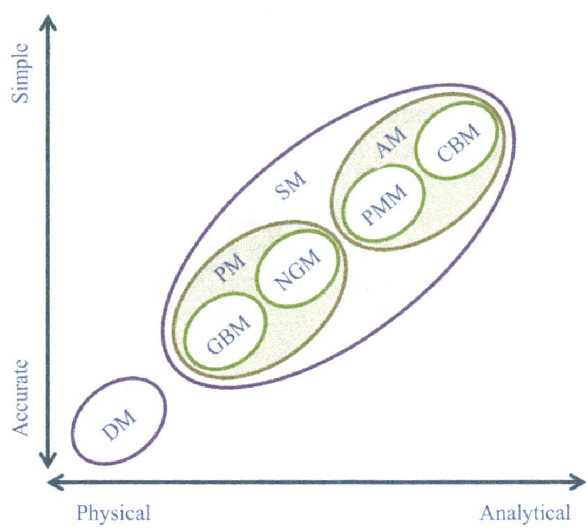

Figure 6.35 Categories of MIMO channel models [12]

Massive MIMO consists of an enhanced MIMO technique using a large number of transmitter and receiver antennas aimed to improve communication features such as reliability, spectral efficiency and energy [12]. As reported in [222–224], massive MIMO channels have some specific characteristics ignored in conventional MIMO channel models provided with relatively small number of antennas [72–74,194,225–227]. Two important ones are described as follows: the spherical wavefront and the cluster appearance and disappearance [224].

For a large number of antennas, the distance D between the cluster of transmitter and receiver elements cannot exceed the Rayleigh distance given by $2L^2/\lambda$ [228,230], with L the dimension of the antenna array and λ the wavelength of the operating frequency [224]. If this criterion is not met, the assumption of conventional plane wavefront [7,73,194,225] is not justified and the effect of spherical wavefront must be considered given its significance [224,230]. As an example, the COST 2100 channel model [74] and the METIS GBSM [229] can incorporate spherical wavefront by means of minor additions [224].

On the array axis can occur cluster appearance and disappearance; therefore, a set of observable clusters must be considered for each antenna [224]. Most GBSMs do not consider this fact including the following [224,230]: extension of the SCM [228], COST 2100 [74], METIS GBSM [229], 3GPP NR [230] and IMT-2020 channel model [231,234].

The two main reviews on massive MIMO channel measurements and models are reported in [224,232]. Most important channel models are described in [231] and [233]. The massive MIMO channel in [231] proposed a 2D confocal ellipse model and a 3D twin-cluster model is described in [233]. Both models applied the spherical wavefront assumption and modified the cluster birth–death process described in [234,235] to model cluster appearance and disappearance for both the time and array axes in [224]. However, those models cannot accommodate propagation in mmWave frequency bands, V2V channels and arbitrary antenna array layouts [224]. Those models did not incorporate directional antennas and polarized antennas [224]. Then, further investigation is required in this direction for massive MIMO channel applied to V2X.

6.4.5 Spatial channel models

According to [236,237], the SCM is a detailed system-level model developed for simulating urban microcell, urban macrocell and suburban macrocell fading environments [238]. The SCM was standardized by 3GPP–3GPP2 for evaluating MIMO system performance under outdoor environments conditions [236,238,239]. It was initially defined for code division multiple access (CDMA) systems operating at 2 GHz with 5 MHz bandwidth [240].

The SCM is based on stochastic modelling, for outdoor moving antennas [241]. It is implemented by ray tracing to characterize the scattering environment by specifying parameters and a procedure for channel modelling [242,243]. The SCM [20,238] approaches practical V2V environments by assuming a random distribution for both of the near and far scatterers along the roadside [241].

Output metrics of SCM for V2V are CIR, PSD, angular power density, ACF and Doppler spread. [241] For including the V2V elevation plane, various 3D channel models have been proposed in [115,244–246], improving the spatial and temporal modelling of the V2V communication links with respect to the 2D channel model approaches.

The 3GPP–3GPP2 ad-hoc group (AHG) provided a report [238] to develop and specify parameters and methods associated with the SCM needs of the 3GPP and 3GPP2 organizations. The SCM important parameters were incorporated, among other: phase, delay, Doppler frequency, AoD, AoA and angle spread to provide a description of MIMO channels. SCM also considers spacing in transmitter and receiver cluster of antennas or arrays to make possible the analysis of the mutual coupling effect.

The scope includes development of specifications for system-level evaluation focused on four areas [238]: physical parameters (e.g. PDPs, angle spreads and dependencies between parameters); system evaluation methodology; antenna arrangements; reference cases and definition of minimum requirements; and some framework (air interface)-dependent parameters. The emphasis of the SCM AHG work is on the first two items [238].

In Table 6.8, it is summarized the physical parameters to be used for link-level modelling. Link-level simulations only provide one snapshot of the channel performance, so they cannot be considered alone in order to compare algorithms. Moreover, link-level simulations either account for system attributes such as scheduling and HARQ. For these reasons, link-level simulations do not allow any conclusions about the typical system performance that can be only achieved by performing SLS.

Therefore, the SCM AHG is focused on SLS for allowing accurate final algorithm comparison [238]. Link-level simulations can be used only for calibration that consists of comparing the performance results obtained from different implementations of a particular algorithm [238].

For the 3GPP-SCM AHG model, the description of the link-level simulation considers the context of a downlink system where a BS transmits to an MS, as shown in Figure 6.36 [238].

In [238] was described a procedure for obtaining the channel coefficients matrix. The link-level 3GPP-SCM model considers N cluster of scatterers, with each cluster corresponding to a resolvable path [236,237]. Each resolvable path (the delay or time difference between successive arriving paths is below a chip period) contains M unresolvable sub-paths (cluster). This implies N received paths with M sub-paths (cluster).

Each resolvable path is described by means of its specific spatial channel parameters: AS, AoD, AoA and power azimuth spectrum (PAzS) [240]. For both sides of the communication link, the properties of resolvable paths (path power, path delay, angular properties) are modelled as random variables [238]. The amplitude of resolvable paths is modelled with a Rayleigh distribution [240], indicating a correlation between consecutive elements of the antenna array so that the amplitude correlation meets the Jakes model [240,247].

Table 6.8 Summary SCM link-level parameters for calibration purposes [238]

Model	Case I		Case II		Case III		Case IV	
Corresponding 3GPP designator*	Case B		Case C		Case D		Case A	
Corresponding3GPP2 designator*	Model A, D, E		Model C		Model B		Model F	
PDP	Modified Pedestrian A		Vehicular A		Pedestrian B		Single path	
# of paths	(1) 4+1 (LOS on, K = 6 dB) (2) 4 (LOS off)		6		6		1	
Relative Path Power (dB) Delay (ns)	(1) 0.0 (2) −Inf	0	0,0	0	0.0	0	0	0
	(1) −6.51 (2) 0.0	0	−1.0	310	−0.9	200		
	(1) −16.21 (2) −9.7	110	−9.0	710	−4.9	800		
	(1) −25.71 (2) −19.2	190	−10.0	1,090	−8.0	1,200		
	(1) −29.31 (2) −22.8	410	−15.0	1,730	−7.8	2,300		
			−20.0	2,510	−23.9	3,700		
Speed (km/h)	(1) 3 (2) 30, 120		3, 30, 120		3, 30, 120		3	

(Continues)

Table 6.8 (Continued)

Model		Case I	Case II	Case III	Case IV
UE/MS	Topology	Reference 0.5λ	Reference 0.5λ	Reference 0.5λ	N/A
	PAS	(1) LOS on: Fixed AoA for LOS component, remaining power has 360-degree uniform PAS (2) LOS off: PAS with a Laplacian distribution, RMS angle spread of 35 degrees per path	RMS angle spread of 35 degrees per path with a Laplacian distribution Or 360-degree uniform PAS.	RMS angle spread of 35 degrees per path with a Laplacian distribution	N/A
	DoT (degrees)	0	22.5	−22.5	N/A
	AoA (degrees)	22.5 (LOS component)67.5 (all other paths)	67.5 (all paths)	22.5 (odd numbered paths),−67.5 (even numbered paths)	N/A
Node B/ BS	Topology	Reference: ULA with 0.5λ-spacing or 4λ-spacing or 10λ-spacing			N/A
	PAS	Laplacian distribution with RMS angle spread of 2 degrees or 5 degrees, per path depending on AoA/AoD			N/A
	AoD/AoA (degrees)	50° for 2° RMS angle spread per path 20° for 5° RMS angle spread per path			N/A

*Designators correspond to channel models previously proposed in 3GPP and 3GPP2 AHGs.

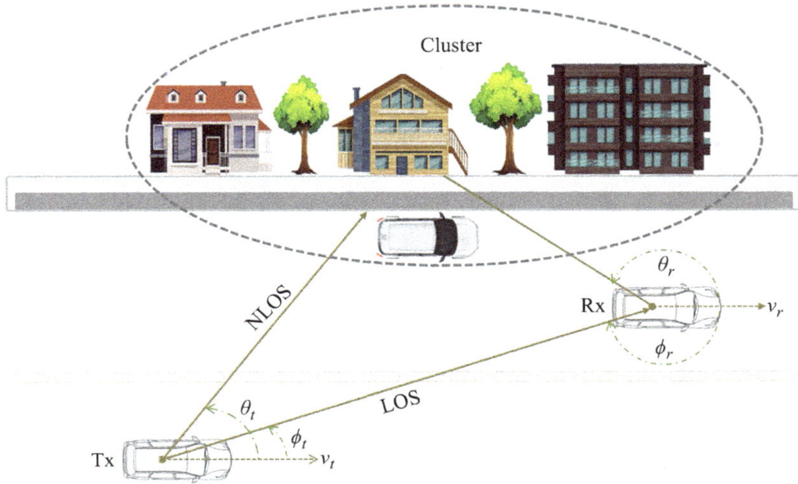

Figure 6.36 Parameters of the 3GPP-SCM for a cluster of scatterers [238]

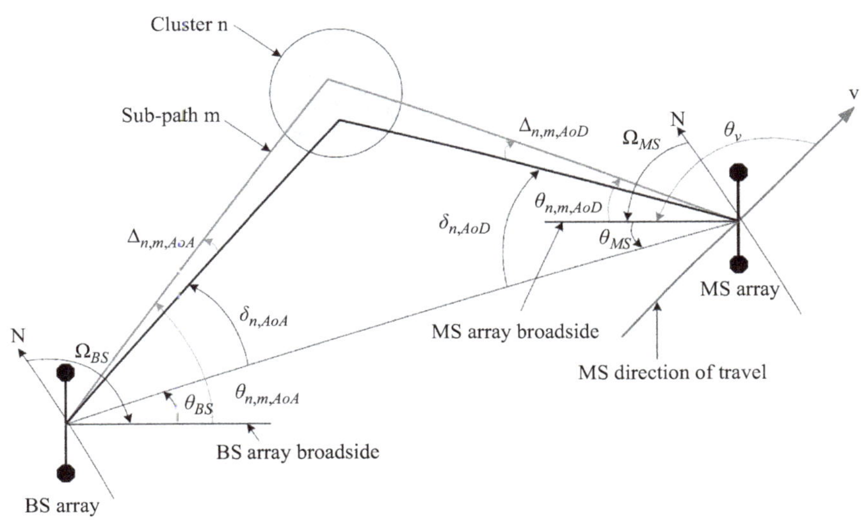

Figure 6.37 Simplified overview of SCM V2V scattering channel model [241]

Figure 6.37 depicts a simplification of SCM V2V scattering channel model with only two paths shown, LOS and NLOS. For the LOS path, ϕ_t and ϕ_r represent, respectively, the AoD and AoA without any obstruction between transmitter and receiver. For the NLOS path, the model provides the estimation of the AoD ϕ_t as the angle between transmitter and scatterers, and the AoA ϕ_r as the angle between scatterers and receiver [241].

In the SCM AHG, for the link-level simulation, a fixed number of $N = 6$ paths are considered in any scenario, with $M = 20$ spatially separated sub-paths composing each resolvable path [238]. The reason behind this choice is that experimental measurements performed in urban environments [240] demonstrated a narrow AS that produces a larger correlation between the antenna elements. To incorporate this effect, the 3GPP model associates $M = 20$ sub-paths to each path (Figure 6.37) to characterize the AS at the MS by its AoA and through AoD for the BS [240]. Other properties of the models are the following:

- The M sub-paths are equally placed around the direction axis of the nominal path [238,240].
- The angular value of each $M = 20$ sub-paths offset is fixed [238,240].
- All the sub-paths have identical powers equal to $P/20$, where P is the power associated to the main path [238,240].
- The resulting PAzS meets a Laplacian distribution with a variance according to the chosen environment: urban macrocell, urban microcell, suburban macrocell [238,240].

Fixing sub-path directions and powers reduces the variability that the model can provide with respect to that of an actual radio channel [240]. Equal power sub-paths require large AS in order to obtaining eligible AS [240]. Then the correlation between MIMO antennas is reduced and the MIMO capacity is overestimated [240].

The SCM AHG to be used in the SLS is also described in [238] for a downlink system with a BS transmitting to an MS; however, the methodology can be also used for the uplink. Unlike the link simulations which only contemplate one downlink transmission between a single BS and a single MS, the SLS consider multiple cells/sectors, BSs and MSs [238].

At system level, the report in [238] defines the methodology and parameters for generating the spatial and temporal channel coefficients for the downlink case. For a linear BS antenna array with N_T element and a linear MS antenna array with N_R element, the channel coefficients for one of the N MPCs are given by an $N_R \times N_T$ matrix of complex amplitudes [236–238].

It is assumed that omnidirectional antenna elements are used at the BS and MS. PL and shadowing are neglected. Finally, the CIR for the lth path between the sth transmitter and uth receiver antenna can be written as [236–238]:

$$h_{u,s,n}(t) = \sqrt{\frac{P_n}{M}} \sum_{m=1}^{M} \left\{ \exp\left[j\left(kd_s \sin\left(\theta_{n,m,AOD}\right) + \phi_{n,m}\right)\right] \right.$$
$$\left. \times \exp\left[jkd_u \sin\left(\theta_{n,m,AOA}\right)\right] \times \exp\left[jkv\cos\left(\theta_{n,m,AOA} - \theta_v\right)t\right] \right\} \tag{6.15}$$

with k the wave number, λ the wavelength in metres, P_n the nth path power, M the number of sub-paths per path, d_s the distance in metres from BS antenna element s to the reference antenna ($s = 1$), d_u the distance in metres from MS antenna element u to the reference antenna ($u=1$), v the MS velocity vector magnitude, $AoA_{n,m}$ the

AoA for the *m*th sub-path of the *n*th path with respect to the MS broadside and $AoD_{n,m}$ the AOD for the *m*th sub-path of the *n*th path with respect to the BS broadside [236,237]. The details of the generation of relevant parameters are given in [238].

The N MPCs considered may be not necessarily time resolvable. The channel matrix for the *n*th MPC ($n = 1, \ldots, N$) is denoted as $H_n(t)$ and becomes a function of time t because the complex amplitudes experience fast fading due to the MS mobility [238].

The details for the general procedure are indicated in [238]. Figure 6.38 provides a roadmap for generating the channel coefficients. The overall procedure comprises three basic steps [238]:

- Specify an environment: suburban macro, urban macro or urban micro.
- Obtain the parameters to be used in simulations, related to the chosen environment.
- Generate the channel coefficients based on the above parameters.

6.4.6 Millimetre wave channel models

In 2014, the 3GPP working groups considered incorporating V2X as an additional feature for LTE-Advanced (LTE-A) [248–251]. 3GPP's Technical Specification Group on Service and System Aspects (TSG-SA), within LTE Release 15, explored the 5G-grade enhanced V2X (eV2X) use cases [252,253]. The envisioned scenarios (platooning, remote driving, smart car sensory, etc.) will require a combination of

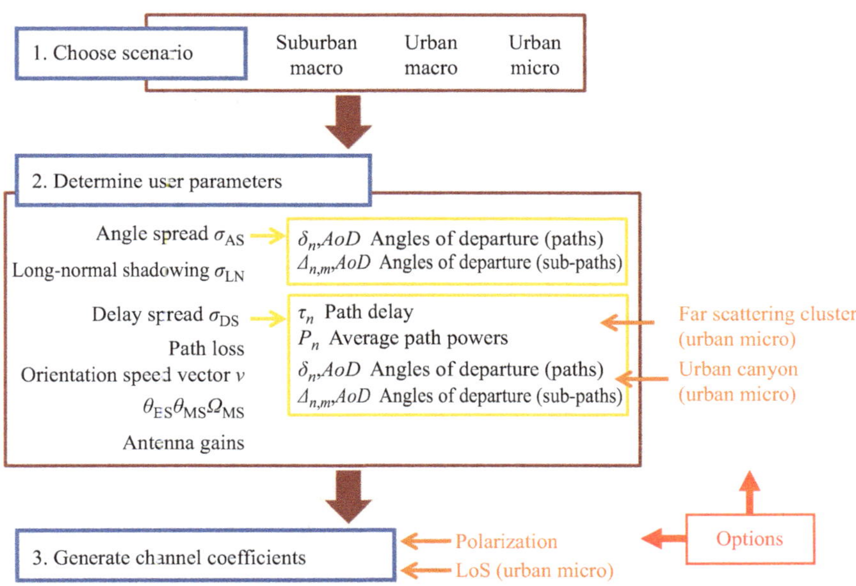

Figure 6.38 Channel model overview for simulations [239]

very high data rates, large reliability levels and low latencies [252–254]. 3GPP introduced LTE-based V2I communication and device-to-device (D2D) communication for the V2V communication [248]. D2D communication is considered as essential for the case where the cellular network coverage is missing or not guaranteed [248].

However, none of those technologies is sufficient to accomplish the stringent requirements demanded by the eV2X or LTE V2X applications. For this reason, 3GPP started developing in Release 19 [250], a standard that will lead to the third wave of V2X communications that will complement LTE by including the 5G New Radio (NR) technologies (NR-V2X) with special attention to the deployment in mmWave bands for vehicular communications [248,252,255]. 5G is intended to support higher mobility, and to improve network reliability, with a packet loss rate of around 10^{-5} valid to offer safety-critical services [248].

5G will operate, among other options, in mmWave spectrum ranging from 30 GHz to 300 GHz. For mmWave-based V2X communications, some interesting frequency portions have been postulated: 60 GHz, 79 GHz, low terahertz (THz) frequencies at 300 GHz and beyond [252,256]. Standardization of NR-V2X will expand LTE V2X with massive deployments of connected vehicles [252].

Among the technologies and standard considered for providing mmWave-based V2X communications, it is found the IEEE 802.11ad standard [257]. It is also known as the Wireless Gigabit Alliance (WiGig), and its specification covers the physical and link layers for communications in the free-licence mmWave band of 60 GHz to provide a gigabit connection between proximate devices [257]. The WiGig employs the 60 GHz spectrum portion achieving a data rate over 4Gb/s with 2 GHz of channel bandwidth [257]. This standard also contemplates indirect communication by using a relay node when a link blockage occurs: a sending station selects a relay node that offers an LOS link to the targeted receiver. The applicability of the IEEE 802.11ad standard to a gigabit-based V2V communication (GiV2V) system in a practical vehicular test bed was achieved in [257,258].

At present, the mmWave band is considered for V2V or V2I communications, required for autonomous or proxy driving in future smart cars [250,257]. The V2X roadmap defined by the 5G Public Private Partnership (5G-PPP) envisioned scenarios that include awareness driving, cooperative driving and synchronized driving with vehicles being autonomously driven [252,253,259,260].

Table 6.9 shows the survey found in [257], summarizing existing research reports regarding the mmWave V2X communications.

6.4.7 V2I channel models

As described in [296], the U.S. Department of Transportation announced the Federal Highway Administration (FHWA) V2I guidance in an attempt to accelerate the development of and deployment of V2I communication systems [297]. Since then, several research projects in the area of ITS [298,299] have offered solutions to three major approaches of the connected vehicles communications: V2V, V2I and V2P [296]. The V2I communication is defined as the bidirectional wireless

Table 6.9 Survey on mmWave-based V2X communications [257]

Reference	Properties
[261–267]	V2X channel models including link blockage, scattering, shadowing and multipath fading. Reflection and diffraction at 60–77 GHz for realistic road surfaces and geometries [262] presents channel models for mmWave V2X communications using ray-tracing simulations [264,267] presents a geometric MIMO channel model for mmWave mobile-to-mobile applications based on the two-ring model
[268–271]	SM for V2X communications including blockage probability from RSU and throughput
[272–274]	Blockage detection and relay/multi-hop routing in V2X communications to avoid obstacles or extend network coverage
[275–281]	Location or situation-based channel estimation, beam direction steering and training. Prior channel information or past measurements are required for each location or situation. Machine learning techniques can be also applied
[282,283]	Configuration of a mmWave link using long-range radar (LRR) coupled to road infrastructures and mounted on vehicles for V2I and V2V communications. Reduction of beam alignment and training overhead
[256,284]	Configuration of a mmWave link assisted by motion and position information of vehicles estimated from vehicular sensors. DSRC beacons carry the sensor information periodically. Reduction of beam alignment and training overhead
[258,285]	Effect of inter-beam interference in V2V networks. Beam alignment and multichannel assignment proposed as solutions
[286–288]	Distributed beam alignment and width selection decision using swarm intelligence based on the channel and queue state information. V2V association and scheduling problem solved in a decentralized manner. Reduction of interference among vehicles solved by a distributed algorithm. Doppler power spectrum in the presence of beamforming for V2I
[289,290]	Multi-connectivity by means of microwave frequencies (DSRC, LTE) considered for both increasing the robustness connectivity (during handover or relay) and reducing the beam tracking overhead
[291–293]	Practical test beds to evaluate the performance of V2I or V2V communications at mmWave band
[294]	Application of IEEE 802.11ad standard (WiGig) for mmWave V2X communications using simulation
[295]	Security techniques of physical layer in mmWave and MIMO systems

exchange of data for control and information between vehicles and RSUs [296]. V2I has been considered for supporting all range of applications, such as safety critical applications [300], stop sign violation warning, railroad crossing violation warning and oversize vehicle warning, but also on-car multimedia streaming [301] and gaming. Other applications taking advantage of V2I communications are traffic information systems, intersection collision warning systems [302], intelligent traffic management systems, vehicle collision prediction [303], parking management systems, [304], cooperative advanced cruise control [305], short-term traffic prediction, platooning, lane detection, obstacle warning, target tracking, etc. The deployment of the V2I applications aimed at the convergence of sensing, infrastructure and information technologies into the ITS depend on achieving a detailed and accurate radio channel characterization as well as the understanding of the involved propagation mechanisms [296].

As aforementioned, in V2I mode the communication takes place between cars and BSs or access points distributed along roadways known as RSUs. The RSUs dedicated to V2I operation mode are similar to the BSs used in cellular radio systems with a small cell size, as picocell. The V2I radio channel properties are similar to the features of the cellular channel, especially with the picocell case: deployment of the RSUs/ BSs along the roadside to define the network layout, and the effect due to vehicle speed. The term V2R is also used to indicate V2I communications [3].

The type and distribution of objects present in the V2I urban environment, such as buildings, vehicles (both static and mobile) and vegetation, as well as their number, size and density, profoundly impacts the characteristics of the radio channel propagation [296]. Some propagation impairments are analogous at the presented in short-range outdoor radio communication systems [306]: reflection from, diffraction around and transmission loss through objects (influence of vegetation, building entry loss, cars, trees, pedestrians, etc.), external environment, which gives rise to issues such as temporal and spatial variation of PL and multipath effects from reflected and diffracted components of the wave [297]. A measurement campaign can be considered as the optimal way to determine and quantify the influence of this phenomenon. Nevertheless, the technological, economical and human resources needed to complete an extensive campaign are not always available. Alternatives such as analytical models, SMs and ray-tracing approaches are considered instead to observe any propagation impairments [296].

A review of V2I channel characterization extracted from [296] is provided as follows:

- In [307] is proposed a statistical channel model for VANET simulation based on the observation of data obtained from real experiments, performed using an IEEE 802.11p/WAVE compliant test bed.
- In [308] is presented a novel approach for modelling the wireless channel of VANETs communications based on discrete-time Markov chain (DTMC) modelling.
- In [309] is provided a realistic analysis of the VANET topology characteristics over time and space for a highway scenario, using various critical metrics of interest, including node degree, neighbour distribution, number of clusters and link duration.
- In [310] is performed a measurement-based analysis of multilink shadowing effects in a V2I communication system with cars as blocking objects [296].
- In [302] is stated that assuming exponential PL in some analytical models is not suitable for scenarios with buildings (large blocking) [296].

Channel characterization by means of SMs offer a macroscopic viewpoint description that might introduce severe deviations from real signal behaviour making security-oriented applications cannot be modelled correctly [296]. V2I channel propagation has been analysed for urban areas using deterministic tools, such as ray launching, recognized as an accurate tool to model propagation phenomena [296,311,312].

Ray-tracing-based models trade off an accurate approximation of realistic propagation and a reasonable computational complexity. Even with a reduced set of details and a limited dielectric characterization of materials, ray-tracing approach can offer an accurate prediction of both geometry and shadow levels [313], offering a tool to simulate radio shadowing environments [296]. An empirical propagation analysis in urban areas for V2I is described by [314]. In [315], an ad-hoc 3D ray-tracing simulation software is used to present a statistical peer-to-peer channel model valid for outdoor urban environments [296]. In [313,316] is reported a deterministic ad-hoc or commercial 3D ray-tracing software used to predict power levels, along with time, frequency and spatial dispersion in an urban V2X scenario.

In [296] is described a deterministic 3D Ray Launching (3D-RL) algorithm, based on geometrical optics and geometrical theory of diffraction [317]. This tool has been previously validated in transportation systems [318] and used to analyse some V2I propagation impairments in an urban environment at 5.9 GHz [296].

The deterministic hybrid simulation technique described in [296] was applied for system-level analysis by simulation of a typical V2I urban environment composed of park, avenues, streets, buildings, roundabout, etc. [296]. The 3D-RL simulation scenario was characterized with high-level detail (objects, materials, geometry of obstacles, electromagnetic properties, etc.) and considers the transmitter–receiver distance, as well as the LOS and NLOS conditions [296]. As outputs, the 3D-RL simulation achieved accurate results offering large-scale PL and multipath metrics [296].

The 3D-RL tool in [296] allowed the combination of physical layer with application-based results for V2I. It was implemented a functional vehicular test bed to analyse a real context-aware environment [296]. The comparison of measurements versus the 3D-RL simulation and analytical models [296] demonstrated an accuracy improvement due to the 3D-RL technique compared to that of the analytical PL models for V2I [296]. The effect of the obstacles in the V2I propagation was studied and it determined larger fitting error for the PL models than for 3D-RL under NLOS conditions [296].

6.4.8 Future trends

To this point, the studies regarding channel models highlights the importance of radio characterization for developing highly reliable applications that use vehicular communications as nominal technology. Far from considering that the demand of radio channel characterization is complete, the results obtained also indicate that there are still many open research questions that are pending response.

One of those questions is the demand of developing more complex channel models that try to reflect as many possible propagation mechanisms and impairments taking place in a vehicular communication. Some promising of those novel channel model approaches are listed as follows:

- MAP-based channel models for link level allow incorporating realistic environments or analysing propagation problems and mechanisms at specific points of a scenario, as that developed by METIS [229].
- System-level channel models: Appropriate for comparing application performance, as the model by 3GPP [228].
- 3D channel models, especially for massive MIMO: Generally reported directional channel models characterize the directional features only in the azimuthal domain, including 3GPP-SCM [239,240], ITU [73] and WINNERII [193] models [217,319,320]. Accordingly, the standardization of MIMO techniques in 3GPP has been also based on 2D channel models [217,323]. Among the reasons for this predominant consideration of 2D models, it is found [217,319,320]:
  - In multiple antenna systems, capacity is directly related to the AS of the MPCs, as seen from the BS. Large AS is achieved in the azimuth domain, whilst in the elevation plane the AS is narrow due to the use of BS antennas with a narrow and non-adaptive beam width [319,321].
  - Massive MIMO systems require considering the elevation domain to obtain an optimal antenna factor by spanning two dimensions [19,53,319,322]. In [323], an extension of the COST 2100 model for massive MIMO system is proposed. However, the best configuration in terms of antennas in the horizontal and vertical domain is still an open issue. Thus, the elevation components become highly important [319].
  - Most of 2D channel models consider that scenario elements such as scatterers, reflectors and transmitter and receiver antennas are located on the same elevation plane [217,320], even 2D moving scattering elements. However, it has been demonstrated that in urban measurements typically 90% of the received power is concentrated in paths with elevations up to 40° [217,320,324], and that the mean elevation of the received power is 1.8° for outdoor–indoor environments, and up to 17.1° for urban macrocell environments [217,320,325]. Assuming a 2D scattering scenario for V2V communications neglects the effect because the transmitter and receiver antennas are located at elevation planes with respect to the height of the surrounding scatterers [217,320].
  - MIMO transmission techniques are capacity limited if only the azimuth dimension of 2D models is provided. This is the case for beamforming, precoding, spatial multiplexing, multi-user MIMO (MU-MIMO) and full-dimension MIMO (FD-MIMO) [217,319,320,326,327]. For the latter case, the transmission is optimally adapted in both elevation and azimuth planes to offer a specific elevation beamforming to particular user equipment [217,319,320,326,327]. 3D channel model including the elevation plane

must be also considered in the vertical cell sectorization that assigns a narrow elevation beam to each vertical sector [217,319,320,326,327].

Developing 3D propagation channel models requires measurements of the elevation features by obtaining the elevation spectrum at the BS, and also at the MS (DoA for the uplink or DoD for the downlink) [319].

- Cooperative models:
  Cooperative MIMO consists of multiple radio devices grouped to form virtual antenna arrays in order to cooperate with each other by exploiting the macro-spatial diversity of mobile fading channels [12,124]. The result is macro-spatial gain for capacity enhancement and interference mitigation [12,124].

Some devices will take on the role of RSs, to enable communication between the BS and the MS in case they are not within reach of each other [12,124].

In cooperative MIMO systems, different types of radio links can be established: BS–BS, BS–RS, RS–RS, RS–MS, BS–MS and MS–MS links [12,124]. The radio link diversity introduces enormous propagation variation due to the different local scattering conditions present around the locations of BS, RS and MS devices [12,124].

Cooperative wireless communication technologies are proposed in the LTE-A systems [328,329], referring cooperative relay [330], coordinated beamforming and joint processing algorithms [331].

Cooperative techniques are of wideband nature that requires further investigation of channel cross-correlation at the small scale for proper modelling of realistic scenarios [12,124].

So far cooperative MIMO measurement campaigns concentrated on the channel characteristics – PL, shadow fading and small-scale fading – of individual links for different outdoor scenarios [12,124]: for static nodes [332,333] and for mobile nodes [334,335]

Another question regards the need to seek new frequency bands and technologies to fully develop applications that use vehicular communications. Following some of them are briefly commented:

- Visible light communications (VLC):
  VLC was recently standardized by the IEEE 802.15.7 group. Among the applications of this LED-based technology, it has been explicitly considered for vehicles and illuminated roadside devices (traffic lights or streetlights) [248,336]. However, the IEEE 802.15.7 proposed in outdoor applications a maximum data rate of 266.6 kb/s that seems insufficient to the requirements of vehicular communications [248].
- 3GPP cellular-V2X (C-V2X):
  3GPP specified specific features for supporting vehicular communication modes generally known as C-V2X technologies based on the use of mobile cellular infrastructure: V2V, V2I and V2R [248]. The basic requirements to support C-V2X PC5 applications were specified by the TSG-SA group in [248,337]. More specifically, LTE release 14 defined V2V use cases in the ITS

5.9 GHz band [248]. The objective is to enable direct and periodic exchange of status information (position, speed, etc.) between vehicles and with everything around. The data exchange could also allow the heading with neighbouring vehicles, pedestrians and road infrastructure nodes. Dissemination of event-driven warning messages is also defined by the C-V2X-based applications [248,338].

In [339,340], it was communicated that the 5G Automotive Association (5GAA) carried out different tests in order to determine the performance C-V2X PC5 radio technologies for delivering V2V safety messages. The global outcomes demonstrated that C-V2X performance as direct communication technology exceeds the features offered by 802.11p/DSRC (ITS-G5 in Europe) [339].

C-V2X demonstrated an improved performance in three main aspects: reliability on large ranges; NLOS performance; and robustness against interference [339]. Those improvements become critical under extreme conditions, as the case of NLOS scenarios (around a corner, highway queue forming), where other technologies (onboard sensors and radars) show performance limitations [339].

C-V2X takes benefits of the low latency and high bandwidth benefits of 5G NR [339] and reduces cost of infrastructure deployment due to the combination of RSUs and 4G/5G small cells, so benefiting from cellular network densification [338].

A comparison of the main characteristics of the C-V2X technologies (releases 14/15/16) with respect to the 802.11p/DSRC provided in [338] are summarized in terms of technology operation in Table 6.10, radio design in Table 6.11 and use cases in Table 6.12.

- THz frequency band:

  Data traffic of up to 1 Tb/h per connected vehicle is predicted to be transmitted over vehicular networks [252,257]. The use of frequencies available in the frequency spectrum bands of high mmWave (beyond 100 GHz) and low THz (beyond 300 GHz) are under study for future allocation of V2X systems [252].

These bands offer data rates of up to 10 Gb/s as demonstrated at 120 GHz for a radio link range up to 850 m [252,341]. The IEEE 802.15.3d Task Group identified the 275–325 GHz band as the one feasible for extremely high access rates of up to 100 Gb/s [252].

In the same line, automotive radars have also been moving at higher frequency bands aiming to improve latencies and image resolution [252], as is the case of the automotive imaging radar operating at 150 GHz described in [252,342].

Some issues strictly related to the band beyond 300 GHz and THz band show up [252], as these ones:

- development of novel low-noise transceivers;
- difficult generation of high-power signals at low THz frequencies, in the so-called THz gap;
- increased radio electric attenuation due to molecular absorption;
- novel antenna systems required of ultra-narrow beamwidth, beam steering and beam tracking implementations.

Table 6.10 C-V2X: comparison of technology operation [338]

Technology operation	802.11p	C-V2X Rel-14/15	C-V2X Rel-16 (expected design)
Specification completed	Completed	Rel-14 in 2016 Rel-15 in 2018	2019
Support for low-latency direct communications	✓	✓ Rel-14–4 ms	✓ ≤ 1 ms
Support for network communications	Limited (via APs only)	✓	✓
Can operate without network assistance		✓	✓
Can operate in ITS 5.9 GHz spectrum		✓	✓
SIM-less operation	✓	✓	✓
Security and privacy on V2V/V2I/V2P	✓ As per IEEE WAVE and ETSI-ITS security services	✓ As per IEEE WAVE and ETSI-ITS security services	✓ As per IEEE WAVE and ETSI-ITS security services
Coexistence in 5.9 GHz	✓ Adjacent channel with 3GPP tech	✓ Adjacent channel with 11p; co-channel coexistence from R14 onwards	✓ Adjacent channel with 11p; co-channel coexistence from R14 onwards &WiFi
Evolution path	✗	✓	✓ Compatible with Rel-14/15

Table 6.13 summarizes the main characteristics of the wireless access technologies for vehicular networks reported in [248].

Finally, it should be mentioned that some propagation mechanisms involved in vehicular radio propagation channel require further investigation. Topics such as modelling of diffraction on scatterers at mmWave and THz, dielectric characterization of road surfaces [343] to achieve total and accurate modelling of the ground reflection are only some examples of the many open research questions

6.5 Channel prediction and estimation

Once channel data sets are obtained from measurement campaigns or software simulation, various data processing methods are applied to obtain the channel model estimation. The common data obtained consist of CIRs or transfer functions. For channel modelling, it is required to determine the directions, delays and amplitudes of the MPCs. The simplest signal processing for parameter extraction is the Fourier analysis, consisting of applying a Fourier transform to the transfer function in order to derive metrics in the delay domain, and a spatial Fourier transform to the signals at the antenna elements to determine their directions. The resolution of those parameter extraction methods is insufficient and might be

Table 6.11 C-V2X: comparison of radio design [338]

Radio design	802.11p	C-V2X Rel-14/15	C-V2X Rel-16 (expected design)
Synchronization	Asynchronous	Synchronous	Synchronous
Channel size	10/20 MHz	Rel-14: 10/20 MHz Rel-15: 10/20/ Nx20 MHz	$\leq$1 ms
Resource multiplexing across vehicles	TDM only	TDM and FDM	TDM and FDM possible
Data channel coding	Convolutional	Turbo	LDPC
HARQ retransmission	No	Rel-14/15: yes Rel-15: ultra-reliable communication possible	Yes, along with ultra-reliable communication
Waveform	OFDM	SC-FDM	Likely OFDMA but many options available
Resource selection	CSMA-CA	Semi-persistent transmission with frequency domain listen-before-talk	Many options available
MIMO support	No support standardized	Rx diversity for two antennas mandatoryTx diversity for two antennas supported	Support up to 8 tx/rx antennasMandatory support for 2tx/rx antennasBoth diversity and spatial multiplexing supported
Modulation support	Up to 64QAM	Up to 64QAM	Up to 256QAM

limited by the measurement bandwidth for delay-based estimations, and by the antenna array size for direction estimations. High-resolution methods are required to provide enough accuracy in the parameter extraction [92].

An example is given by the high-resolution parameter extraction (HRPE) algorithms described in [92] that achieve an order of magnitude better resolution. HRPE is based on a parametric model of the wireless propagation channel with bidirectional CIRs composed of a sum of MPCs. Each MPC is characterized by its corresponding amplitude, delay and direction at transmitter and receiver sides [92]. The estimation algorithm provides the parameters of the MPCs from the measurement data sets obtained from a channel sounder: ToA, DoD, DoA, Doppler shift and signal power for strong discrete MPCs, statistics for diffuse MPCs and the continuous time-varying receiver angular power spectrum [92].

The first class of HRPE algorithm is the iterative maximum-likelihood (ML) estimator, called the RiMax algorithm [342,345], that consists of an improved version of the SAGE algorithm [92]. As result of the implementation of the joint parameter optimization, the RiMAX algorithm convergence is considerably faster than that of the SAGE for a case of multiple signals with similar parameters [92].

Table 6.12 C-V2X: comparison of use cases and performance [338]

Use cases	**802.11p**	**C-V2XRel-14/15**	**C-V2X Rel-16 (expected design)**
Target use cases	Target use cases	Day 1 safety and enhanced safety use cases	Advanced use cases to assist in autonomous driving including, ranging assisted positioning, high throughput sensor sharing and local 3D HD map updates
Performance			
High density support	Packet loss at high densities ✓	Can guarantee no packet loss at high densities ✓	Can guarantee no packet loss at high densities ✓
High mobility support	Up to relative speeds of 500 km/h with advanced receiver implementation ✓	Up to relative speeds of 500 km/h as minimum requirement ✓	Up to relative speeds of 500 km/h as minimum requirement ✓
Transmission range @ 90% error, 280 km/h relative speed	Up to ~225 m	Over 450 m using direct modeVery large via cellular infrastructure	Over 450 m using direct modeVery large via cellular infrastructure
Typical transmission frequency for periodic traffic	Once every 100 ms (50 ms is also possible)	Once every 100 ms (20 ms is also possible)	Supports packet periodicities of few milliseconds

A HRPE algorithm based on the EKF can be used for estimating the channel time evolution by means of the correlation between parameters of MPCs [92]. The estimation accuracy is further improved with this algorithm. The outcome provides the MPCs time tracking [92].

High-resolution methods are more commonly used for channel measurement data processing rather than conventional methods [12]. Some of these high-resolution estimation methods are based on the single specular path model and include two classes [12]:

- *Subspace-based approaches*:
  - MUSIC algorithm [346]
  - ESPRIT [347]
- *Approximations of the ML methods*:
  - EM algorithm [348]
  - SAGE algorithm [349,350]
  - RiMAX algorithm [344,345]

Table 6.13 Main characteristics of wireless access for vehicular networks [248]

	Wave IEEE802.11p	**3GPP LTE-A**	**IEEE802.15.7 (VLC)**	**5G**
Frequency	5.9 GHz	400 MHz–3.5 GHz	380–800 THz	New bands 700 MHz–100 GHz
Data rate	27 Mb/s	3 Gb/s (downlink) 1.5 Gb/s (up-link)	11.67 kb/s–96 Mb/s	Up to 1,000 times greater
Communication range	<1,000 m	Ubiquitous	<100 m	Ubiquitous
Latency	<100 ms	<100 ms	<50 ms	<10 ms
Mobility speed	<300 km/h	<300 km/h	N/A	<500 km/h
V2I communication	Supported after deployment of RSUs	Supported	May use the available roads lights	Networks cooperation
V2V communication	Supported	Supported from Release 14 in 5.9 GHz	Supported	Networks cooperation
Deployment	Requiring RSUs set-up	May use the available eNodes B	May use the available LEDs	Upper layers cooperation and management

These methods are applied to extract the parameters of the specular path components in multiple dimensions [67,351–354].

The above-listed estimation methods can be categorized into two groups according to [12,68]:

- *Spectral-based approaches*: Methods in this group estimate the channel parameters by determination of maxima, or minima, of spectrum-like functions of the dispersion parameters [12]. These techniques exhibit a reasonable computational complexity as the maxima searching can be performed in one dimension for all paths [12]. In cases with coherent multipath signals, spectral-based approaches usually fail or perform badly [12].
- *Parametric approaches*: Techniques in this category estimate the parameters of an underlying parametric model that characterizes the effect of the propagation channel on the transmitted signal from the signals at the array output [12]. These techniques exhibit better estimation accuracy and higher resolution than spectral-based techniques at expenses of a higher computational complexity due to the multidimensional searching required [12]. Parametric approaches are applicable in cases of coherent multipath signals with accurate results [12].

The above classes of algorithms and methods make use of a deterministic model that describes the channel composition in instantaneous snapshots [12].

For this reason, they are considered deterministic channel parameter estimation methods [12].

6.6 Channel model selection

Understanding of the radio propagation channel is of great importance for accurately evaluating vehicular protocols and applications as a first step to the actual system deployment in different scenarios.

Selecting the most suitable channel model is essential, and some criteria must be chosen in order to assess the usability of a model. First of all, usability of the models can be decided at scale level [14]:

- *Link-level*: If specific issues or channel statistics need to be considered prior to the actual deployment of a vehicular communication system, geographic information should be available.
- *System-level*: Only the global behaviour is aimed, whether accurate geographical information is available or not.

For the link-level category, a model channel should offer information regarding critical characteristics that will determine the usability of the model. Among other aspects that a channel model must describe, the following ones should be emphasized, as described per [14]:

- *Spatial and temporal dependency*:

  A channel model should properly describe the spatial and temporal dependency of the communication links occurring within a given environment: spatial correlation is created by the environment surrounding objects, static and dynamic, whereas the time dependency is due to the vehicle mobility and VTD variation. The spatial and temporal statistics allow determining the small-scale fading.
- *Temporal variance and NS*:

  Vehicular channels exhibit NS derived from the discontinuous interruption of the LOS component due to the static and mobile surrounding objects present in the environment, especially high-speed moving vehicles.
- *Extensibility to different environments*:

  Not all channel models allow its general extension to other propagation environments beyond those initially considered to develop the model.

  The channel models derived from site-specific measurements can give no accurate results when applied to locations with different characteristics from those of the locations initially considered. It should be needed to model a large set of specific locations to give guarantees of extensibility.

  Conversely, channel models based on geometry-specific information of the environment can model different environments. The complexity of implementation is the far side of these models.
- *Applicability*:

  This feature indicates the ability of a model to adapt to scenarios of specific application. For example, instead of generally analysing highways, only one highway lane may need to be evaluated. Critical applications, like road safety warning, require dedicated channel models; however, it is valuable that models support a wide range of application cases.

- *Antenna configuration*:

  Channel model should support different antenna settings, especially MIMO antenna configuration in case of requiring assessing its benefits to counteract the signal impairments introduced by small-scale fading.
- *Scalability*:

  The scalability grade of a channel model is related to its efficiency for evaluating detailed and accurate channel statistics for a particular vehicular application. The efficiency is measured as the ability of a channel model support large-scale simulations. The efficiency level of the model to predict results will depend on the complexity of the underlying signal propagation mechanisms considered for estimating the channel statistics. Generally, ray-tracing-based models provide good accuracy but show constraints to scale a scenario, e.g. due to the computational load or processing speed. Geometric-based (GB) models trade off scalability and accuracy.
- *Complexity of implementation*:

  The practical implementation of channel models offers various trade-offs between complexity and other features, like accuracy, efficiency, scalability and processing speed. Time constraints may be imposed with respect to the simulation execution.
- *Data input availability*:

  The practical implementation of channel models may require the user to enter a significant amount of input data, some difficult to find, for example, the GB data of the environment, or the number and features of the surrounding objects, static or mobile.

  Non-geometric or stochastic models (NG) derived of measurements not requiring environment information reduce the complexity, keeping a high scalability and achieving lower accurate results.
- *Estimation accuracy*:

  If GB models are considered but the actual locations for surrounding objects or the communicating vehicles are unknown, the estimation accuracy of non-LOS links will be reduced.

In measurement-based models like the log-distance fit, if the PL exponent is unknown or the number of slopes that the model fits is obviated, the simulation or manual estimation will result in unrealistic power-fading values.

Channel parameters should be set as close as possible to the values derived from the measurement data. The estimation accuracy increases measurement data for a given location available.

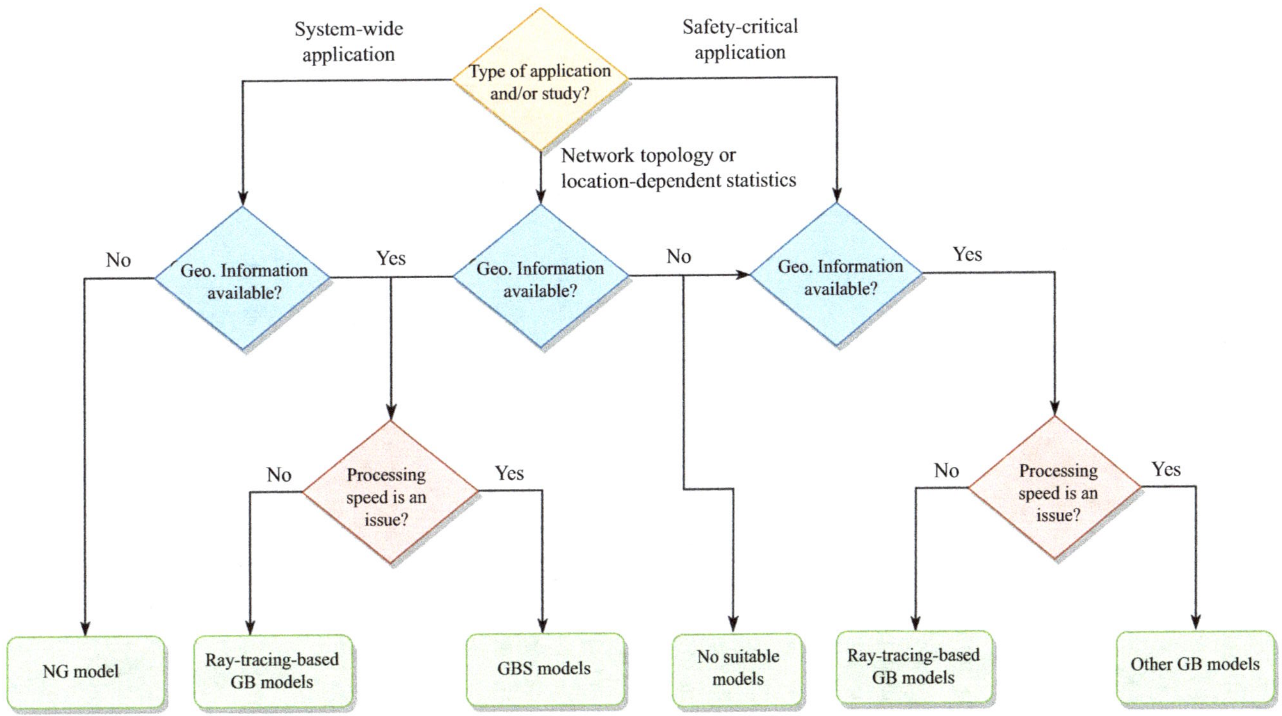

Figure 6.39 Suitability flow chart for channel model selection

Generally, if the geographical information is available, then the GBDMs are a preferable choice in terms of estimation accuracy.

In Figure 6.39, it has been rearranged the flow chart presented in [14] that depicts a guideline for selecting a suitable channel model according three choice levels. The first step to select a suitable model consists of determining the category of the targeted vehicular application [14]. Three application categories are possible:

- For system-wide performance analysis any type of model might be suitable.
- For estimation of network topology statistics or location dependent statistics, GB models are suitable to model dynamic link transitions and small-scale variations.
- For safety-critical applications, deterministic GB models are the best choice.

The second step after the application category is identified will be to determine the need for geographic information. The third level of choice regards the processing capability of the channel model. One of four types of suitable models can be finally selected:

- Non-geometric-based models: SMs without requiring environment information; simple implementation and high scalability with lower accuracy.
- Ray-tracing-based GB models: Complete geographic information is required, and processing speed is not a constraint; these offer maximum accuracy.
- Simplified GB models (GBS): Limited information about the propagation environment is needed but the processing speed is critical. These achieve reasonable accuracy-scalability trade-off, enough accuracy and simple implementation.
- Other GB models: These only require the qualitative type of simulated environment to estimate the channel statistics – highly scalable with environment-specific channel statistics.

Table 6.14 qualitatively summarizes from [3,12–14] a basic state of the art in radio channel modelling of vehicular communications. The following acronyms have been used:

- *Model type:* NG, non-geometry-based model; GBS, geometry-based stochastic model; GBD, geometry-based deterministic model.
- *Channel statistics:* PDP, power delay profile; DD, Doppler-delay; PSD, power spectrum density; STF, space–time–frequency; CF, correlation function; LCR, level crossing rate; SDF, space-Doppler-frequency; PDF, probability density function; PL, path loss; CDF, cumulative distribution function; CT, coherence time.

Table 6.14 Basic V2X radio channel models

Reference	Environment	Model	Propagation scale	Properties
Sun *et al.* [355]	Parking garage	GBD	LS: Log distance	$\sigma_\tau = 109$ ns (mean, NLOS)
Fayziyev *et al.* [356]	Tunnel	GBD	LS: Measurement-fitted impulse response	
Abbas *et al.* [51]	Highway	GBS	LS: Log Distance	
Sen and Matolak [43]	Urban canyon, highway	NGSISOWideband	SM: Weibull (amplitude PDF)	5 GHz Spatial-temporal dependence Non-stationarity (CF) High and low traffic density Scatterer region: 2D non-isotropic $\sigma_\tau = 125.8$ ns (urban mean) $\sigma_\tau = 200$ ns (90th perc. highway)
Wang *et al.* [117]	All	GBS	SM: Rician	Spatial–temporal dependence Non-stationarity
Cheng *et al.* [16]	Suburban	NGNarrowbandSISO	LS: Dual slopeLog distanceSM: Nakagami	5.9 GHz Low traffic density PL, CT, amplitude CDF Doppler PSD Stationary 2D isotropic (two-ring)
Mangel *et al.* [357]	Intersections	GBD	LS: Log distance SM: Nakagami (LOS) Normal (NLOS)	Spatial–temporal dependence
Karedal *et al.* [39]	Rural, highway	RS-GBSWidebandMI-MO	LS: Log distanceSM: Simpli-fiedRay tracing	Spatial–temporal dependenceNon-stationarity2D non-isotropic (randomly)
Boban *et al.* [45]	All	GBD	LS: Two-ray (LOS) Log Dist. (NLOS) SM: Normal	Spatial–temporal dependence Non-stationarity

(Continues)

Table 6.14 (Continued)

Reference	Environment	Model	Propagation scale	Properties
Pilosu *et al.* [358]	All	GBD	LS and SM: Pre-processed Ray tracing	Spatial–temporal dependence Non-stationarity
Maurer *et al.* [47]	All	GBD Wideband SISO	LS: Ray tracing SM: Weibull	Spatial–temporal dependence Non-stationarity
Maurer *et al.* [110]	All	GBD Wideband MIMO		Non-stationary Scatterer region: 3D non-isotropic High and low traffic density
Acosta-Marum and Ingram [17]	Urban canyon highway	NG Wideband SISO		5.9 GHz Non-stationary Scatterer region: 2D non-isotropic Low traffic density PDP, Doppler PSD Traffic direction: oncoming $\sigma_\tau = 40.2$ ns
Acosta, Tokuda and Ingram [41]	Urban, highway	NG Wideband SISO		2.4 GHz, low traffic density PDP, Doppler PSD traffic direction: same
Kunisch and Pamp [359]	Urban, highway	NG		$\sigma_\tau = 14$ ns (highway mean) $\sigma_\tau = 22$ ns (urban highway)
Renaudin *et al.* [360]	Urban, highway	NG		$\sigma_\tau = 165$ ns (highway mean) $\sigma_\tau = 373$ ns (urban hwy)
Paschalidis *et al.* [361]	Highway	NG		Traffic direction: oncoming $\sigma_\tau = 54$ ns (highway mean)
Bernado *et al.*[362]	Tunnel	NG		$\sigma_\tau = 60$ ns (tunnel mean)
Zajic *et al.* [42]	Urban canyon, highway	RS-GBS Wideband MIMO	Scattering region: 3D non-isotropic (two concentric-cylinder)	2.4 GHz Stationary Low traffic density STF CF, LCR, SDF, PSD Traffic direction: same
Zajic *et al.* [197]	SimulationAll	RS-GBS	Scattering region: 2D non-	Stationary High and low traffic density 2D

(Continues)

		Narrowband MIMO	isotropic (two-ring+ellipse) M-to-M Ricean fading channels	isotropic (two-ring)
Paier *et al.* [41]	Highway	NG Wideband MIMO		5.2 GHz traffic direction: oncoming low traffic density PL, PDP, DD power profile
Akki *et al.* [112]	Urban, suburban	RS-GBS Narrowband SISO	Scattering region: 2D isotropic (two-ring)	Stationary Amplitude PSD Spatial-time CF
Cheng *et al.* [182]	Macro-, micro- and picocells	RS-GBS Narrowband MIMO	Scattering region: 2D non-isotropic (two-ring+ellipse)	Stationary high and low traffic density Space–time CF, space Doppler PSD
Cheng *et al.* [38]	Urban	RS-GBS Wideband MIMO	Scattering region: 2D non-isotropic(two-ring+multiple confocal ellipses)	Stationary High and low traffic density TAP model Space–time CF, space Doppler PSD
Azpilicueta *et al.* [311]	Simulation Urban	GBD Narrowband SISO		stationary low traffic density PDP, Doppler PSD
Jiang *et al.* [363]	Simulation	GBS Wideband MIMO	Scattering region: 2D non-isotropic (two-ring+ellipse)	Non-stationary time-variant geometric statistics, VTD impact
Jiang *et al.* [364]	Simulation UAV for A2G	GBS Wideband MIMO		3D, non-stationary AoD, AoA for azimuthal and elevation planes Spatial CCF, temporal ACF, Doppler PSD, PDP
Ruisi *et al.* [365]	Simulation	GBS Wideband MIMO		Non-stationary Gauss–Markov mobility Doppler PSD, MPC cluster Cluster-based two-ring Time-variant time frequency CF

References

[1] Bello, P.A., 'Characterization of randomly time-variant linear channels', *IEEE Trans. Commun. Syst.*, 1963, 11(4): 360–393.

[2] Belhabib, M., 'Investigation on radio channel over the air emulation by multi-probe setup'. *PhD Doctoral Thesis*, Université Rennes, 2017.

[3] Matolak, D.W., 'Modeling the vehicle-to-vehicle propagation channel: a review', *Radio Science*, 2014, 49(9): 721–736.

[4] Michelson, D.G., and Ghassemzadeh, S.S., 'Measurement and modeling of wireless channels', In: Tarokh, V. (ed.), *New Directions in Wireless Communications Research*, Springer, Boston, MA, pp. 1–27, 2009.

[5] Kerr, D.E., *Propagation of Short Radio Waves*, Vol. 13 of the MIT Radiation Laboratory Series, McGraw-Hill: New York, NY, 1951.

[6] Okumura, Y., E. Ohmori, T. Kawano, and K. Fukuda, 'Field strength and its variability in VHF and UHF land-mobile radio service', *Rev. Elec. Commun. Lab.*, 1968, 16(9-10): 825–873.

[7] Clarke, R.H., 'A statistical theory of mobile radio reception', *Bell Sys. Tech. J.*, 1968, 47: 957–1000.

[8] Cox, D.C., 'Delay Doppler characteristics of multipath propagation at 910 MHz in a suburban mobile radio environment', *IEEE Trans. Antennas Propag.*, 1972, 20(5): 625–635.

[9] European Cooperation in Science and Technology, COST Actions, https://www.cost.eu/cost-actions/

[10] Parsons, J.D., *The Mobile Radio Propagation Channel*, Halsted Press: New York, NY, 1992.

[11] Bernadó, L., Zemen, T., Tufvesson, F., Molisch, A.F., and Mecklenbräuker, C.F., 'The (in-)validity of the WSSUS assumption in vehicular radio channels', *IEEE 23rd International Symposium on Personal, Indoor and Mobile Radio Communications*, Sydney, NSW, Australia, 9–12 September 2012.

[12] Yin, X., and Cheng, X, *Propagation Channel Characterization, Parameter Estimation and Modelling for Wireless Communications*, John Wiley & Sons: Singapore, 2016.

[13] Wang, C.-X., Cheng, X., and Laurenson, D.I., 'Vehicle-to-vehicle channel modeling and measurements: recent advances and future challenges', *IEEE Communications*, 2009, 47(11): 96–103.

[14] Viriyasitavat, W., Boban, M., Tsai, H.-M., and Vasilakos, A.V., 'Vehicular communications: survey and challenges of channel and propagation models', *IEEE Vehicular Technology Magazine*, 2015, 10(2): 55–66.

[15] Dhar, S., Bera, R., Giri, R.B., Anand, S., Nath, D., and Kumar, S., 'An overview of V2V communication channel modeling', *International Symposium on Devices MEMS, Intelligent Systems & Communication*, pp. 24–31, 2011.

[16] Cheng, L., Henty, B.E., Stancil, D.D., Bai, F., and Mudalige, P., 'Mobile vehicle-to-vehicle narrowband channel measurement and characterization of the 5.9 GHz dedicated short range communication (DSRC) frequency band', *IEEE Journal on Selected Areas in Communications*, 2007, 25(8): 1501–1516.

[17] Acosta-Marum, G., and Ingram, M.A., 'Six time-and frequency-selective empirical channel models for vehicular wireless LANs', *IEEE Vehic. Tech. Mag.*, 2007, 2(5): 4–11.

[18] Matolak, D.W., 'Channel modeling for vehicle-to-vehicle communications', *IEEE Communications Magazine*, 2008, 46(5): 76–83.

[19] IEEE P802.11p/D2.01, 'Standard for wireless local area networks providing wireless communications while in vehicular environment', Mar. 2007.

[20] Molisch, A., Tufvesson, F., Karedal, J., and Mecklenbräuker, C., 'A survey on vehicle-to-vehicle propagation channels', *IEEE Wireless Communications*, 2009, 16(6): 12–22.

[21] Davis, J.S., and Linnartz, M.G., 'Measurements of vehicle-to-vehicle propagation', *Proc. Asilomar Conference*, Monterey, Calif., 31 Oct.–1 Nov 1994.

[22] Wada, T., Maeda, M., Okada, M., Tsukamoto, K., and Komaki, S., 'Theoretical analysis of propagation characteristics in millimeter-wave intervehicle communication system', *Electron. Commun. Jpn., Part 2*, 2000, 83(11): 33–43.

[23] Azpilicueta, L., Aguirre-Gallego, E., Falcone, F., Lopez-Iturri, P., and Alejos, A.V., 'Radio channel characterization of vehicle-to-infrastructure communications at 60GHz', *IEEE Antennas and Propagation in Wireless Communications*, Torino (Italy), 7–11 September 2015.

[24] ETSI TR 138 901 (V14.0. 0) (2017-03): '5G; Study on channel model for frequencies from 0.5 to 100 GHz (3GPP TR 38.901 version 14.0.0 Release14)'.

[25] Giordani, M., Shimizu, T., Zanella, A., Higuchi, T., Altintas, O., and Zorzi, M., 'Path loss models for V2V mmWave communication: performance evaluation and open challenges', *2nd IEEE Connected and Automated Vehicles Symposium (CAVS)*, September 2019.

[26] Kato, A., Sato, K., and Fujise, M., 'ITS wireless transmission technology. Technologies of millimeter-wave inter-vehicle communications: propagation characteristics', *Journal of the Communications Research Laboratory*, 2001, 48: 99–110.

[27] Takahashi, S., Kato, A., Sato K., and Fujise, M., 'Distance dependence of path loss for millimeter wave inter-vehicle communications', *IEEE 58th Vehicular Technology Conference (VTC-Fall)*, October 2003.

[28] Yamamoto, A., Ogawa, K., Horimatsu, T., Kato, A., and Fujise, M., 'Path-loss prediction models for intervehicle communication at 60 GHz', *IEEE Trans. on Vehicular Technology*, 2008, 57(1): 65–78.

[29] European Automobile Manufacturers Association, ACEA Position Paper. Frequency Bands for V2X, November 2018, https://www.acea.be/uploads/publications/ACEA_position_paper-Frequency_bands_for_V2X.pdf

[30] Alejos, A.V., Sánchez, M.G., and Cuiñas I., 'Measurement and analysis of propagation mechanisms at 40 GHz: viability of site shielding forced by obstacles', *IEEE Transactions on Vehicular Technology*, 2008, 57(6): 3369–3380.

[31] Committee on Science and Technology (COST) (2007), 231 Final Report, http://www.lx.it.pt/cost231/final_report.htm, November 2013

[32] Matolak, D.W., 'V2V communication channels: state of knowledge, new results, and what's next', *Nets4Cars/Nets4Trains: 5th Int. Workshop on Communications Tech. for Vehicles*, Lille, France, 14–15 May 2013.

[33] Molisch, A.F., Asplund, H., Heddergott, R., Steinbauer, M., and Zwick, T., 'The COST259 directional channel model – Part I: Overview and methodology', *IEEE Trans. Wireless Commun.*, 2006, 5(12): 3421–3433.

[34] Asplund, H., Glazunov, A.A., Molisch, A.F., Pedersen, K.I., and Steinbauer, M., 'The COST259 directional channel model – Part II: Macrocells', *IEEE Trans. Wireless Commun.*, 2006, 5(12): 3434–3450.

[35] De Cerio, D.P.-D., and Valenzuela, J.L., 'Provisioning vehicular services and communications based on a bluetooth sensor network deployment', *Sensors*, 2015, 15: 12765–12781.

[36] Arena, F., and Pau, G., 'An overview of vehicular communications', *Future Internet*, 2019, 11(27): 1–12.

[37] Level, F., 'Improving self-driving car safety and reliability with V2X protocols', July 2018, https://fifthlevel.ai/archives/2506

[38] Cheng, X., Wang, C.-X., and Laurenson, D.I., 'A geometry-based stochastic model for wideband MIMO mobile-to-mobile channels', *Proc. IEEE GLOBECOM '09*, Hawaii, USA, Nov.–Dec. 2009.

[39] Karedal, J., Tufvesson, F., Czink, N., *et al.*, 'A geometry-based stochastic MIMO model for vehicle-to-vehicle communications', *IEEE Trans. Wireless Commun.*, 2009, 8(7): 3646–3657.

[40] Anaya, J.J., Merdrignac, P., Shagdar, O., Nashashibi, F., and Naranjo, J.E., 'Vehicle to pedestrian communications for protection of vulnerable road users', *IEEE Intelligent Vehicles Symposium*, pp. 1037–1042, 2014.

[41] Acosta, G., Tokuda, K., and Ingram, M.A., 'Measured joint Doppler-delay power profiles for vehicle-to-vehicle communications at 2.4 GHz', *Proc. IEEE GLOBECOM*, Dallas, TX, pp. 3813–3817, November 2004.

[42] Zajic, A.G., Stuber, G.L., Pratt, T.G., and Nguyen, S.T. 'Wideband MIMO mobile-to-mobile channels: geometry-based statistical modeling with experimental verification', *IEEE Trans. Vehic. Tech.*, 2009, 58(2): 517–534.

[43] Sen, I., and Matolak, D.W., 'Vehicle-vehicle channel models for the 5-GHz band', *IEEE Trans. Intelligent Transportation Sys.*, 2008, 9(2): 235–245.

[44] Paier, A., Karedal, J., Czink, N., *et al.*, 'Characterization of vehicle-to-vehicle radio channels from measurements at 5.2 GHz', *Wireless Personal Communications*, 2009, 50(1): 19–32, http://dx.doi.org/10.1007/s11277-008-9546-6

[45] Boban, M., Barros, J., and Tonguz, O., 'Geometry-based vehicle-to-vehicle channel modeling for large-scale simulation', *IEEE Transactions on Vehicular Technology*, 2014, 63(9): 4146–4164.

[46] Vlastaras, D., Abbas, T., Nilsson, M., Whiton, R., Olback, M., and Tufvesson, F., 'Impact of a truck as an obstacle on vehicle-to-vehicle communications in

rural and highway scenarios', *IEEE 6th International Symposium on Wireless Vehicular Communications*, pp. 1–6, 2014.

[47] Maurer, J., Fugen, T., Schafer, T., and Wiesbeck, W., 'A new inter-vehicle communications (IVC) channel model', *IEEE Vehicular Technology Conference (VTC-Fall)*, pp. 9–13, September 2004.

[48] New Connected Car Standards Put Europe into Top Gear, February 2014, https://europa.eu/rapid/press-release_IP-14-141_en.htm

[49] National Highway Traffic Safety Administration, V2V Statement Attributable to the Department of Transportation, November 2017, https://www.nhtsa.gov/press-releases/v2v-statement

[50] Whaiduzzaman, M., Sookhak, M., Gani, A., and Buyya, R., 'A survey on vehicular cloud computing', *Journal of Network and Computer Applications*, 2014, 40(1): 325–344.

[51] Abbas, T., Sjöberg, K., Karedal, J., and Tufvesson, F., 'A measurement based shadow fading model for vehicle-to-vehicle network simulations', *International Journal of Antennas and Propagation*, 2015, 2015, article ID 190607, 12 pages.

[52] Cheng, L., Henty, B.E., Cooper, R., and Stancil, D.D., 'A measurement study of time-scaled 802.11a waveforms over the mobile-to-mobile vehicular channel at 5.9 GHz', *IEEE Commun. Mag.*, 2008, 46(5): 84–91.

[53] Gallagher, B., and Akatsuka, H., 'Wireless communications for vehicle safety: radio link performance and wireless connectivity methods', *IEEE Veh. Technol. Mag.*, 2006, 1: 4–16.

[54] Lee, W.C.Y., and Yeh, Y.S., 'On the estimation of the second-order statistics of log normal fading in mobile radio environment', *IEEE Trans. Comm.*, 1974, 22: 869–873.

[55] Bush, J., 'The significance of Rayleigh fading in coverage measurements and 40 Lambda criteria', Berkeley Varitronics Systems, http://www.bvsystems.com, November 2002.

[56] Hammoudeh, A.M., Sánchez, M.G., and Grindrod, E., 'Experimental analysis of propagation at 62 GHz in suburban mobile radio microcells', *IEEE Transactions on Vehicular Technology*, 1999, 48(2): 576–588.

[57] Sánchez, M.G., Hammoudeh, A.M., Grindrod, E., and Kermoal, J.-P., 'Coherence bandwidth characterization in an urban microcell at 62.4 GHz', *IEEE Transactions on Vehicular Technology*, 2000, 49(2): 607–613.

[58] Bello, P.A., and Nelin, B.D., 'The effect of frequency-selective fading on the binary error probabilities of incoherent and differentially-coherent matched filter receivers', *IRE Trans. Commun. Syst.*, 1963, COM-11: 170–186.

[59] Anderson, C.R., 'Design and implementation of an ultrabroadband millimeter-wavelength vector sliding correlator channel sounder and in-building measurements at 2.5 & 60 GHz', Master's Thesis, Virginia Polytechnic Institute and State University, May 2002, http://scholar.lib.vt.edu/theses/index.html

[60] Bultitude, R.J.C., 'Estimating frequency correlation functions from propagation measurements on fading radio channels: a critical review', *IEEE Journal on Selected Areas in Communications*, 2002, 20(6): 1133–1143.

[61] Alejos, A.V., Sánchez, M.G., Cuiñas, I, and Dawood, M, 'Wideband noise radar based in phase coded sequences', *Radar Technology*, Guy Kouemou, IntechOpen, 1 January 2010, doi: 10.5772/7173, https://www.intechopen.com/books/radar-technology/wideband-noise-radar-based-in-phase-coded-sequences

[62] Alejos, A.V., Dawood, M., and Ur-Rahman Mohammed, H.U.-R., 'Estimation of sidelobe level variations of phased codes in presence of random interference for bistatic wideband noise radar', *International Journal of Antennas and Propagation*, 2015, 2015, article ID 297823, 11 pages.

[63] Salous, S., 'FMCW channel sounder with digital processing for measuring the coherence of wideband HF radio links', *Proc. IEE*, 1986, 133: 456–462.

[64] Peter Aberl, P., 'An efficient and safe approach of radar signal capturing and processing', Texas Instruments, June 2015, https://www.mwee.com/content/efficient-and-safe-approach-radar-signal-capturing-and-processing-1/page/0/1

[65] Parsons, J.D., Demery, D.A., and Turkamani, A.M.D., 'Sounding techniques for wideband mobile radio channels: a review', *IEE Proceedings*, 1992, 138 (5): 437–446.

[66] Affandi, A., El Zein, G., and Citerne, J., 'Investigation on frequency dependence of indoor radio propagation parameters', *IEEE 50th Vehicular Technology Conference*, vol. 4, pp. 1988–1992, 1999.

[67] Steinbauer, M., Molisch, A.F., and Bonek, E., 'The double-directional radio channel', *IEEE Antennas Propag. Mag.*, 2001, 43(4): 51–63.

[68] Krim, H., and Viberg, M., 'Two decades of array signal processing research: the parametric approach', *IEEE Transactions on Signal Processing* 1996, 13: 67–94.

[69] Steinbauer, M., Hampicke, D., Sommerkorn, G., *et al.*, 'Array measurement of the double-directional mobile radio channel', *Proceedings of IEEE 51st Vehicular Technology Conference (VTC-Spring)*, vol. 3, pp. 1656–1662, 2000.

[70] Wallace, J., Jensen, M., Swindlehurst, A., and Jeffs, B., 'Experimental characterization of the MIMO wireless channel: data acquisition and analysis', *IEEE Transactions on Wireless Communications*, 2003, 2(2): 335–343.

[71] MacCartney, G., and Rappaport, T., '73 GHz millimeter wave propagation measurements for outdoor urban mobile and backhaul communications in New York City', *Proceedings of IEEE International Conference on Communications (ICC)*, pp. 4862–4867, Sydney, 2014.

[72] Kyösti, P., Meinila, J., Hentila, L., *et al.*, 'WINNER II Channel Models D1.1.2 v1.1', 2007.

[73] ITU, 'ITU-R M.2135-1: guidelines for evaluation of radio interface technologies for IMT-Advanced', Technical Report, 12/2009.

[74] Liu, L., Oestges, C., Poutanen, J., *et al.*, 'The COST 2100 MIMO channel model', *IEEE Transactions on Wireless Communications*, 2012, 19(6): 92–99.

[75] Rappaport, T., Sun, S., Mayzus, R., *et al.*, 'Millimeter wave mobile communications for 5G cellular: it will work!', *IEEE Access*, 2013, 1: 335–349.

[76] Xu, H., Rappaport, T., Boyle, R., and Schaffner, J., '38 GHz wideband point-to-multipoint radio wave propagation study for a campus environment', *IEEE 49th Vehicular Technology Conference*, vol. 2, pp. 1575–1579, 1999.

[77] Xu, H., Rappaport, T., Boyle, R., and Schaffner, J., 'Measurements and models for 38-GHz point-to-multipoint radiowave propagation', *IEEE Journal on Selected Areas in Communications*, 2000, 18(3): 310–321.

[78] Yin, X., Ling, C., and Kim, M.D., 'Experimental multipath cluster characteristics of 28 GHz propagation channel in office environments', *IEEE Access*, 2015, 3: 3138–3150.

[79] Sánchez, M.G., Cid, E.L., and Alejos, A.V., 'Empirical modeling of radiowave angular power distributions in different propagation environments at 60 GHz for 5G', *Electronics*, 2018, 7(2): 365.

[80] Cid, E.L., Taboas, M.P., Sánchez, M.G., and Alejos, A.V., 'Microcellular radio channel characterization at 60 GHz for 5G communications', *IEEE Antennas and Wireless Propagation Letters*, 2016, 16: 1476–1479.

[81] Sánchez, M.G., Alejos, A.V., and Cuinas, I., 'Comparison of space diversity performance in indoor radio channels at 40 GHz and 60 GHz', *European Conference on Wireless Technology*, pp. 170–173, Amsterdam (The Netherlands), 2008.

[82] Cid, E.L., Sanchez, M.G., and Alejos, A.V., 'High speed transmission at 60 GHz for 5G communications', *IEEE APS-URSI*, Vancouver (Canada), July 2015.

[83] Czink, N., Bonek, E., Yin, X., and Fleury, B.H., 'Cluster angular spreads in a MIMO indoor propagation environment', *Proceedings of the 16th IEEE International Symposium on Personal, Indoor and Mobile Radio Communications (PIMRC'05)*, vol. 1, pp. 664–668, Berlin, Germany, 2005.

[84] Stucki, A., 'PropSound system specifications document: concept and specifications'. Technical Report, Elektrobit AG, Switzerland, 2001.

[85] Salous, S., Fillipides, P., and Hawkins, I., 'Architecture of multichannel sounder for multiple antenna applications', Technical Report TD-02-002, COST-273, 2002.

[86] Zetik, R., Thomä, R., and Sachs J., 'Ultra-wideband real-time channel sounder design and application', Technical Report TD-03-201, COST 273, 2003.

[87] Pedersen, T., Pedersen, C., Yin, X., and Fleury, B., 'Optimization of spatiotemporal apertures in channel sounding', *IEEE Transactions on Signal Processing*, 56(10): 4810–4824.

[88] Pedersen T, Pedersen C, Yin X, *et al.*,'Joint estimation of Doppler frequency and directions in channel sounding using switched Tx and Rx arrays', *Proceedings of IEEE Global Telecommunications Conference (Globecom)*, vol. 4, pp. 2354–2360, 2004.

[89] Yin, X., Fleury, B.H., Jourdan, P., and Stucki, A., 'Doppler frequency estimation for channel sounding using switched multiple transmit and receive antennae', *Proceedings of the IEEE Global Communications Conference (Globecom'03)*, vol. 4, pp. 2177–2181, 2003.

[90] Mahler, K., Paschalidis, P., Kortke, A., Peter, M., and Keusgen, W., 'Realistic IEEE 802.11p transmission simulations based on channel sounder measurement data', *Vehicular Technology Conference (VTC Fall), 2013 IEEE 78th*, Las Vegas, NV, 2013.

[91] Paschalidis, P., Wisotzki, M., Kortke, A., Keusgen, W., and Peter, M., 'A wideband channel sounder for car-to-car radio channel measurements at 5.7 GHz and results for an urban scenario', *Vehicular Technology Conference, 2008. VTC 2008-Fall. IEEE 68th*, pp. 1–5, September 2008.

[92] Wireless Devices and Systems Group, Propagation Channels, University of South California, https://wides.usc.edu/research.html

[93] Devasirvatham, D.M.J., Krain, M.J., Rappaport, D.A., and Banerjee, C., 'Radio propagation measurements at 850 MHz, 1.7 GHz and 4 GHz inside two dissimilar office buildings', *Electronics Letters*, 1990, 26(7): 445–447.

[94] Newhall, W.G., Saldanha, K.J., and Rappaport, T.S.. 'Propagation time delay spread measurements at 915 MHz in a large train yard', *Vehicular Technology Conference, 1996. Mobile Technology for the Human Race., IEEE 46th*, vol. 2, pp. 864–868. IEEE, 1996.

[95] Kivinen, J., Korhonen, T.O., Aikio, P., Gruber, R., Vainikainen, P., and Haggman, S.-G., 'Wideband radio channel measurement system at 2 GHz', *IEEE transactions on Instrumentation and Measurement*, 1999, 48(1):39–44.

[96] Nielsen, J.O., Andersen, J.B., Eggers, P.C.F, Pedersen, G.F., Olesen, K., and Suda, H.,. 'Measurements of indoor 16×32 wideband mimo channels at 5.8 GHz', *Spread Spectrum Techniques and Applications, 2004 IEEE Eighth International Symposium on*, pp. 864–868, IEEE, 2004.

[97] Salous, S.,, Filippidis, P., Lewenz, R., Hawkins, I., Razavi-Ghods, N., and Abdallah, M., 'Parallel receiver channel sounder for spatial and mimo characterization of the mobile radio channel', *IEE Proceedings-Communications*, 2005, 152(6): 912–918.

[98] Ranvier, S., Kyro, M., Haneda, K., Mustonen, T., Icheln, C., and Vainikainen, P., 'VNA-based wideband 60 GHz mimo channel sounder with 3-d arrays', *Radio and Wireless Symposium, 2009. RWS'09. IEEE*, pp. 308–311, IEEE, 2009.

[99] Felhauer, T., Baier, P. W., Konig, W., and Werner, M., 'Optimized wideband system for unbiased mobile radio channel sounding with periodic spread spectrum signals', *IEICE Transactions on Communications*, 1993, 76 (8):1016–1029.

[100] Heddergott, R., Truffer, P., Tschudin, M., and Nold, M., 'Comparison of high resolution channel parameter measurements with ray tracing simulations in a multipath environment', In: *EPMCC*, vol. 99, pp. 9–11, Citeseer, 1999.

[101] MEDAV GmbH, Rusk Channel Sounder, https://www.tu-ilmenau.de/de/ei-ms-csp/mscsp-labs/mimo-rusk-channel-sounder/ (accessed 1 October 2019).

[102] Trautwein, U., Schneider, C., Sommerkorn, G., Hampicke, D., Thoma, R., and Wirnitzer, W., 'Measurement data for propagation modeling and wireless system evaluation', COST273, MCM, TD-03-021, Barcelona, 2003.

[103] Hentila, L., Kyösti, P., Ylitalo, J., Zhao, X., Meinila, J., and Nuutinen, J.P., 'Experimental characterization of multi-dimensional parameters at 2.45 and 5.25 GHz indoor channels', In: *WPMC*, vol. 5, pp. 18–22, 2005.

[104] Cullen, P.J., Fannin, P.C., and Molina, A., 'Wide-band measurement and analysis techniques for the mobile radio channel', *IEEE Transactions on Vehicular Technology*, 1993, 42(4): 589–603.

[105] Zhang, L., 'Channel measurement and modeling in complex environments', *PhD Dissertation*, School of Telecommunications Systems and Engineering, Universidad Politécnica de Madrid, September 2016.

[106] Molisch, A. F., *Wireless Communications*, New York, USA: John Wiley & Sons, 2007.

[107] Wu, X., Miucic, R., Yang, S., *et al.*, 'Cars talk to phones: a DSRC based vehicle-pedestrian safety system', *IEEE Vehicular Technology Conference (VTC Fall)*, pp. 1–7, 2014.

[108] Parsons, J.D., *The Mobile Radio Propagation Channel*, New York, USA: John Wiley & Sons, 2000.

[109] Tan, I., Tang, W., Laberteaux, K., and Bahai, A., 'Measurement and analysis of wireless channel impairments in DSRC vehicular communications', *IEEE International Conference on Communications (ICC)*, pp. 4882–4888, May 2008.

[110] Maurer, J., Fuegen, T., Porebska, M., Zwick, T., and Wiesbeck, W., 'A ray-optical channel model for mobile to mobile communications', 4th MCM COST 2100, Wroclaw, Poland, 6–8 February 2008.

[111] Akki, A., 'Statistical properties of mobile-to-mobile land communication channels', *IEEE Transactions on Vehicular Technology*, 1994, 43(4): 826–831.

[112] Akki, A.S., and Haber, F., 'A statistical model for mobile-to-mobile land communication channel', *IEEE Trans. Vehic. Tech.*, 1986, 35(1): 2–10.

[113] Pätzold, M., Hogstad, B., and Youssef, N., 'Modeling, analysis, and simulation of MIMO mobile-to-mobile fading channels', *IEEE Transactions on Wireless Communications*, 2008, 7(2), 510–520.

[114] Zajic, A.G., and Stuber, G.L., 'Three-dimensional modeling, simulation, and capacity analysis of space-time correlated mobile-to-mobile channels', *IEEE Trans. Veh. Technol.*, 2008, 57(4): 2042–2054.

[115] Zajic, A., and Stuber, G., 'Three-dimensional modeling and simulation of wideband MIMO mobile-to-mobile channels', *IEEE Transactions on Wireless Communications*, 2009, 8(3): 1260–1275.

[116] ETSI Technical Report TR 103 257-1 v1.1.1, 'Intelligent Transport Systems (ITS); Access Layer; Part 1: Channel Models for the 5,9 GHz frequency band', May 2019.

[117] Wang, X., Anderson, E., Steenkiste, P., and Bai, F., 'Improving the accuracy of environment-specific vehicular channel modeling', *Proceedings of the Seventh ACM International Workshop on Wireless Network Testbeds, Experimental Evaluation and Characterization, WiNTECH '12*, pp. 43–50, New York, NY, USA: ACM, 2012.

[118] Website of GEMV^2 Channel Model, http://vehicle2x.net/

[119] Boban, M., Viriyasitavat, W., and Tonguz, O.K., 'Modeling vehicle-to-vehicle line of sight channels and its impact on application-layer performance', *Proceeding of the 10th ACM International Workshop on Vehicular Inter-networking, Systems, and Applications (VANET 13)*, pp. 91–94, Taipei, Taiwan, June 2013.

[120] Karedal, J., Czink, N., Paier, A., Tufvesson, F. and Molisch, A., 'Path loss modelling for vehicle-to-vehicle communications', *IEEE Transactions on Vehicular Technology*, 2011, 60(1): 323–328.

[121] Mangel, T., Klemp, O., and Hartenstein, H., '5.9 GHz inter-vehicle communication at intersections: a validated non-line-of-sight path-loss and fading model', *EURASIP Journal on Wireless Communications and Networking 2011*, 2011,182, doi: 10.1186/1687-1499-2011-182

[122] Perez Fontan, F., and Espineira, P.M., *Modeling the Wireless Propagation Channel: A Simulation Approach with MATLAB*, John Wiley: West Sussex, UK, 2008.

[123] Pätzold, M., *Mobile Fading Channels*, John Wiley: New York, NY, 2002.

[124] Alexander, P., Haley, D., and Grant, A., 'Cooperative intelligent transport systems: 5.9 GHz field trials', *Proceedings of the IEEE*, 2011, 99(7): 1213–1235.

[125] Bernadó, L., Zemen, T., Tufvesson, F., Molisch, A.F., and Mecklenbräuker, C.F., 'Delay and Doppler spreads of non-stationary vehicular channels for safety relevant scenarios', *IEEE Transactions on Vehicular Technology*, 2014, 63(1): 82–93.

[126] Silvennoinen, A., 'Multipath propagation and LTV channel model', Communications Laboratory, Postgraduate Course in Radio Communications, Autumn 2004, http://www.comlab.hut.fi/opetus/333/2004_2005_slides/Multipath_propagation_text.pdf

[127] Matz, G., and Hlawatsch, F., 'Fundamentals of time-varying communication channels', In: Hlawatsch, F., and Matz, G. (eds.), *Wireless Communications over Rapidly Time-varying Channels*, Academic Press, Amsterdam, The Netherlands, Chapter 1, pp. 1–63, 2011.

[128] Fleury, B., 'First- and second-order characterization of direction dispersion and space selectivity in the radio channel', *IEEE Transactions on Information Theory*, 2000, 46(6): 2027–2044.

[129] Deng, Y., Burr, A., and White, G., 'Performance of MIMO systems with combined polarization multiplexing and transmit diversity', *Proceedings of the 61st Vehicular Technology Conference (VTC2005-Spring)*, vol. 2, pp. 869–873, 2005.

[130] Kyritsi, P., and Cox, D., 'Effect of element polarization on the capacity of a MIMO system', *Wireless Communications and Networking Conference, WCNC2002*, vol. 2, pp. 892–896, 2002.

[131] Pedersen, K., and Mogensen, P., 'Simulation of dual-polarized propagation environments for adaptive antennas', *IEEE VTS 50th Vehicular Technology Conference, VTC 1999 - Fall*, vol. 1, pp. 62–66, 1999.

[132] Vaughan, R., 'Polarization diversity in mobile communications', *IEEE Transactions on Vehicular Technology*, 1990, 39(3): 177–186.

[133] Yin, X., Fleury, B.H., Jourdan, P., and Stucki, A., 'Polarization estimation of individual propagation paths using the SAGE algorithm', *Proceedings of the IEEE International Symposium on Personal, Indoor and Mobile Radio Communications (PIMRC)*, Beijing, China, pp. 1795–1799, 2003.

[134] Acosta-Marum, G., Walkenhorst, B.T., and Baxley, R.J., 'An empirical doubly-selective dual-polarization vehicular MIMO channel model', *2010 IEEE 71st Vehicular Technology Conference (VTC 2010-Spring)*, pp. 1–5, 2010.

[135] Degli-Esposti, V., Kolmonen, V.M., Vitucci, E., Fuschini, F., and Vainikainen, P., 'Analysis and ray tracing modelling of co- and cross-polarization radio propagation in urban environment', *The Second European Conference on Antennas and Propagation*, pp. 1–4, 2007.

[136] Hämäläinen, J., Nuutinen, J.P., Wichman, R., Ylitalo, J., and Jämsä, T., 'Analysis and measurements for indoor polarization MIMO in 5.25 GHz band', *Proc. IEEE Vehicular Technology Conference, VTC 2002 Spring*, pp. 252–256, 2005.

[137] Kwon, S.C., and Stüber, G., '3-D geometry-based statistical modeling of cross-polarization discrimination in wireless communication channels', *2010 IEEE 71st Vehicular Technology Conference (VTC 2010-Spring)*, pp. 1–5, 2010.

[138] Almers, P., Wyne, S., Tufvesson, F., and Molisch A., 'Effect of random walk phase noise on MIMO measurements', *Vehicular Technology Conference, 2005, VTC 2005-Spring, 2005 IEEE 61st*, vol. 1, pp. 141–145, 2005.

[139] MEDAV GmbH 2001 ManualW701W1.096: RUSK MIMO: Broadband Vector Channel Sounder for MIMO Channels.

[140] Kim, M.D., Park, J.J., Chung, H.K., and Yin, X., 'Cross-correlation characteristics of multi-link channel based on channel measurements at 3.7 GHz', *14th International Conference on Advanced Communication Technology (ICACT)*, pp. 351–355, 2012.

[141] Park, J.J., Kim, M.D., Kwon, H.K., Chung, H.K., Yin, X., and Fu, Y., 'Measurement-based stochastic cross-correlation models of a multilink channel in cooperative communication environments', *ETRI Journal on Wired and Wireless Telecommunication Technologies* 34(6), 858–868.

[142] Yin, X., Liang, J., Chen, J., Park, J., Kim, M., and Chung, H., 'Empirical models of cross-correlation for small-scale fading in co-existing channels', *Proceedings of Asia-Pacific Communication Conference 2012*, pp. 327–332, Jeju, Korea, 2012.

[143] Yin, X., Liang, J., Fu, Y., *et al.*, 'Measurement-based stochastic modeling for co-existing propagation channels in cooperative relay scenarios', *Future Network Mobile Summit (FutureNetw), 2012*, pp. 1–8, 2012.

[144] Yin, X., Liang, J., Fu, Y., *et al.*, 'Measurement-based stochastic models for the cross-correlation of multi-link small-scale fading in cooperative relay

environments', *Antennas and Propagation (EUCAP), 2012 6th European Conference on*, pp. 1–5, 2012.

[145] Tan, C., Beach, M., and Nix, A., 'Enhanced-SAGE algorithm for use in distributed-source environments', *Electronics Letters,* 39(8), 697–698.

[146] Bengtsson, M., and Ottersten, B., 'Low-complexity estimators for distributed sources', *IEEE Transactions on Signal Processing*, 2000, 48(8), 2185–2194.

[147] Valaee, S., Champagne, B., and Kabal, P., 'Parametric localization of distributed sources', *IEEE Transactions on Signal Processing*, 1995, 43, 2144–2153.

[148] Meng, Y., Stoica, P., and Wong, K., 'Estimation of the directions of arrival of spatially dispersed signals in array processing', *IEE Proceedings Radar, Sonar and Navigation*, 1996, 143(1), 1–9.

[149] Trump, T., and Ottersten, B., 'Estimation of nominal direction of arrival and angular spread using an array of sensors', *Signal Processing*, 1996, 50, 57–69.

[150] Czink, N., Tian, R., Wyne, S., *et al.*, 'Tracking time-variant cluster parameters in MIMO channel measurements: algorithm and results', *Proceedings of International Conference on Communications and Networking in China (ChinaCOM)*, pp. 1147–1151, Shanghai, China, 2007.

[151] Kwakkernaat, M., and Herben, M., 'Analysis of clustered multipath estimates in physically nonstationary radio channels', *Proceedings of the 17th IEEE International Symposium on Personal, Indoor and Mobile Radio Communications (PIMRC)*, pp. 1–5, Athens, Greece, 2007.

[152] Chung, P., and Böhme, J.F., 'Recursive EM and SAGE-inspired algorithms with application to DOA estimation', *IEEE Transactions on Signal Processing,* 53(8), 2664–2677.

[153] Richter, A., Salmi, J., and Koivunen, V., 'An algorithm for estimation and tracking of distributed diffuse scattering in mobile radio channels', *Proceedings of the 7th IEEE International Workshop on Signal Processing Advances for Wireless Communications (SPAWC'06)*, pp. 1–5, 2006.

[154] Salmi J., Richter, A., and Koivunen, V., 'Enhanced tracking of radio propagation path parameters using state-space modeling', *Proceedings of the 2006 European Signal Processing Conference (EUSIPCO)*, pp. 1–5, Florence, Italy, 2006.

[155] Correia, L. (ed.), *Wireless Flexible Personalised Communication (COST 259: European Co-operation in Mobile Radio Research)*, New York, USA: John Wiley & Sons, 2001.

[156] Medbo, J., Riback, M., and Berg, J., 'Validation of 3GPP spatial channel model including WINNER wideband extension using measurements', *Proceedings of IEEE 64th Vehicular Technology Conference, (VTC2006-Fall)*, pp. 1–5, Montréal, Canada, 2006.

[157] Bengtsson, M., and Völcker, B., On the estimation of azimuth distributions and azimuth spectra', *Proceedings of the 54th IEEE Vehicular Technology Conference (VTC2001-Fall)*, vol. 3, pp. 1612–1615, Atlantic City, USA, 2001.

[158] Besson, O., and Stoica, P., 'Decoupled estimation of DoA and angular spread for spatially distribute sources', *IEEE Transactions on Signal Processing,* 1999, 49: 1872–1882.

[159] Betlehem, T., Abhayapala, T.D., and Lamahewa, T.A., 'Space-time MIMO channel modelling using angular power distributions', *Proceedings of the 7th Australian Communications Theory Workshop*, pp. 165–170, Perth, Australia, 2006.

[160] Ribeiro, C.B., Ollila, E., and Koivunen, V., 'Stochastic maximum likelihood method for propagation parameter estimation', *Proceedings of the 15th IEEE International Symposium on Personal, Indoor and Mobile Radio Communications (PIMRC'06)*, vol. 3, pp. 1839–1843, Helsinki, Finland, 2004.

[161] Chizhik, D., Foschini, G., and Valenzuela, R., 'Capacities of multi-element transmit and receive antennas: correlations and keyholes', *Electronics Letters,* 2000, 36(13): 1099–1100.

[162] Shin, H., and Lee, J.H., 'Capacity of multiple-antenna fading channels: spatial fading correlation, double scattering, and keyhole', *IEEE Transactions on Information Theory,* 2003, 49(10): 2636–2647.

[163] Chizhik, D., Foschini, G., Gans, M., and Valenzuela, R., 'Keyholes, correlations, and capacities of multielement transmit and receive antennas', *IEEE Transactions on Wireless Communications,* 2002, 1(2): 361–368.

[164] Almers, P., Tufvesson, F., and Molisch, A., 'Measurement of keyhole effect in a wireless multiple-input multiple-output (MIMO) channel', *IEEE Communications Letters*, 2003, 7(8): 373–375.

[165] Gesbert, D., Bolcskei, H., Gore, D., and Paulraj, A., 'Outdoor MIMO wireless channels: models and performance prediction', *IEEE Transactions on Communications*, 2002, 50(12): 1926–1934.

[166] Karasawa, Y., Tsuruta, M., and Taniguchi, T., 'Multi-keyhole model for MIMO radio-relay systems', *Antennas and Propagation, 2007. EuCAP 2007. The Second European Conference on*, pp. 1–6, 2007.

[167] Levin, G., and Loyka, S., 'On the outage capacity distribution of correlated keyhole MIMO channels', *IEEE Transactions on Information Theory,* 2008, 54(7): 3232–3245.

[168] Yin, X., Zhou, Y., and Liu, F., 'A generic wideband channel model for keyhole propagation scenarios and experimental evaluation', *Communications and Networking in China, 2009. ChinaCOM 2009. Fourth International Conference on*, pp. 1–5, 2009.

[169] Molisch, A.F., 'A generic model for MIMO wireless propagation channels in macro- and microcells', *IEEE Transactions on Signal Processing*, 2004, 52(1): 61–71.

[170] Oestges, C., Erceg, V., and Paulraj, A.J., 'A physical scattering model for MIMO macrocellular broadband wireless channels', *IEEE Journal on Selected Areas in Communications*, 2003, 21(5): 721–729.

[171] Almers, P., Bonek, E., Burr, A., *et al.*, 'Survey of channel and radio propagation models for wireless MIMO systems', *EURASIP Journal on Wireless Communications and Networking*, 2007, 2007.

[172] Shiu, D.S., Foschini, G., Gans, M., and Kahn, J., 'Fading correlation and its effect on the capacity of multielement antenna systems', *IEEE Transactions on Communications*, 2000, 48(3): 502–513.

[173] Abdi, A., and Kaveh, M., 'A space-time correlation model for multielement antenna systems in mobile fading channels', *IEEE Journal on Selected Areas in Communications*, 2002, 20(3): 550–560.

[174] Wu, S.-H., 'MIMO wireless communications', Lecture Notes, Department of Communication Engineering, NCTU, 2008, http://140.113.144.123/MIMOWC/MIMO%20Channel%20Modeling.pdf

[175] Anioke, C.L., Nnamani, C.O., and Ani, C.I., 'Review of wireless MIMO channel models', *Nigerian Journal of Technology*, 2016, 35(2): 381 – 391, https://www.ajol.info/index.php/njt/article/viewFile/140618/130358

[176] Jensen, M.A., and Wallace, J.W., *Space-time Processing for MIMO Communications*, New York, USA: John Wiley and Sons, 2005.

[177] Tsoulos, G., *MIMO System Technology for Wireless Communications*, Taylor and Francis: London, 2006.

[178] Pätzold, M., and Hogstad, B.O., 'A space-time channel simulator for MIMO channels based on the geometrical one-ring scattering model', *IEEE 60th Vehicular Technology Conference*, pp. 144–149, 26–29 September, 2004.

[179] Ozcelik, H., Czink, N., and Bonek E., 'What makes a good MIMO channel model?', *IEEE 61st Vehicular Technology Conference*, pp. 156–160, 30 May–1 June 2006.

[180] Kermoal, J.P., Schumacher, L., and Pedersen, K.I. 'A stochastic MIMO radio channel model with experimental validation', *IEEE Journal on Selected Areas in Communications,* 2002, 20(6): 1211–1226.

[181] Liberti, J.C., and Rappaport, T.S., 'A geometrically based model for line-of sight multipath radio channels', *IEEE 46th Vehicular Technology Conference*, pp. 844–848, 28 April–1 May 1996.

[182] Cheng, X., Wang, C.-X., Laurenson, D. I., Salous, S., and Vasilakos, A.V., 'An adaptive geometry-based stochastic model for non-isotropic MIMO mobile-to-mobile channels', *IEEE Transactions on Wireless Communications*, 2009, 8: 4824–4835.

[183] Veeravalli, V., Liang, Y., and Sayeed, A.M. 'Correlated MIMO wireless channels: capacity, optimal signalling and asymptotics', *IEEE Transactions on Information Theory*, 2005, 51(6): 2058–2072.

[184] Borhani, A., and Pätzold, M., 'A non-stationary one-ring scattering model', *IEEE Wireless Communications and Networking Conference*, Shanghai, China, 7–10 April 2013.

[185] Krajzewicz, D., Hertkorn, G., Rössel, C., and Wagner, P., 'SUMO (simulation of urban mobility)', *Proc. of the 4th Middle East Symposium on Simulation and Modelling*, pp. 183–187, 2002.

[186] Simulation of Vehicular Mobility SUMO Website, http://sumo.sourceforge.net/

[187] Haklay, M., and Weber, P., 'OpenStreetMap: user-generated street maps', *IEEE Pervasive Computing*, 2008, 7(4): 12–18.

[188] Open Street Map Project Website, https://www.openstreetmap.org

[189] NASA World Wind Project Website, https://worldwind.arc.nasa.gov/

[190] Guttman, A., 'R-trees: a dynamic index structure for spatial searching', *SIGMOD Rec.*, 1984, 14: 47–57.

[191] Boban, M., Vinhoza, T.T.V., Ferreira, M., Barros, J., and Tonguz, O.K., 'Impact of vehicles as obstacles in vehicular ad hoc networks', *IEEE Journal on Selected Areas in Communications*, 2011, 29(1): 15–28.

[192] WINNER Project Board, D5.4 v1.4 – Final Report on Link Level and System Level Channel Models, 18 November 2005.

[193] WINNER Project Board, D1.1.2 V1.2 – WINNER II Channel Models, 30 09 2007, https://www.cept.org/files/8339/winner2%20-%20final%20report.pdf

[194] WINNER Project Board, D5.3 – WINNER+ Final Channel Models, 30 June 2010, https://www.researchgate.net/publication/261467821_CP5-026_WINNER_D53_v10_WINNER_Final_Channel_Models

[195] 3GPP TR 37.885 (2018-05): 'Study on evaluation methodology of new Vehicle-to-Everything V2X use cases for LTE and NR'.

[196] Recommendation ITU-R P.526: 'Propagation by diffraction', International Telecommunication Union Radiocommunication Sector, Geneva, November 2013.

[197] Zajic, A.G., and Stüber, G.L., 'Space-time correlated mobile-to-mobile channels: modelling and simulation', *IEEE Trans. Vehic. Tech.*, 2008, 57 (2): 715–726.

[198] Matolak, D.W., 'Channel modeling for V2V communications', *Proc. 2nd Int. Workshop on Vehicle-to-Vehicle Communications, IEEE/ACM MobiQuitous*, San Jose, Calif., 21 July 2006.

[199] Wang, B., Sen, I., and Matolak, D.W., 'Performance evaluation of 802.16e in vehicle to vehicle channels', *Proc. IEEE Fall VTC*, Baltimore, MD, 1–3 Oct. 2007

[200] Cohen, L., *Time-frequency Analysis*, Prentice-Hall: Upper Saddle River, NJ, 1995.

[201] Matz, G., 'On non-WSSUS wireless fading channels', *IEEE Trans. Wireless Commun.*, 2005, 4(5): 2465–2478.

[202] Renaudin, O., Kolmonen, V.-M., Vainikainen, P., and Oestges, C., 'Non-stationary narrowband MIMO inter-vehicle channel characterization in the 5 GHz band', *IEEE Trans. Veh. Technol.*, 2010, 59(4): 2007–2015.

[203] Paier, A., Zemen, T., Bernadó, L., *et al.*, 'Non-WSSUS channel characterization in highway and urban scenarios at 5.2 GHz using the local scattering function', *Proc. Int. Workshop on Smart Antennas*, Helsinki, Finland, 26–27 February 2008.

[204] Pätzold, M., and Borhani, A., 'A non-stationary multipath fading channel model incorporating the effect of velocity variations of the mobile station'. *Proceedings of the Wireless Communication and Network Conference*, pp. 194–199, Istanbul, Turkey, 6–9 April 2014.

[205] Zemen, T., and Molisch, A.F, 'Adaptive reduced-rank estimation of non-stationary time-variant channels using subspace selection', *IEEE Trans. Veh. Technol.*, 2012, 61: 4042–4056.

[206] Chong, C.-C., Tan, C.-M., Laurenson, D. I., McLaughlin, S., Beach, M.A., and Nix, A.R., 'A novel wideband dynamic directional indoor channel model based on a Markov process', *IEEE Trans. Wireless Commun.*, 2005, 4(4): 1539–1552.

[207] Matolak, D.W., and Wu, Q., 'Markov models for vehicle-to-vehicle channel multipath persistence processes', *Proc. 1st IEEE Veh. Tech. Society Wireless Access in Veh. Env. (WAVE) Conf.*, Dearborn, MI, 8–9 December 2008.

[208] Paschalidis, P., Mahler, K., Kortke, A., Peter, M., and Keusgen, W., 'Statistical evaluation of multipath component lifetime in the car-to-car channel at urban street intersections based on geometrical tracking', *Proc. IEEE Spring Veh. Tech. Conf.*, Yokohama, Japan, 6–9 May, 2012.

[209] Paier, A., Bernadó, L., Karedal, J., Klemp, O., and Kwoczek, A., 'Overview of vehicle-to-vehicle radio channel measurements for collision avoidance applications', *Proc. IEEE Spring Veh. Tech. Conf.*, Taipei, Taiwan, 16–19 May 2010.

[210] Renaudin, O., Kolmonen, V.-M., Vainikainen, P., and Oestges, C., 'Wideband measurement-based modeling of inter-vehicle channels in the 5 GHz band', *IEEE Trans. Veh. Technol.*, 2013, 62(8): 3531–3540.

[211] Reichardt, L., Fugen, T., and Zwick, T., 'Influence of antennas placement on car to car communications channel', *Proc. European Conf. on Antennas & Propagation*', Berlin, Germany, 23–27 March.

[212] Schack, M., Kornek, D., Slottke, E., and Kuerner, T., 'Analysis of channel parameters for different antenna configurations in vehicular environments', *Proc. IEEE Fall Veh. Tech. Conf.*, Ottawa, Ont., Canada, 6–9 September 2010.

[213] Cheng, X., Yao, Q., Wang, C.-X., *et al.*, 'An improved parameter computation method for a MIMO V2V Rayleigh fading channel simulator under non-isotropic scattering environments', *IEEE Commun. Lett.*, 2013, 17(2): 265–268.

[214] Paschalidis, P., Mahler, K., Kortke, A., Wisotzki, M., Peter, M., and Keusgen, W., '2 X 2 MIMO measurements of the wideband car-to-car channel at 5.7 GHz on urban street intersections', *Proc. IEEE Fall VTC*, San Francisco, CA, 5–8 September 2011.

[215] Ohlmer, E., Fettweis, G., and Plettemeier, D., 'MIMO system design and field tests for terminals with confined space – impact on automotive communication', *Proc. European Conf. on Antennas & Propagation*, Rome, Italy, 11–15 April 2011.

[216] Nuckelt, J., and Kuerner, T., 'MRC performance benefit in V2V communication systems in urban traffic scenarios', *Proc. European Conf. on Antennas & Propagation*, Prague, Czech Republic, 26–30 March 2012.

[217] Du, D., Zeng, X., Jian, X., Yang, F., and Sun, M., '3-D V2V MIMO channel modeling in different roadway scenarios with moving scatterers', *Progress in Electromagnetics Research M*, 2018, 64: 43–54.

[218] Chelli, A., and Pätzold, M., 'The impact of fixed and moving scatterers on the statistics of MIMO vehicle-to-vehicle channels', *Proc. IEEE VTC Spring*, pp. 1–6, Barcelona, Spain, April 2009.

[219] Chelli, A., and Pätzold, M., 'A dynamic MIMO vehicle-to-vehicle channel model derived from the geometrical street model', Proc. IEEE VTC Fall, pp. 1–6, San Francisco, CA, USA, Sep. 2011.

[220] Liang, X., Zhao, X., Li, S., Wang, Q., and Li, J., 'A dynamic geometry-based scattering model for street vehicle-to-vehicle wideband MIMO channels', Proc. IEEE 26th Annual International PIMRC, 2239–2243, Hong Kong, Sep. 2015.

[221] Zhao, X., Liang, X., Li, S., and Ai, B., 'Two-cylinder and multi-ring GBSSM for realizing and modeling of vehicle-to-vehicle wideband MIMO channels', *IEEE Trans. Intell. Trans. Syst.*, 2016, 17(10): 2787–2799.

[222] Payami, S., and Tufvesson, F., 'Channel measurements and analysis for very large array systems at 2.6 GHz', *Proc. EuCAP*, pp. 433–437, Prague, Czech Republic, Mar. 2012.

[223] Gao, X., Tufvesson, F., Edfors, O., and Rusek, F., 'Measured propagation characteristics for very-large MIMO at 2.6 GHz', *Proc. ASILOMAR*, pp. 295–299, Pacific Grove, CA, USA, Nov. 2012.

[224] Wu, S., Wang, C.-X., Aggoune, E.-H.M., Alwakeel, M.M., and You, X., 'A general 3-D non-stationary 5G wireless channel model', *IEEE Trans. on Communications*, 2018, 66(7): 3065–3078.

[225] Study on 3D Channel Model for LTE, Version 12.2.0, document 3GPP T.R. 36.873, June 2015.

[226] Zhu, M., Eriksson, G., and Tufvesson, F., 'The COST 2100 channel model: parameterization and validation based on outdoor MIMO measurements at 300 MHz', *IEEE Trans. Wireless Commun.*, 2013, 12(2): 888–897.

[227] Verdone, R., and Zanella, A. (eds.), *Pervasive Mobile and Ambient Wireless Communications: COST Action 2100,* Springer: London, UK, 2012.

[228] Baum, D.S., Hansen, J., and Salo, J., 'An interim channel model for beyond-3G systems: Extending the 3GPP spatial channel model (SCM)', *Proc. IEEE VTC-Spring*, pp. 3132–3136, vol. 5. Stockholm, Sweden, May 2005.

[229] Nurmela, V., Karttunen, A., Roivainen, A., *et al.*, METIS Channel Models, ICT-317669-METIS DeliverableD1.4 V3, July 2015, https://www.metis2020.com/wp-content/uploads/METIS_D1.4_v3.pdf

[230] Study on Channel Model for Frequencies From 0.5 to 100 GHz, Version 14.2.0, Document 3GPP T.R. 38.901, September 2017.

[231] Wu, S., Wang, C.-X., Haas, H., Aggoune, E.-H.M., Alwakeel, M.M., and Ai, B., 'A non-stationary wideband channel model for massive MIMO communication systems', *IEEE Trans. Wireless Commun.*, 2015, 14(3): 1434–1446.

[232] Wang, C.-X., Wu, S., Bai, L., You, X., Wang, J., and I, C.-L., 'Recent advances and future challenges for massive MIMO channel measurements and models', *Sci. China Inf. Sci.*, 2016, 59(2): 1–16.

[233] Wu, S., Wang, C.-X., Aggoune, E.-H. M., Alwakeel, M.M., and He, Y., 'A non-stationary 3-D wideband twin-cluster model for 5G massive MIMO channels', *IEEE J. Sel. Areas Commun.*, 2014, 32(6): 1207–1218.

[234] Zwick, T., Fischer, C., Didascalou, D., and Wiesbeck, W., 'A stochastic spatial channel model based on wave-propagation modeling', *IEEE J. Sel. Areas Commun.*, 2000, 18(1): 6–15.

[235] Zwick, T., Fischer, C., and Wiesbeck, W., 'A stochastic multipath channel model including path directions for indoor environments', *IEEE J. Sel. Areas Commun.*, 2002, 20(6): 1178–1192.

[236] Pan, S., Durrani, S., and Bialkowski, M. E., 'MIMO capacity for spatial channel model scenarios', *8th Australian Communications Theory Workshop*, pp. 25–29, Adelaide, Australia, 31 January–2 February, 2007.

[237] Pan, S., 'Faculty of engineering and information technology', *Master Thesis*, 2006, Australian National University, https://pdfs.semanticscholar.org/cdc1/b48dcae7680648d9806f0885c3796fa0993b.pdf

[238] 3rd Generation Partnership Project (3GPP), 'Spatial channel model for multiple input multiple output (MIMO) simulations (3GPP TR 25.996 version 6.1.0 release 6)', ETSI, Technical Specification Group Radio Access Network, September 2003.

[239] Salo, J., Del Galdo, G., Salmi, J., *et al.*, 'MATLAB implementation of the 3GPP spatial channel model (3GPP TR 25.996)', Jan. 2005, http://www.tkk.fi/Units/Radio/scm/

[240] Mucchi, L., Staderini, C., Ylitalo, J., and Kyösti, P., 'Modified spatial channel model for MIMO wireless systems', *EURASIP Journal on Wireless Communications and Networking*, 2007, 2007, Article ID 68512, 7 pages, doi:10.1155/2007/68512

[241] Chen, X., Fang, Y., Xiang, W., and Zhou, L., 'Research on spatial channel model for vehicle-to-vehicle communication channel in roadside scattering environment', *International Journal of Antennas and Propagation*, 2007, 2007, Article ID 3098198, 12 pages, https://doi.org/10.1155/2017/3098198

[242] Al-Khalil, A.B., Turner, S., and Al-Sherbaz, A., 'Utilising SCM – MIMO channel model based on V-BLAST channel coding in V2V communication', In: *Communication Technologies for Vehicles*, vol. 9066 of *Lecture Notes in Computer Science*, pp. 3–11, Springer International Publishing, Sousse, Tunisia, 2015.

[243] Jaeckel, S., Borner, K., Thiele, L., and Jungnickel, V., 'A geometric polarization rotation model for the 3-D spatial channel model', *IEEE Transactions on Antennas and Propagation*, 2012, 60(12):5966–5977.

[244] Nawaz, S.J., Riaz, M., Khan, N.M., and Wyne, S., 'Temporal analysis of a 3D ellipsoid channel model for the vehicle-to-vehicle communication environments', *Wireless Personal Communications*, 2015, 82 (3): 1337–1350.

[245] Yuan, Y., Wang, C.-X., He, Y., Alwakeel, M. M., and Aggoune, E.-H. M., '3D wideband non-stationary geometry-based stochastic models for non-isotropic MIMO vehicle-to-vehicle channels', *IEEE Transactions on Wireless Communications*, 2015, 14(12): 6883–6895.

[246] Li, Y., He, R.S., Lin, S., *et al.*, 'Cluster-based non-stationary channel modeling for vehicle-to-vehicle communications', *IEEE Antennas and Wireless Propagation Letters*, 2017, 16: 408–411.

[247] Gesbert, D., Ekman, T., and Christophersen, N., 'Capacity limits of dense palm-sized MIMO arrays', *Proceedings of the IEEE Global Telecommunications Conference (GLOBECOM '02)*, vol. 2, pp. 1187–1191, Taipei, Taiwan, November 2002.

[248] Silva, C.M., Masini, B.M., Ferrari, G., and Thibault, I., 'A survey on infrastructure-based vehicular networks', *Mobile Information Systems*, 2007, 2007, Article ID 6123868, 28 pages, https://doi.org/10.1155/2017/6123868

[249] 3GPP, Study on LTE support for Vehicle to Everything (V2X) Services. TR22.885, 2016.

[250] 3GPP, Study on LTE-based V2X Services, TR36.885, 2016.

[251] Bazzi, A., Masini, B.M., and Zanella, A., 'Performance analysis of V2v beaconing using LTE in direct mode with full duplex radios', *IEEE Wireless Communications Letters*, 2015, 40(6): 685–688.

[252] Petrov, V., Fodor, G., Kokkoniemi, J., Moltchanov, D., and Lehtomäki, J., 'On unified vehicular communications and radar sensing in millimeter-wave and low terahertz bands', *IEEE Wireless Communications*, 2019, 26 (3): 146–153.

[253] Kwoczek, A., Rech, B., Hehn, T., *et al.*, '5G automotive vision', 5G Infrastructure Public Private Partnership (5G PPP), White Paper, Oct. 2015.

[254] Petrov, V., Lema, M.A., Gapeyenko, M., *et al.*, 'Achieving end-to-end reliability of mission-critical traffic in softwarized 5G networks', *IEEE J. on Sel. Areas in Commun.*, 2018, 36(3): 485–501.

[255] Chen, S., Hu, J., Shi, Y., *et al.*, 'Vehicle-to-everything (V2X) services supported by LTE-based systems and 5G', *IEEE Commun. Stand. Mag.*, 2017, 1(2): 70–76.

[256] Choi, J., Va, V., Gonzalez-Prelcic, N., Daniels, R., Bhat, C.R., and Heath, R.W., 'Millimeter-wave vehicular communication to support massive automotive sensing', *IEEE Commun. Mag.*, 2016, 54(12): 160–167.

[257] Kim, W., 'Experimental demonstration of mmWave vehicle-to-vehicle communications using IEEE 802.11ad', *Sensors*, 2019, 19: 2057–2079, doi:10.3390/s19092057.

[258] Kim, W., 'MC-GiV2V: multichannel allocation in mmWave-based vehicular ad hoc networks', *Wirel. Commun. Mob. Comput.*, 2018, 2018.

[259] 5G-PPP Project, 5G Automotive Vision, December 2015, https://5g-ppp.eu/wp-content/uploads/2014/02/5G-PPP-White-Paper-on-Automotive-Vertical-Sectors.pdf

[260] Masini, B.M., Bazzi, A., and Zanella, A., 'A survey on the roadmap to mandate on board connectivity and enable V2V-based vehicular sensor networks', *Sensors*, 18(7): 2207, 2018.

[261] Va, V., Shimizu, T., Bansal, G., and Heath, R.W., Jr., 'Millimeter wave vehicular communications: a survey', *Found. Trends Netw.*, 2016, 10: 1–113.

[262] Antonescu, B., Moayyed, M.T., and Basagni, S., 'mmWave channel propagation modeling for V2X communication systems', *Proceedings of the 2017 IEEE 28th Annual International Symposium on Personal, Indoor, and Mobile Radio Communications (PIMRC)*, pp. 1–6, Montreal, QC, Canada, 8–13 October 2017.

[263] Wang, Q., Matolak, D.W., and Ai, B., 'Shadowing characterization for 5 GHz vehicle-to-vehicle channels', *IEEE Trans. Veh. Technol.*, 2017, 67: 1855–1866.

[264] He, R., Ai, B., Stüber, G.L., Wang, G., and Zhong, Z., 'Geometrical-based modeling for millimeter-wave MIMO mobile-to-mobile channels', *IEEE Trans. Veh. Technol.*, 2018, 67: 2848–2863.

[265] Schneider, R., Didascalou, D., and Wiesbeck,W., 'Impact of road surfaces on millimeter-wave propagation', *IEEE Trans. Veh. Technol.*, 2000, 49: 1314–1320.

[266] Karasawa, Y., 'Multipath fading due to road surface reflection and fading reduction by means of space diversity in ITS vehicle-to-vehicle communications at 60 GHz', *Electron. Commun. Jpn. Part I Commun.*, 2002, 85: 35–42.

[267] He, R., Ai, B., Stuber, G.L., Wang, G., and Zhong, Z., 'A cluster based geometrical model for millimeter wave mobile-to-mobile channels', *Proceedings of the 2017 IEEE/CIC International Conference on Communications in China (ICCC)*, pp. 1–6, Qingdao, China, 22–24 October 2017.

[268] Tassi, A., Egan, M., Piechocki, R.J., and Nix, A. 'Modeling and design of millimeter-wave networks for highway vehicular communication', *IEEE Trans. Veh. Technol.*, 2017, 66: 10676–10691.

[269] Lorca, J., Hunukumbure, M., and Wang, Y., 'On overcoming the impact of Doppler spectrum in millimeter-wave V2I communications', *Proceedings of the 2017 IEEE GlobecomWorkshops (GC Wkshps)*, pp. 1–6, Singapore, 4–8 December 2017.

[270] Wang, Y., Venugopal, K., Molisch, A.F., and Heath, R.W., 'MmWave vehicle-to-infrastructure communication: analysis of urban microcellular networks', *IEEE Trans. Veh. Technol.*, 2018, 67: 7086–7100.

[271] Wang, Y., Venugopal, K., Molisch, A.F., and Heath, R.W., 'Blockage and coverage analysis with mmwave cross street BSs near urban intersections', *Proceedings of the 2017 IEEE International Conference on Communications (ICC)*, pp. 1–6, Paris, France, 21–25 May 2017.

[272] Eltayeb, M.E., Al-Naffouri, T.Y., and Heath, R.W., 'Compressive sensing for blockage detection in vehicular millimeter wave antenna arrays', *Proceedings of the 2016 IEEE Global Communications Conference (GLOBECOM)*, pp. 1–6, Washington, DC, USA, 4–8 December 2016.

[273] He, Z., Mao, S., Kompella, S., and Swami, A., 'On link scheduling in dual-hop 60-GHz mmWave networks', *IEEE Trans. Veh. Technol.*, 2017, 66, 11180–11192.

[274] Taya, A., Nishio, T., Morikura, M., and Yamamoto, K., 'Coverage expansion through dynamic relay vehicle deployment in mmWave V2I communications', *Proceedings of the 2018 IEEE 87th Vehicular Technology Conference (VTC Spring)*, pp. 1–5, Porto, Portugal, 3–6 June 2018.

[275] Va, V., Zhang, X., and Heath, R.W., 'Beam switching for millimeter wave communication to support high speed trains', *Proceedings of the 2015 IEEE 82nd Vehicular Technology Conference (VTC2015-Fall)*, pp. 1–5, Boston, MA, USA, 6–9 September 2015.

[276] Va, V., Shimizu, T., Bansal, G., and Heath, R.W., 'Beam design for beam switching based millimeter wave vehicle-to-infrastructure communications', *Proceedings of the 2016 IEEE International Conference on Communications*, pp. 1–6, Kuala Lumpur, Malaysia, 22–27 May 2016.

[277] Garcia, N., Wymeersch, H., Ström, E.G., and Slock, D., 'Location-aided mm-wave channel estimation for vehicular communication', *Proceedings of the 2016 IEEE 17th International Workshop on Signal Processing Advances in Wireless Communications*, pp. 1–5, Edinburgh, UK, 3–6 July 2016.

[278] Maschietti, F., Gesbert, D., de Kerret, P., and Wymeersch, H., 'Robust location-aided beam alignment in millimeter wave massive MIMO', *Proceedings of the GLOBECOM 2017 – 2017 IEEE Global Communications Conference*, pp. 1–6, Singapore, 4–8 December 2017.

[279] Va, V., Shimizu, T., Bansal, G., and Heath, R.W., 'Position-aided millimeter wave V2I beam alignment: a learning-to-rank approach', *Proceedings of the 2017 IEEE 28th Annual International Symposium on Personal, Indoor, and Mobile Radio Communications (PIMRC)*, pp. 1–5, Montreal, QC, Canada, 8–13 October 2017.

[280] Va, V., Choi, J., Shimizu, T., Bansal, G., and Heath, R.W., 'Inverse multipath fingerprinting for millimeter wave V2I beam alignment', *IEEE Trans. Veh. Technol.*, 2018, 67: 4042–4058.

[281] Wang, Y., Klautau, A., Ribero, M., Narasimha, M., and Heath, R.W., 'MmWave vehicular beam training with situational awareness by machine learning', *Proceedings of the 2018 IEEE Globecom Workshops (GC Wkshps)*, pp. 1–6, Abu Dhabi, UAE, 9–13 December 2018.

[282] Kumari, P., Choi, J., Prelcic, N.G., and Heath, R.W., 'IEEE 802.11 ad-based radar: an approach to joint vehicular communication-radar system', *IEEE Trans. Veh. Technol.*, 2017, 67: 3012–3027.

[283] González-Prelcic, N., Méndez-Rial, R., and Heath, R.W., 'Radar aided beam alignment in mmwave V2I communications supporting antenna diversity', *Proceedings of the 2016 Information Theory and Applications Workshop (ITA)*, pp. 1–7, La Jolla, CA, USA, 31 January–5 February 2016.

[284] Mavromatis, I., Tassi, A., Piechocki, R.J., and Nix, A., 'Beam alignment for millimetre wave links with motion prediction of autonomous vehicles',

2017, https://arxiv.org/pdf/1702.04264.pdf (accessed on 23 September 2019).

[285] Petrov, V., Kokkoniemi, J., Moltchanov, D., Lehtomäki, J., Juntti, M., and Koucheryavy, Y., 'The impact of interference from the side lanes on mmWave/THz band V2V communication systems with directional antennas', *IEEE Trans. Veh. Technol.*, 2018, 67: 5028–5041.

[286] Perfecto, C., Del Ser, J., and Bennis, M., 'On the interplay between scheduling interval and beamwidth selection for low-latency and reliable V2V mmWave communications', *Proceedings of the 2017 20th Conference on Innovations in Clouds, Internet and Networks (ICIN)*, pp. 1–8, Paris, France, 7–9 March 2017.

[287] Perfecto, C., Del Ser, J., and Bennis, M., 'Millimeter-wave V2V communications: distributed association and beam alignment', *IEEE J. Sel. Areas Commun.*, 2017, 35: 2148–2162.

[288] Va, V., Choi, J., and Heath, R.W., 'The impact of beamwidth on temporal channel variation in vehicular channels and its implications', *IEEE Trans. Veh. Technol.*, 2017, 66: 5014–5029.

[289] Giordani, M., Mezzavilla, M., Rangan, S., and Zorzi, M., 'Multi-connectivity in 5G mmWave cellular networks', *Proceedings of the 2016 Mediterranean Ad Hoc Networking Workshop (Med-Hoc-Net)*, pp. 1–7, Vilanova i la Geltru, Spain, 20–22 June 2016.

[290] Giordani, M., Mezzavilla, M., Rangan, S., and Zorzi, M., 'An efficient uplink multi-connectivity scheme for 5G millimeter-wave control plane applications', *IEEE Trans. Wirel. Commun.*, 2018, 17: 6806–6821.

[291] Loch, A., Asadi, A., Sim, G.H., Widmer, J., and Hollick, M., 'mm-Wave on wheels: practical 60 GHz vehicular communication without beam training', *Proceedings of the 2017 9th International Conference on Communication Systems and Networks (COMSNETS)*, pp. 1–8, Bengaluru, India, 4–8 January 2017.

[292] Park, J.J., Lee, J., Liang, J., Kim, K.W., Lee, K.C., and Kim, M.D., 'Millimeter wave vehicular blockage characteristics based on 28 GHz measurements', *Proceedings of the 2017 IEEE 86th Vehicular Technology Conference (VTC-Fall)*, pp. 1–5, Toronto, ON, Canada, 24–27 September 2017.

[293] Ben-Dor, E., Rappaport, T.S., Qiao, Y., and Lauffenburger, S.J., 'Millimeter-wave 60 GHz outdoor and vehicle AOA propagation measurements using a broadband channel sounder', *Proceedings of the 2011 IEEE Global Telecommunications Conference-GLOBECOM 2011*, pp. 1–6, Houston, TX, USA, 5–9 December 2011.

[294] Coll-Perales, B., Gruteser, M., and Gozalvez, J. 'Evaluation of IEEE 802.11 ad for mmWave V2V communications', *Proceedings of the Wireless Communications and Networking Conference Workshops (WCNCW)*, pp. 290–295, Barcelona, Spain, 15–18 April 2018.

[295] Eltayeb, M.E., Choi, J., Al-Naffouri, T.Y., and Heath, R.W., 'Enhancing secrecy with multiantenna transmission in millimeter wave vehicular communication systems', *IEEE Trans. Veh. Technol.*, 2017, 66: 8139–8151.

[296] Azpilicueta, F.G.L., Vargas-Rosales, C., Lopez-Iturri, P., *et al.*, 'Spatial characterization of radio propagation channel in urban vehicle-to-infrastructure environments to support WSNs deployment', *Sensors*, 2017, 17(6): 1313–1335.

[297] United States Department of Transportation (USDOT), Vehicle-to-Infrastructure (V2I) Communications for Safety, http://www.its.dot.gov/research_archives/safety/v2i_comm_safety.htm (accessed on October 2019).

[298] massachusetts Institute of Technology – News SMART Automation, http://news.mit.edu/2017/smart-automation-daniela-rus-0119 (accessed on October 2019).

[299] University of Michigan Transportation Research Institute-ITS Publications, http://www.umtri.umich.edu/our-focus/intelligent-transportation-systems (accessed on October 2019).

[300] U.S. Department of Transportation (USDOT), Vehicle-to-Infrastructure (V2I) Safety Applications. Concept of Operation Document, USDOT: Washington, DC, USA, 2013.

[301] Belyaev, E., Vinel, A., Surak, A., Gabbouj, M.. Jonsson, M., and Egiazarian, K., 'Robust vehicle-to-infrastructure video transmission for road surveillance applications', *IEEE Trans. Veh. Technol.*, 2014, 64: 1.

[302] Sommer, C., and Dressler, F., *Vehicular Networking*, Cambridge University Press: Cambridge, UK, 2015.

[303] Raut, S.B., and Malik, L.G., 'Survey on vehicle collision prediction in VANET', *Proceedings of the 2014 IEEE International Conference on Computational Intelligence and Computing Research (IEEE ICCIC)*, pp. 735–751, Tamilnadu, India, 18–20 December 2014.

[304] Lee, S., Yoon, D., and Ghosh, A., 'Intelligent parking lot application using wireless sensor networks', *Proceedings of the 2008 International Symposium on Collaborative Technologies and Systems*, pp. 48–57, Irvine, CA, USA, 19–23 May 2008.

[305] Milanes, V., Shladover, S.E., Spring, J., Nowakowski, C., Kawazoe, H., and Nakamura, M., 'Cooperative adaptive cruise control in real traffic situations', *IEEE Trans. Intell. Transp. Syst.*, 2014, 15: 296–305.

[306] ITU-R. Recommendation ITU-R P.1238–8: Propagation Data And Prediction Methods for the Planning of Indoor Radiocommunication systems and Radio Local Area Networks in the Frequency Range 300 MHz to 100 GHz, ITU-R: Geneva, Switzerland, 2015, p. 22.

[307] Cardote, A., Neves, F., Sargento, S., and Steenkiste, P., 'A statistical channel model for realistic simulation in VANET', *Proceedings of the 2012 IEEE Vehicular Networking Conference (VNC)*, pp. 48–55, Seoul, Korea, 14–16 November 2012.

[308] Fazio, P., Tropea, M., Sottile, C., and Lupia, A., 'Vehicular networking and channel modeling: a new Markovian approach', *Proceedings of the 12th Annual IEEE Consumer Communications & Networking Conference (CCNC)*, pp. 702–707, Las Vegas, NV, USA, 9–12 January 2015.

[309] Akhtar, N., Coleri Ergen, S., and Ozkasap, O., 'Vehicle mobility and communication channel models for realistic and efficient highway VANET simulation', *IEEE Trans. Veh. Technol.*, 2015, 64: 248–262.

[310] Nilsson, M., Gustafson, C., Abbas, T., and Tufvesson, F., 'A measurement based multilink shadowing model for V2V network simulations of highway scenarios', *IEEE Trans. Veh. Technol.*, 2017.

[311] Azpilicueta, L., López-Iturri, P., Aguirre, E., *et al.*, 'Evaluation of deployment challenges of wireless sensor networks at signalized intersections', *Sensors*, 2016, 16.

[312] Souley, A.H., and Cherkaoui, S., 'Realistic urban scenarios simulation for ad hoc networks', *Proceedings of the Second International Conference on Innovations in Information Technology*, pp. 1–10, Dubai, UAE, 26–28 September 2005.

[313] Rodriguez, I., Almeida, E., Lauridsen, M., *et al.*, 'Measurement-based evaluation of the impact of large vehicle shadowing on V2X communications', *Proceedings of the European Wireless Conference*, pp. 142–149, Oulu, Finland, 18–20 May 2016.

[314] Shemshaki, M., Lasser, G., Ekiz, L., and Mecklenbrauker, C., 'Empirical path loss model fit from measurements from a vehicle-to-infrastructure network in Munich at 5.9 GHz', *Proceedings of the IEEE International Symposium on Personal, Indoor, and Mobile Radio Communications (PIMRC)*, pp. 181–185, Hong Kong, China, 30 August–2 September 2015.

[315] Wang, Z., Tameh, E., and Nix, A., 'Statistical peer-to-peer channel models for outdoor urban environments at 2GHz and 5GHz', *Proceedings of the IEEE Vehicular Technology Conference, VTC2004-Fall 2004*, vol. 10, pp. 5101–5105, Los Angeles, CA, USA, 26–29 September 2004.

[316] Campolo, C., Cozzetti, H.A., Molinaro, A., and Scopigno, R., 'Vehicular connectivity in urban scenarios: effectiveness and potential of roadside, moving WAVE providers and hybrid solutions', *EURASIP J. Wirel. Commun. Netw.*, 2011, 1: 10.

[317] Azpilicueta, L., Rawat, M., Rawat, K., Ghannouchi, F., and Falcone, F., 'Convergence analysis in deterministic 3D ray launching radio channel estimation in complex environments', *Appl. Comput. Electromagn. Soc. J.*, 2014, 29: 256–271.

[318] Azpilicueta, L., Vargas-Rosales, C., and Falcone, F., 'Deterministic propagation prediction in transportation systems', *IEEE Veh. Technol. Mag.*, 2016, 11: 29–37.

[319] Molisch, A.F., and Tufvesson, A., 'Propagation channel models for next-generation wireless communications systems', *IEICE Trans. Commun.*, 2014, E97–B(10): 2022–.

[320] Mondal, B., Thomas, T.A., Visotsky, E., *et al.*, '3D channel model in 3GPP', *IEEE Communications Magazine*, 2015, 53(3): 16–23.

[321] Cardona, N., *Cooperative Radio Communications for Green Smart Environments*, River Publishers Series in Communications, Gistrup, Denmark: River Publishers, 2016.

[322] F. Rusek, Persson, D., Lau, B.K., *et al.*, 'Scaling up MIMO: opportunities and challenges with very large arrays', *IEEE Signal Process. Mag.*, 2013, 30(1): 40–60.

[323] Gao, X., Tufvesson, F., and Edfors, O., 'Massive MIMO channels – measurements and models', *47th Annual Asilomar Conference on Signals, Systems, and Computers*, November 2013.

[324] Fuhl, J., Rossi, J.-P., and Bonek, E., 'High-resolution 3-D direction-of-arrival determination for urban mobile radio', *IEEE Trans. Antennas Propag.*, 1997, 45(4): 672–682.

[325] Kalliola, K., Sulonen, K., Laitinen, H., Kivelas, O., Krogerus, J., and Vainikainen, P., 'Angular power distribution and mean effective gain of mobile antenna in different propagation environments', *IEEE Trans. Veh. Technol.*, 2002, 51(5): 823–838.

[326] Jiang, H., Zhang, Z., Dang, J., and Wu, L., 'A Novel 3D massive MIMO channel model for vehicle-to-vehicle communication environments', *IEEE Transactions on Communications*, 2018, 66(1): 79 – 90.

[327] Nam, Y.-H., Ng, B.L., Sayana, K., *et al.*, 'Full-dimension MIMO (FD-MIMO) for next generation cellular technology', *IEEE Communications Magazine*, 2013, 51(6).

[328] 3GPP 2008 TS 36.300 v8.7.0, Evolved Universal Terrestrial Radio Access (EUTRA) and Evolved Universal Terrestrial Radio Access Network (EUTRAN), Overall Description, Stage 2.

[329] 3GPP 2009 TR 36.814 v9.0.0, Further Advancements for Evolved Universal Terrestrial Radio Access (EUTRA) Physical Layer Aspects (Release 9).

[330] Beres, E., and Adve, R., 'Selection cooperation in multi-source cooperative networks', *IEEE Transactions on Wireless Communications*, 2008, 7(1): 118–127.

[331] Boldi, M., Botella, C., Boccardi, F., *et al.*, 'D1.8 intermediate report on CoMP', Technical Report, Wireless World Initiative New Radio (WINNER+) Document Identifier: Protocols, Measurements and Signalling, 2009.

[332] Soma, P., Baum, D., Erceg, V., Krishnamoorthy, R., and Paulraj, A., 'Analysis and modeling of multiple-input multiple-output (MIMO) radio channel based on outdoor measurements conducted at 2.5 GHz for fixed BWA applications', *Communications, 2002. ICC 2002. IEEE International Conference on*, vol. 1, pp. 272–276, 2002.

[333] Ahumada, L., Feick, R., Valenzuela, R., and Morales, C., 'Measurement and characterization of the temporal behavior of fixed wireless links', *IEEE Transactions on Vehicular Technology*, 2005, 54(6): 1913–1922.

[334] Kaltenberger, F., Gesbert, D., Knopp, R., and Kountouris, M., 'Correlation and capacity of measured multi-user MIMO channels', *Personal, Indoor and Mobile Radio Communications, 2008. PIMRC 2008. IEEE 19th International Symposium on*, pp. 1–5, IEEE, 2008.

[335] Zhang, J., Dong, D., Nie, X., Liu, G., and Dong, W., 'Propagation characteristics of wideband relay channels in urban environment', *Proceedings of The 3rd International Workshop on Broadband MIMO Channel Measurement and Modeling*, August 25, pp. 1–6, 2009.
[336] 'IEEE standard for local and metropolitan area networks-part 15.7: short-range wireless optical communication using visible light', *IEEE Std*, pp. 1–309, IEEE, 2011.
[337] 3GPP, Service Requirements for V2X Services, TS 22.185, 2016.
[338] Qualcomm, Connecting Vehicles to Everything with C-V2X, 2017, https://www.qualcomm.com/invention/5g/cellular-v2x
[339] Europa Press, 'Connected car technology: cellular V2X outperforms DSRC/ITS-G5 in comprehensive tests as mobility industry moves towards', 9 November 2018, https://www.europapress.es/comunicados/internacional-00907/noticia-comunicado-connected-car-technology-cellular-v2x-outperforms-dsrc-its-g5-in-comprehensive-tests-as-mobility-industry-moves-towards-20181109151200.html
[340] 5G Automotive Association, TR P-180106, 'V2X functional and performance test report; test procedures and results', March 2018, https://5gaa.org/wp-content/uploads/2018/11/P-180106-V2X-Functional-and-Performance-Test-Report_Final_051118.pdf
[341] Kallfass, I., Boes, F., Messinger, T., *et al.*, '64 Gbit/s transmission over 850 m fixed wireless link at 240 GHz carrier frequency', *J. of Infrared, Millim., and Ter. Waves*, 2015, 36(2): 221–233.
[342] Jasteh, D., Gashinova, M., Hoare, E.G., Tran, T.-Y., Clarke, N., and Cherniakov, M., 'Low-THz imaging radar for outdoor applications', *Proc. of IRS*, June 2015.
[343] González-Domínguez, M., Bao, D., Dadín, S., Cuiñas, I., and García-Sánchez, M., 'Measuring in-situ reflection coefficients due to road pavements at 5.9 GHz', *IET Electronics Letters*, 2016, 52(25): 2072–2074.
[344] Richter, A., 'RIMAX – a flexible algorithm for channel parameter estimation from channel sounding measurements', Technical Report TD-04-045, COST 273, Athens, Greece.
[345] Richter, A., Landmann, M., and Thomä, R.S., 'Maximum likelihood channel parameter estimation from multidimensional channel sounding measurements', *Proceedings of the 57th IEEE Semiannual Vehicular Technology Conference (VTC)*, vol. 2, pp. 1056–1060, 2003.
[346] Schmidt, R.O., 'Multiple emitter location and signal parameter estimation', *IEEE Transactions on Antennas and Propagation*, 1986, AP-34(3): 276–280.
[347] Roy, R., and Kailath, T., 'ESPRIT – estimation of signal parameters via rotational invariance techniques', *IEEE Transactions on Acoustics, Speech, and Signal Processing*, 1989, 37(7): 984–995.
[348] Moon, T., 'The expectation-maximization algorithm', *IEEE Signal Processing Magazine*, 1997, 13(6): 47–60.

[349] Fessler, J.A., and Hero, A.O., 'Space-alternating generalized expectation-maximization algorithm', *IEEE Transactions on Signal Processing*, 42(10): 2664–2677.

[350] Fleury, B.H., Tschudin, M., Heddergott, R., Dahlhaus, D., and Pedersen, K. L., 'Channel parameter estimation in mobile radio environments using the SAGE algorithm', *IEEE Journal on Selected Areas in Communications*, 17 (3): 434–450.

[351] Fleury, B.H., Yin, X., Rohbrandt, K.G., Jourdan, P., and Stucki, A., 'Performance of a high-resolution scheme for joint estimation of delay and bidirection dispersion in the radio channel', *Proceedings of the IEEE Vehicular Technology Conference (VTC-Spring)*, vol. 1, pp. 522–526, 2002.

[352] Heneda, K., Takada, J., and Kobayashi, T., 'Double directional ultra wideband channel characterization in a line-of-sight home environment' *IEICE – Transactions on Fundamentals of Electronics, Communications and Computer Sciences*, 2005, E88(9): 2264–2271.

[353] Zwick, T., Hampicke, D., Richter, A., Sommerkorn, G., Thoma, R., and Wiesbeck, W., 'A novel antenna concept for double-directional channel measurements', *IEEE Transactions on Vehicular Technology*, 2004, 53(2): 527–537.

[354] Dawood, M., Quraishi, N., and Alejos, A.V., 'Super-resolution velocity estimation technique for multiple moving targets within a range bin of a UWB random noise radar', *IEEE Transactions on Aerospace and Electronic Systems*, 2013, 49(1): 325 – 340,.

[355] Sun, R., Matolak, D.W., and Liu, P., 'Parking garage channel characteristics at 5 GHz for V2V applications', *IEEE 78th Vehicular Technology Conference (VTC Fall)*, pp. 1–5, 2013.

[356] Fayziyev, A., Pätzold, M., Masson, E., Cocheril, Y., and Berbineau, M., 'A measurement-based channel model for vehicular communications in tunnels', *IEEE Wireless Communications and Network Conference (WCNC)*, pp. 128–133, 2014.

[357] Mangel, T., Klemp, O., and Hartenstein, H., 'A validated 5.9 GHz nonline-of-sight path-loss and fading model for inter-vehicle communication', *11th International Conference on ITS Telecommunications (ITST)*, pp. 75 –80, August 2011.

[358] Pilosu, L., Fileppo, F., and Scopigno, R., 'RADII: a computationally affordable method to summarize urban ray-tracing data for VANETs', *International Conference on Wireless Communications, Networking and Mobile Computing (WiCOM)*, pp. 1–6, 2011.

[359] Kunisch, J., and Pamp, J., 'Wideband car-to-car radio channel measurements and model at 5.9 GHz', *Proc. IEEE Fall Veh. Tech. Conf.*, Calgary, Alberta, Canada, 21–24 September 2008.

[360] Renaudin, O., Kolmonen, V.-M., Vainikainen, P., and Oestges, C., 'Wideband MIMO car-to-car radio channel measurements at 5.3 GHz', *Proc. IEEE Fall Veh. Tech. Conf.*, Calgary, Alberta, Canada, 21–24 September 2008.

[361] Paschalidis, P., Wisotzki, M., Kortke, A., Peter, M., and Keusgen, W., 'Wideband car-to-car MIMO radio channel measurements at 5.7 GHz and issues concerning application-oriented systems', *Proc. 1st IEEE Veh. Tech. Society Wireless Access in Veh. Env. (WAVE) Conf.*, Dearborn, MI, 8–9 December 2008.

[362] Bernadó, L., Roma, A., Paier, A., *et al.*, 'In-Tunnel vehicular radio channel characterization', *Proc. IEEE Spring VTC*, Budapest, Hungary, 15–18 May 2011.

[363] Jiang, H., Zhang, Z., Wu, L., and Dang, J., 'A non-stationary geometry-based scattering vehicle-to-vehicle MIMO channel model', *Communications Letters IEEE*, 2018, 22(7): 1510–1513.

[364] Jiang, H., Zhang, Z., and Gui, G., 'Three-dimensional non-stationary wideband geometry-based UAV channel model for A2G communication environments', *Access IEEE*, 2019, 7: 26116–26122.

[365] He, R., Ai, B., Stüber, G.L., and Zhong, Z., 'Non-stationary mobile-to-mobile channel modeling using the Gauss-Markov mobility model', *Wireless Communications and Signal Processing (WCSP) 2017 9th International Conference on*, pp. 1–6, 2017.

[366] Saunders, S.R., and Aragón-Zavala, A., *Antennas and Propagation for Wireless Communication Systems*, 2nd ed., Wiley: West Sussex, UK, 2007.

[367] Preliminary Draft New Report ITU-R M. [IMT-2020.EVAL], Channel Model (Part), Version 1, Document ITU-R R15-WP5D-170613-TD-0332, June 2017.

Chapter 7
Wireless communication system optimization

In this chapter, we integrate the propagation models with channel models to carry out analysis of link budget, coverage and network design that are used for network design, planning, dimensioning and optimization. Different types of vehicular channels (intra-vehicular and inter-vehicular) and environments will be considered, in order to gain insight on the impact in coverage/capacity relations.

7.1 Radio planning for vehicular environments

Wireless communication systems provide inherent advantages owing to high levels of mobility and ubiquity, given by radio channel propagation conditions. In this sense, radio channels provide flexibility but are also vulnerable, due to large potential losses in transmitter to receiver link balance, rapid signal variation and impact of different interference sources. In parallel, wireless communication systems are evolving in order to adapt to more restrictive specifications in terms of increasing simultaneous active connections, higher transmission bit rates and lower transmission delays. In the specific case of vehicular communications, further requirements are imposed in terms of maximum tolerable delay and overall system availability. These parameters have direct impact on quality of service (QoS) and quality of experience (QoE) values, which can be further detailed to specific key performance indicators (KPIs) of the system. These KPIs provide information in terms of network access conditions, level of mobility, delay or achievable transmission bit rate, among others. As a function of the requirements of the network, planning and optimization tasks must be performed in order to deploy the network and, as function of KPI analysis, to optimize network configuration so it can meet the foreseen operation specifications. In this way, the number of network nodes, their location and configuration are determined for initial deployment and the network is subsequently modified with progressive and continuous optimization. A schematic description of the network planning and optimization process is depicted in Figure 7.1.

One of the fundamental aspects in network planning and optimization is to establish coverage/capacity relations, which determine maximum transmitter–receiver distance ranges as a function of receiver sensitivity thresholds. These coverage/capacity relations in case of wireless systems are going to be determined by link path loss (as a function of distance and surrounding environment),

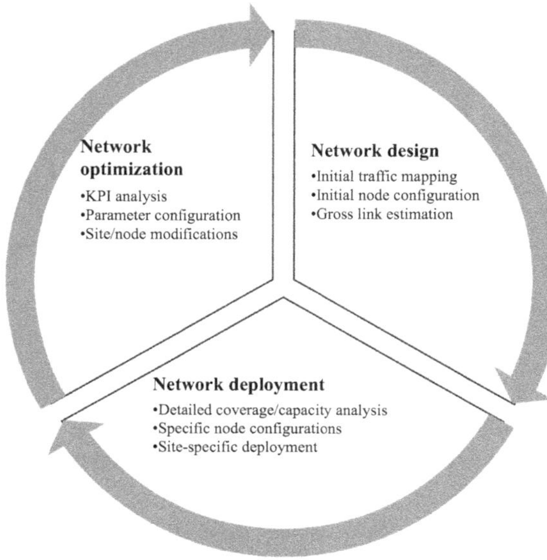

Figure 7.1 Schematic description of network planning and optimization process

variations owing to mobility (with greater contribution from fast-fading components) and interference (given by user/traffic distribution and characteristics). Moreover, variations in receiver sensitivity levels are given by the use of adaptive modulation and coding schemes, in addition to other radio link-level functionalities, such as handovers or diversity schemes. Multiple system operation and the use of cooperative networking provide the grounds for heterogenous network operation, in which different coverage/capacity relations are employed in order to maximize QoS/QoE parameters.

Radio network planning and optimization tasks are multi-goal optimization problems, in which different elements such as range extension, energy consumption, access rate, transmission rate and delay can be optimized. Multiple approaches have been proposed in order to provide location of base stations or relay stations [1–7]. Different case studies and conditions have also been considered, such as use of multiple input multiple output (MIMO) systems, network optimization supported by call trace information or the consideration of ultra-dense network operation [8–27]. More recently, wireless network planning has taken advantage of elements such as spatial characterization, machine learning optimization or the consideration of ultra-dense node density, typical within Internet of Things (IoT) and machine-type communications. Vehicular environments pose additional challenges in terms or network planning and optimization tasks, given by high channel variability, presence of vehicles, specific low-delay requirements and large bandwidth requirements [28–30]. In this sense, channel characterization towards accurate coverage/capacity estimations is compulsory in order to provide adequate estimations for node range, location and configuration.

In this chapter, radio network analysis results will be obtained for different types of network conditions, considering intra-vehicular as well as inter-vehicular communication links. Different communication systems have been taken into account for both cases, in order to consider variable coverage/capacity conditions. The parameters employed are given in Table 7.1 (intra-vehicular case) and Tables 7.2 and 7.3 (inter-vehicular case), employing deterministic three-dimensional (3D) ray-launching (3D-RL) algorithm within urban scenario definitions previously presented in Chapters 4 and 5. Simulation parameters have been set up in order to optimize the performance of the 3D-RL algorithm, in order to reduce computational complexity. Results have been obtained for the complete simulation volume, which enables to explore transceiver locations in any infrastructure or vehicular element of interest within the scenario.

Views of the scenario under analysis are depicted in Figure 7.2, considering full topological complexity within a complex urban scenario. Different locations for vehicle as well as infrastructure transceivers have been considered and are

Table 7.1 3D-RL simulation parameters for intra-vehicular case

System	Parameters	Values
Bluetooth Class 2	Transmitter power	4 dBm
	Frequency / Bit rate	2.4 GHz / 1 Mb/s
	Antenna radiation pattern (RX, TX) / Gain	Omnidirectional / 0 dB
	3D-RL: Angular resolution / Reflections / Diffraction / Cuboids size	1 degree / 6 / Yes / 0.1 m
	Receiver sensitivity	−82 dBm
ZigBee 868	Transmitter power	13 dBm
	Frequency / Bit rate	868 MHz / 80 kb/s–
	Antenna radiation pattern (RX, TX) / Gain	Omnidirectional / 0 dB
	3D-RL: Angular resolution / Reflections / Diffraction / Cuboids size	1 degree / 6 / Yes / 0.1 m
	Receiver sensitivity	−106 dBm
ZigBee 915	Transmitter power	13 dBm
	Frequency / Bit rate	915 MHz / 250 kb/s
	Antenna radiation pattern (RX, TX) / Gain	Omnidirectional / 0 dB
	3D-RL: Angular resolution / Reflections / Diffraction / Cuboids size	1 degree / 6 / Yes / 0.1 m
	Receiver sensitivity	−103 dBm
ZigBee 2400	Transmitter power	8 dBm
	Frequency / Bit rate	2.4 GHz / 250 kb/s
	Antenna radiation pattern (RX, TX) / Gain	Omnidirectional / 0 dB
	3D-RL: Angular resolution / Reflections / Diffraction / Cuboids size	1 degree / 6 / Yes / 0.1 m
	Receiver sensitivity	−100 dBm

Table 7.2 3D-RL simulation parameters for the I2V link

I2V link	Parameters	Values
802.11p	Transmitter power / Position	0 dBm/street light
	Frequency / Bit rate	5.9 GHz/1–27 Mb/s
	Antenna radiation pattern (RX, TX) / Gain	Omnidirectional/0 dB
	3D-RL: Angular resolution / Reflections / Diffraction / Cuboids size	1 degree/6/Yes/1 m
	Receiver sensitivity	−100 dBm
5G-FR1	Transmitter power / Position	0 dBm /street light
	Frequency / Bit rate	3.7 GHz / 1 Mb/s
	Antenna radiation pattern (RX, TX) / Gain	Omnidirectional / 0 dB
	3D-RL: Angular resolution / Reflections / Diffraction / Cuboids size	1 degree / 6 / Yes / 1 m
	Receiver sensitivity	−100 dBm

Table 7.3 3D-RL simulation parameters for the V2V link

V2V link	Parameters	Values
802.11p	Transmitter power / Position	0 dBm / Rooftop vehicle
	Frequency / Bit rate	5.9 GHz / 27 Mb/s
	Antenna radiation pattern (RX, TX) / Gain	Omnidirectional / 0 dB
	3D-RL: Angular resolution / Reflections / Diffraction / Cuboids size	1 degree / 6 / Yes / 1 m
	Receiver sensitivity	−100 dBm

detailed in Table 7.4. Two different conditions have been taken into account: urban scenario and suburban scenario, in which clutter in terms of building density has been reduced.

7.2 Coverage/capacity inter-vehicular analysis

In order to perform coverage/capacity estimations, received power-level estimations have been obtained for both types of vehicular links. The simulation scenario configuration is depicted in Figure 7.2, considering an average vehicle density of 1 vehicle/528 m^2, corresponding to 50 vehicles in the scenario under test. Results have been obtained considering infrastructure to vehicle (I2V) communication links, from a transmitter located within a lamp post in the vicinity of the street termed as AV-2, for the cases of 802.11p as well as for future 5G systems operating within FR1 (below 6 GHz spectrum allocation). The comparison of linear distribution of received power levels for both systems is presented in Figure 7.3. The

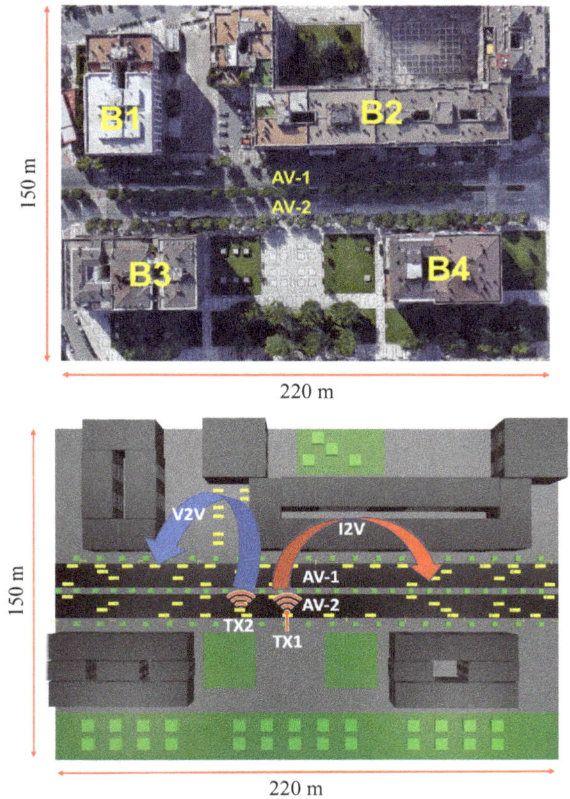

Figure 7.2 Urban scenario: Google Map aerial view (top); 2D-map schematic aerial view with the transmitter's positions (down)

Table 7.4 Scenario reference

Reference	Abbreviation	Coordinates (x, y, z) m
Main avenues	AV-1/AV-2	(x, 83, 0)/(x, 69, 0)
Street light transmitter	TX1	(101.5, 63, 3.5)
Car transmitter	TX2	(80.5, 67.25, 1.5)
Buildings	B1, B2, B3, B4	Not applicable

results exhibit higher received power levels in case of 5G New Radio (5G-NR), owing to higher transmit power as compared to the case of 802.11p. In general, power levels are above sensitivity levels in excess of 20 dB. In case of 802.11p, there is one specific location in which values are within thresholds limits, given by higher shadowing. In general, values in case of AV-2 are higher than those in case of AV-1, given by the fact that line-of-sight conditions are given more frequently in

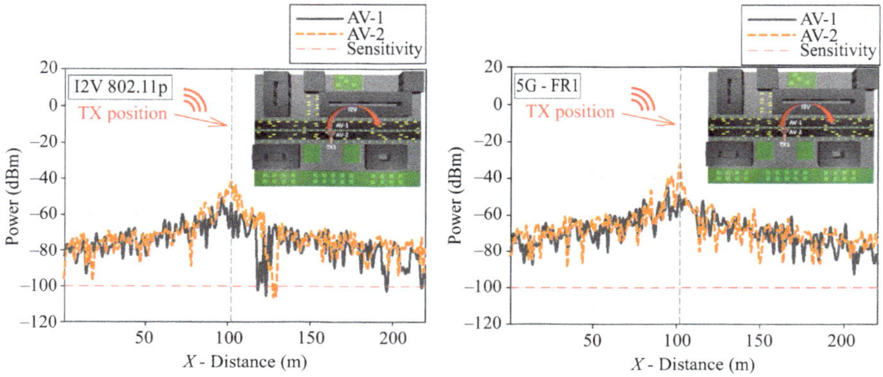

Figure 7.3 Comparison of linear distribution of received power levels for the different considered systems for the I2V link within the urban environment, in case of 802.11p and 5G FR1

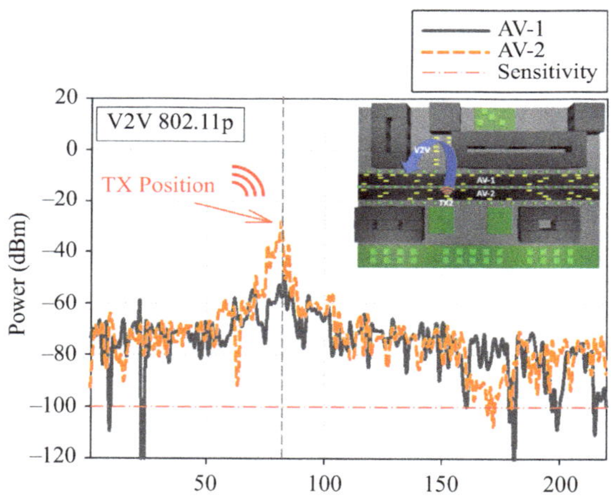

Figure 7.4 Linear distribution of V2V link received power, urban environment

that case, since static transmitter positions are closer to the vehicles within AV-2 than in case of AV-1.

Estimations of received power levels have also been obtained for the case of vehicle to vehicle (V2V) links, following the network schematic depicted in Figure 7.2. The results for the case of the urban environment scenario are shown in Figure 7.4. It can be seen that within an average distance of 70 m within the location of the transmitter, received power levels are well above sensitivity thresholds. Specific points exhibit sharp fading, owing mainly to non-line-of-sight conditions and out-of-phase conditions in multipath components.

In order to gain insight in terms of impact on QoS given by the variations in received power levels inherent to vehicular channel conditions, bit error rate (BER) estimations have been obtained as a function of transceiver location, for different linear radials within the scenario. Interference conditions have been established by providing background noise spectral power density values of −100 dBm/Hz. Modulation has been set to 16 quadrature amplitude modulation (QAM), with an initial transmission speed of 3 Mb/s. Results for I2V as well as for V2V links in case of 802.11p are depicted in Figure 7.5. As expected, BER values decrease as received power levels increase, a condition that can clearly be seen in the vicinity of the transmitter locations. In case of adaptive modulation and coding schemes,

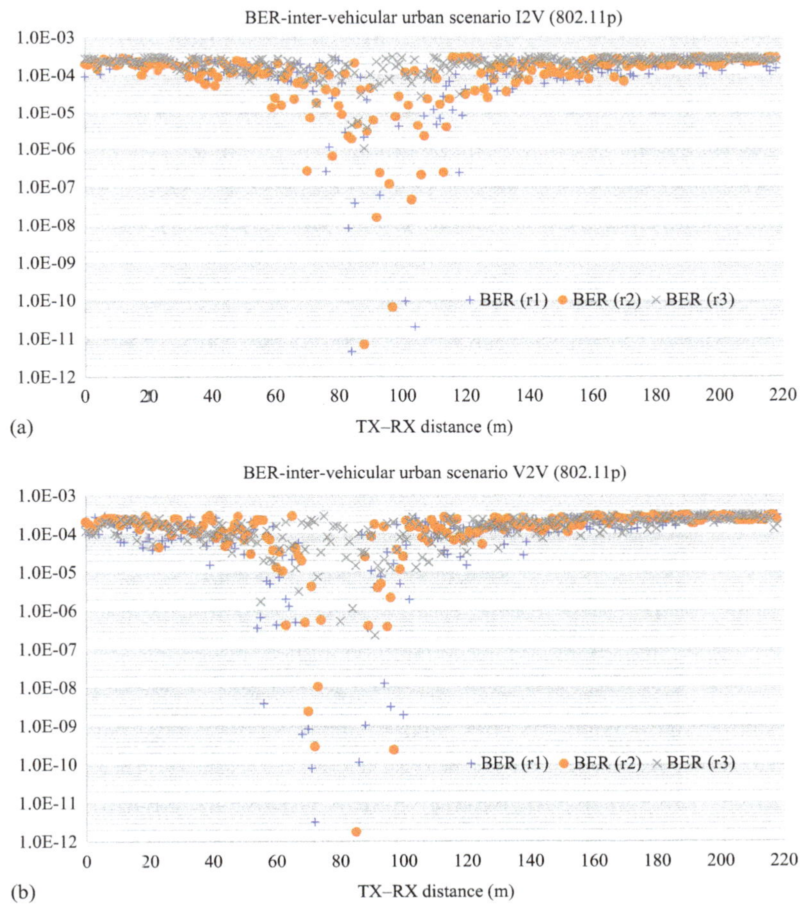

Figure 7.5 Estimation of BER for different radial distributions in case of urban scenario conditions employing 802.11p standard (a) I2V communication links (b) V2V communication links

further reduction requirements in BER will lead to lowering modulation order, as we will describe further in Section 7.3.

The distribution of received power levels has also been obtained for I2V and V2V links in case of suburban scenario conditions, depicted in Figure 7.6. The results for I2V links are depicted in Figure 7.7, for the case of 802.11p and 5G-FR1 systems. As in the previous case, received power levels are in general above receiver sensitivity with a power excess margin above 20 dB for the edge of the simulation scenario. Received power levels are higher in case of 5G-FR1, owing to lower operation frequency and configuration setup. Fading dips in received power levels can be observed at specific distances (10, 120, 155 and 200 m). owing to phase-counter phase summation in the distribution of multipath components in the scenario. Received power levels are in average higher again for the case of AV-2 in relation with AV-1, given again by the fact that line-of-sight conditions have a higher probability of occurrence.

Simulation results for inter-vehicular links for V2V communications in case of suburban scenario are depicted in Figure 7.8. The results follow a similar trend in the case of the results previously shown in Figure 7.4, with slightly higher average values in received power levels, owing to lower clutter levels in the scenario. In any case, it is worth noting that fast-fading components are very relevant in this scenario, given by the fact that the presence of vehicles (location of transceiver antennas and metallic structure) plays a key role in propagation conditions of V2V communication links.

The impact on QoS metrics can be seen with the estimation of BER values as a function of transceiver location. The results, depicted in Figure 7.9, have been obtained for three different linear radials within the scenario, for both I2V and V2V communication links. The simulations consider background noise levels of

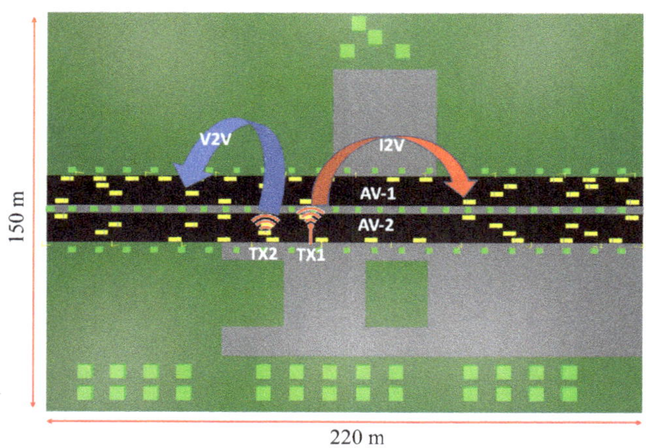

Figure 7.6 Suburban scenario: 2D-map schematic aerial view with the transmitters positions

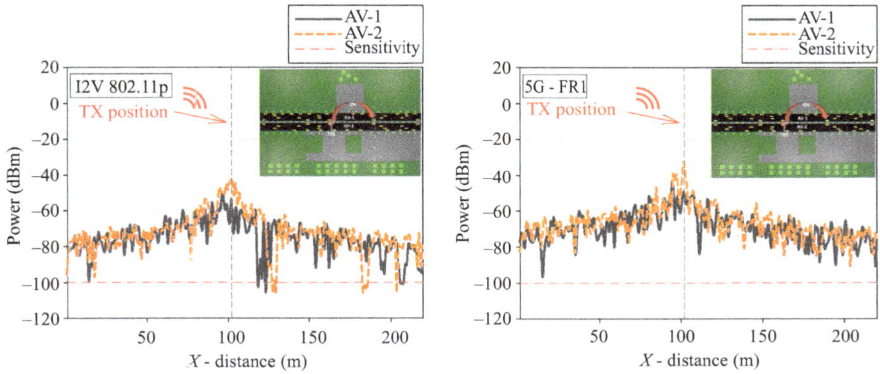

Figure 7.7 Comparison of linear distribution of received power levels for the different considered systems for the I2V link within the suburban environment, in case of 802.11p and 5G-FR1

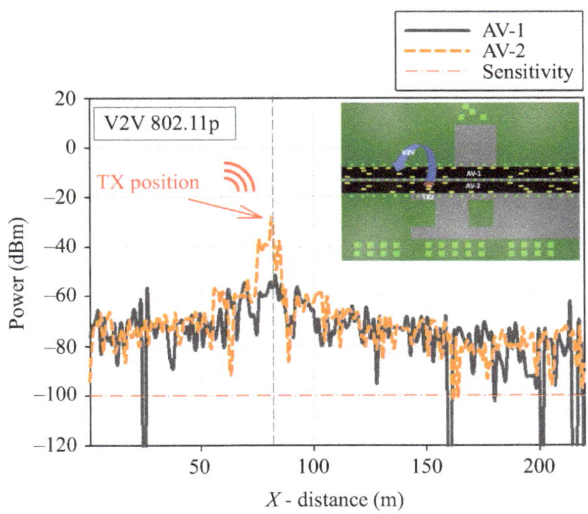

Figure 7.8 Linear distribution of received power 802.11p V2V link within suburban environment conditions

−100 dBm/Hz, with a 16-QAM modulation scheme, showing the BER values are lower in the vicinity of the transmitter node, a direct consequence of higher signal to noise ratios in those locations. BER values are in average lower in case of V2V links, an expected result considering that overall received power levels are higher given more obstruction and non-line-of-sight propagation conditions as compared with the I2V case.

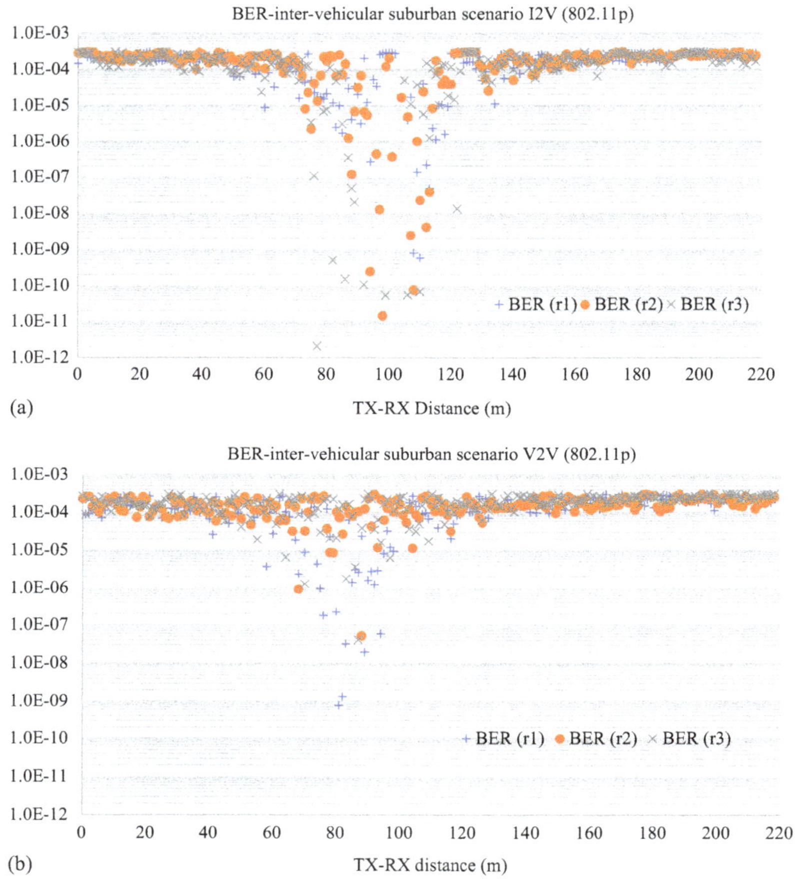

Figure 7.9 Estimation of BER for different radial distributions in case of suburban scenario conditions employing 802.11p standard (a) I2V communication links and (b) V2V communication links

Variations in coverage/capacity relations are given mainly by changes in fading conditions, as well as by an increase in interference levels. This has a direct impact on the values of signal-to-noise ratios (SNR), which in turn impacts maximum achievable transmission speed, given by receiver sensitivity thresholds. In case of vehicular communications, the specific conditions given by the presence of multiple scatterers (vehicles, surrounding buildings, infrastructure elements or vegetation, among others) and vehicular dynamics imposes higher constraints to communication system performance. In order to gain insight in the impact of the scenario in wireless links, SNR distributions for urban as well as suburban scenario are presented. The results have been obtained by deterministic

3D-RL simulations, considering the full complexity in terms of volume and frequency dispersive material parameters of the elements in the scenarios depicted in Figures 7.2 and 7.6. The results have been obtained for all potential transceiver locations, although they have been particularized to specific cut planes for the sake of clarity. Figure 7.10 shows the results for SNR distribution in case of urban scenario conditions, for I2V communication 802.11p links. Variable interference conditions have been considered, with background interferences spectral power density values of −70, −90 and −110 dBm/Hz, uniformly distributed, considering maximum likelihood reception and additive white Gaussian noise. As it can be seen, as a function of losses for the transceiver location and background interference conditions, SNR values tend to improve in the vicinity of the transmitter, with general values higher in case of the road, owing to higher percentage of line-of-sight conditions. Similar results have been obtained in case of I2V communication 5G-FR1 communication links. As in the 802.11p, different background noise conditions have been considered, with spectral power density values of −70, −90 and −110 dBm/Hz, uniformly distributed, considering maximum likelihood reception and additive white Gaussian noise. The results for each one of these interference conditions is depicted in Figure 7.11, showing higher values of SNR in the scenario, as a consequence of higher average received power levels previously described.

The simulation results have been obtained for the complete simulation scenario. In this way, the impact of changes in transceiver height location can also be addressed. Different vertical planes have been extracted, specifically for the middle position of roads AV-1 and AV-2, following the schematic depicted in Figure 7.12. The results for the both 802.11p and 5G-FR1 are depicted in Figures 7.13 and 7.14, respectively. As a function of the corresponding height, values of SNR vary, with higher values in the vicinity of the transmitter location. It is worth noting that relevant variations, in the order of 5–10 dB, occur as a function of spatial shifts in the location of the transceivers, which can play a relevant role in the election of potential locations for infrastructure nodes.

Inter-vehicular V2V communication links have also been considered, in which the values of background interference spectral density levels vary from −70, −90 and −110 dBm/Hz. The results for 2D SNR distributions for a constant height cut plane at 1.5 m is depicted in Figure 7.15, whereas vertical planes for the middle position of AV-1 and AV-2 are depicted in Figure 7.16. It can be observed from the results for horizontal as well as for vertical planes that the relative position of the transceivers as well as the corresponding background interference levels greatly influence SNR distributions and hence, the expected system performance. It is worth noting that the proposed methodology is flexible in terms of locating multiple transmitting point sources, as well as in applying modifications in radiation diagrams of the transmitting and receiving sources. In this way, it is possible to consider different operation conditions, such as the use of sector-type antennas or the impact of beamforming techniques.

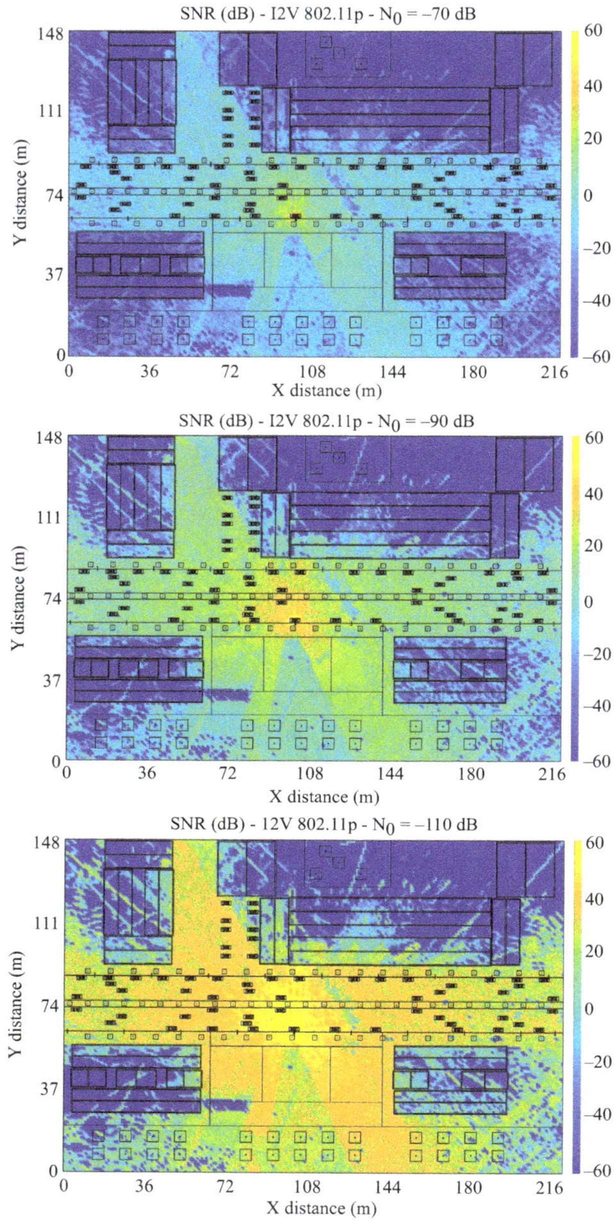

Figure 7.10 2D distribution of SNR for horizontal cut planes at a height of h = 1.5 m. Results have been obtained considering variable background noise levels, of −70, −90 and −110 dBm/Hz, for the case of I2V 802.11p communication links

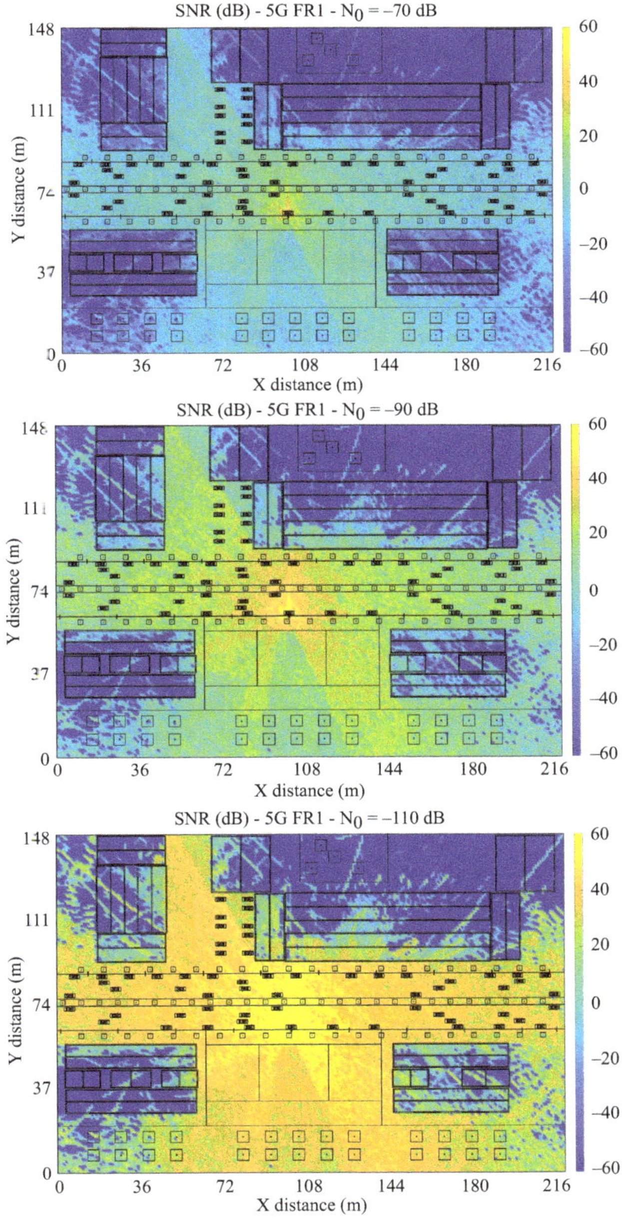

Figure 7.11 2D distribution of SNR for horizontal cut planes at a height of $h = 1.5$ m. Results have been obtained considering variable background noise levels, of −70, −90 and −110 dBm/Hz, for the case of I2V 5G-FR1 communication links

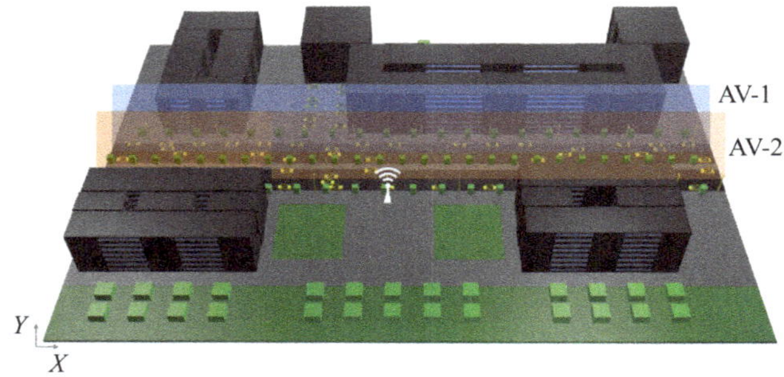

Figure 7.12 Vertical planes representation in urban scenario with transmitter position in the street light

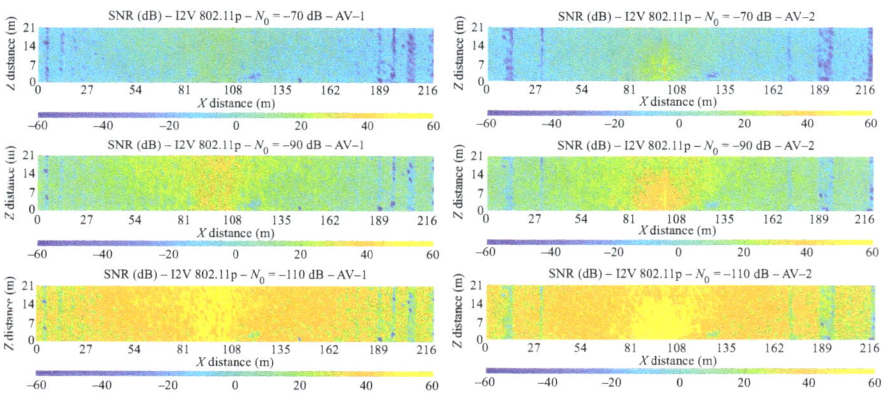

Figure 7.13 2D distribution of SNR for vertical cut planes located at the middle positions of AV-1 and AV-2. Results have been obtained considering variable background noise levels, of −70, −90 and −110 dBm/Hz, for the case of I2V 802.11p communication links

7.3 Coverage/capacity intra-vehicular analysis

In order to provide vehicular context-aware environments, specific intra-vehicle communication networks can be deployed, mainly based on the use of wireless sensor network standards and wireless local area network standards. In this case, it's compulsory to consider the specific characteristics inside the vehicle, such as interior morphology and the materials employed in vehicle construction. Moreover, by applying full volumetric simulation techniques, adequate transceiver location can be determined, in order to increase coverage/capacity relations.

To gain insight in intra-vehicular communication characteristics, different vehicle types and wireless systems have been simulated by means of 3D-RL

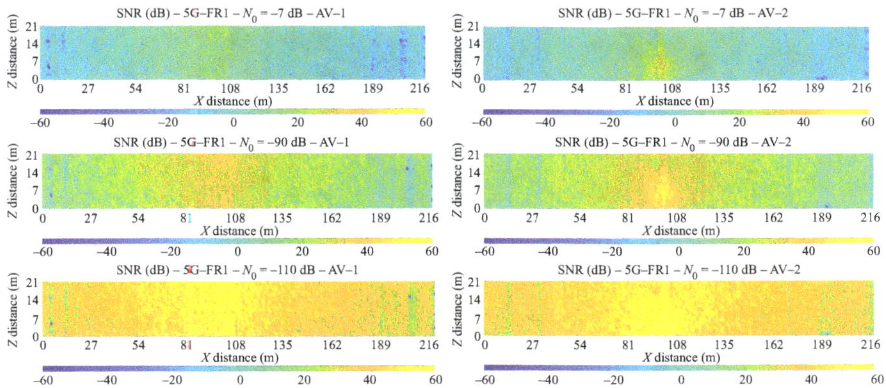

Figure 7.14 2D distribution of SNR for vertical cut planes located at the middle positions of AV-1 and AV-2. Results have been obtained considering variable background noise levels, of −70, −90 and −110 dBm/Hz, for the case of I2V 5G FR1 communication links

technique. Full vehicle models have been implemented, considering the specific characteristics of their indoor environment, such as location of seats, handrails, doors, windows or dashboards, among others. Wireless systems taken into consideration are ZigBee (@868 MHz, @915 MHz, @2.4 GHz) and Bluetooth.

The first vehicle analysed is a conventional sedan, in which an on-board transmitter is located within the upper part of the dashboard. The full vehicle has been recreated, including two human body models in the front seats. The implemented model as well as the linear received power radials are shown in Figure 7.17. The results for received power levels for each one of the linear radials, located at relative $Y = 1.2$ m and $Y = 2.8$ m are represented in Figure 7.18 for ZigBee as well as Bluetooth Class 2 devices.

The results show that there are relevant losses in received power levels in the region below 1 m, corresponding to the external part of the vehicle, from the front to the dashboard. The results for both radials are also different, which is given by the fact that the scenario isn't symmetric, providing different distribution of scatterers and elements that obstruct the wireless link. The vehicle model also considers non-visible elements, such as the motor block, which is more distributed on the driver side of the implemented vehicle model. In terms of received power levels, systems operating at 868 and 915 MHz exhibit higher values compared with both ZigBee and Bluetooth at 2.4 GHz, as expected from the use of lower frequency bands. Within the vehicle cockpit, received power levels are above receiver sensitivity in excess of 40 dB in case of ZigBee and in the order of 20 dB in case of Bluetooth.

The same vehicular sedan environment has been simulated, considering in this case that occupancy is increased, with two users in the front part of the vehicle and two users in the rear part, as schematically depicted in Figure 7.19. Results for

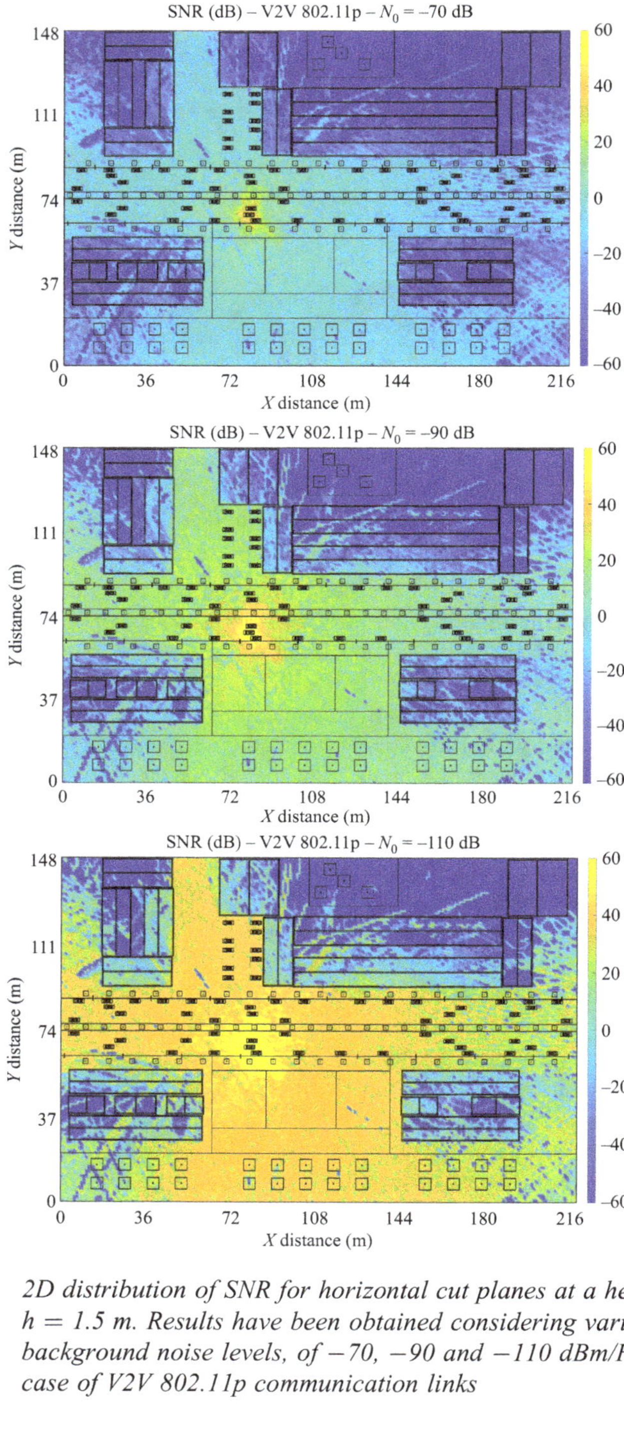

Figure 7.15 2D distribution of SNR for horizontal cut planes at a height of h = 1.5 m. Results have been obtained considering variable background noise levels, of −70, −90 and −110 dBm/Hz, for the case of V2V 802.11p communication links

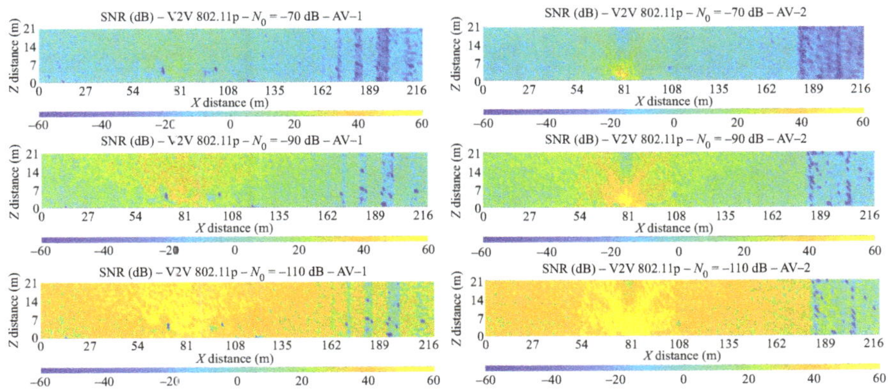

Figure 7.16 2D distribution of SNR for vertical cut planes located at the middle positions of AV-1 and AV-2. Results have been obtained considering variable background noise levels, of −70, −90 and −110 dBm/Hz, for the case of V2V 802.11p communication links

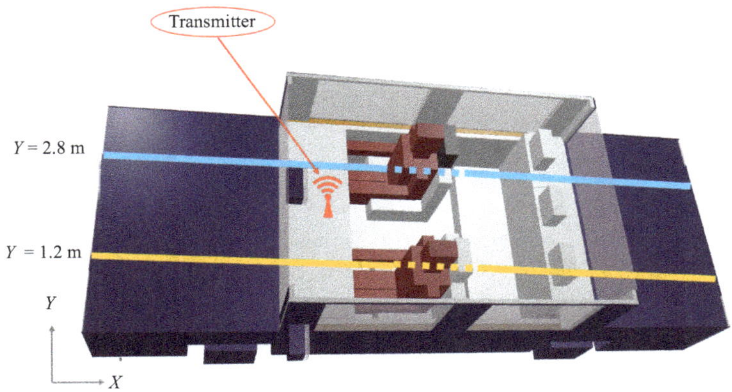

Figure 7.17 Simulation model for a conventional sedan, including two human body models inside. Radial lines considered for the car with two people located in the front seats are depicted for positions Y = 1.2 m and Y = 2.8 m

radials $Y = 1.2$ m and $Y = 2.8$ m for Bluetooth and ZigBee devices are shown in Figure 7.20. The results follow a similar trend as in case of Figure 7.18, with slightly higher losses owing to higher user density within the vehicle cockpit.

The next scenario that has been simulated corresponds to urban transport buses. Two different models of buses, rigid (single cabin) and articulated (dual cabin), have been considered. The implemented simulation models are depicted in Figure 7.21. In both cases, the radial line considered is located in the central

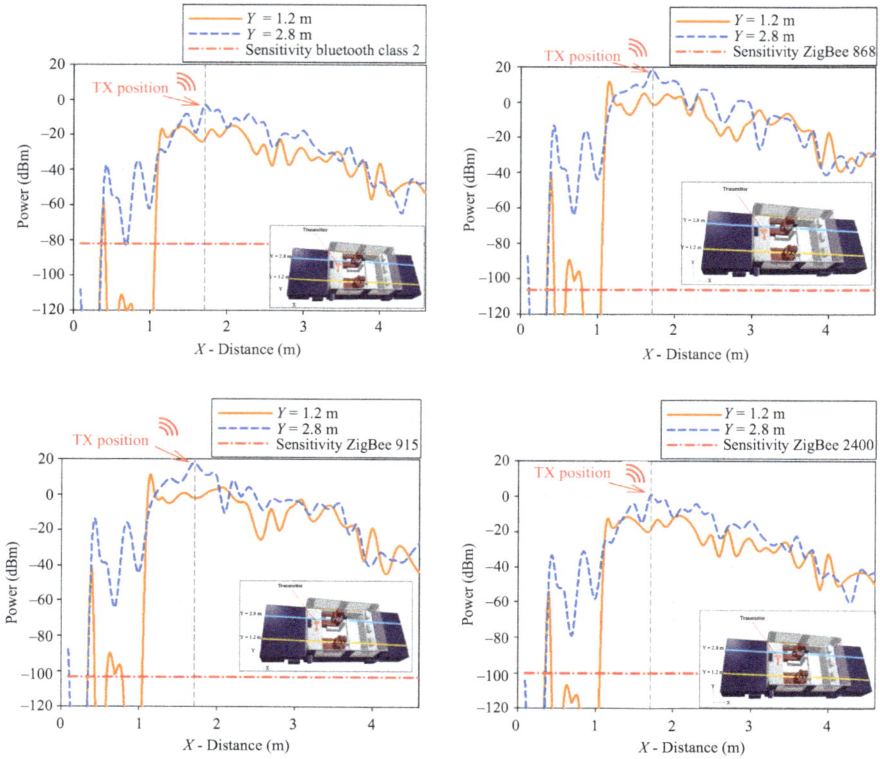

Figure 7.18 Estimation of received power levels within the sedan simulation scenario, with two users in the front seats, for Bluetooth Class 2 devices and ZigBee @ 868 MHz, @ 915 MHz and @2.4 GHz

position of the bus, corresponding to relative coordinate $Y = 1.4$ m. Human body models have been included inside the buses, considering low-density and high-density occupancy conditions, as previously described in Chapter 4. The simulation results for ZigBee and Bluetooth devices and the different user density conditions are depicted in Figure 7.22 for the rigid bus and in Figure 7.23 for the articulated. It can be observed that user density plays a key role in the distribution of received power levels within the bus environment, as well as the bus structure. In case of high user density, non-line-of-sight propagation conditions is more predominant, increasing shadowing and, hence, path loss values. This gives rise to locations within the bus in which coverage/capacity relations can't be satisfied, hence QoS will be degraded. The difference between low user density and high user density in terms of path loss estimations exceeds 40 dB in these points, which clearly indicates the need to consider the presence of users and perform radio planning analysis in order to deploy the network nodes considering this worst-case scenario. In case of the articulated bus, losses are higher, owing to larger vehicle dimensions and the

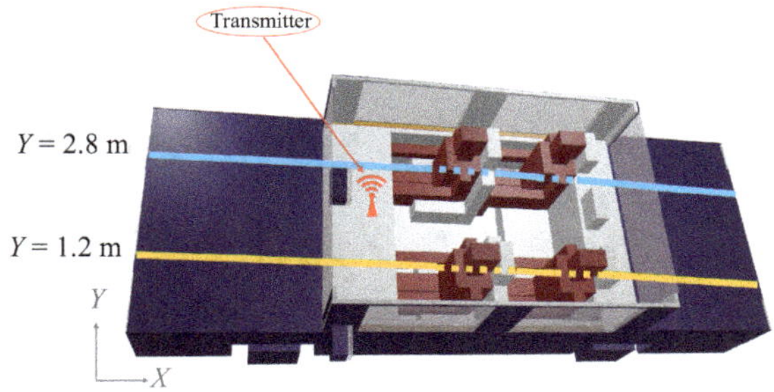

Figure 7.19 *Simulation model for a conventional sedan, including two human body models inside. Radial lines considered for the car with four people, located in the front and rear seats, are depicted for positions Y = 1.2 m and Y = 2.8 m*

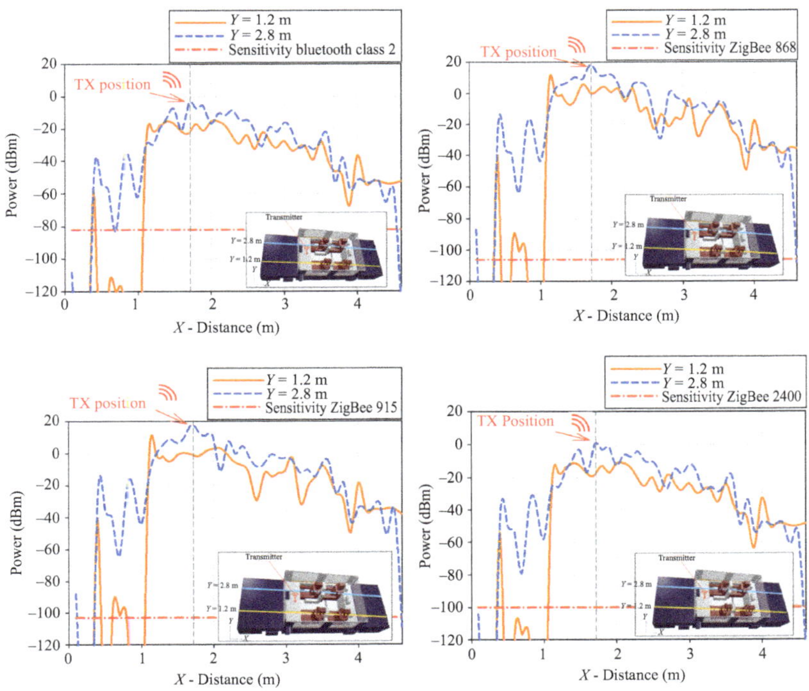

Figure 7.20 *Estimation of received power levels within the sedan simulation scenario, with four users (two in the front seats and two in the rear seats), for Bluetooth Class 2 devices and ZigBee @ 868 MHz, @ 915 MHz and @2.4 GHz*

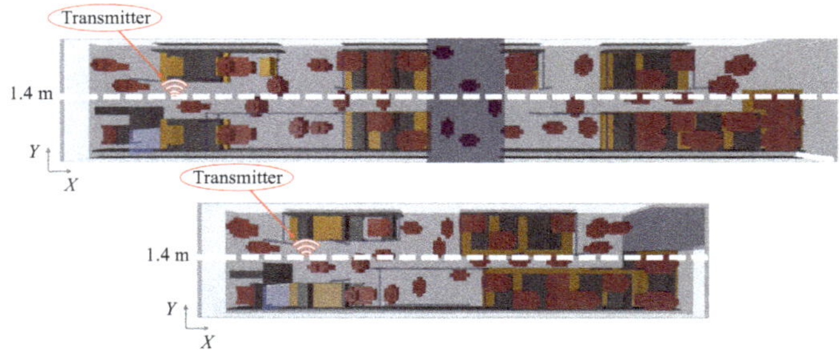

Figure 7.21 Simulation models for the articulated (top) and rigid (bottom) bus. The radial lines considered in both cases is Y = 1.4 m. The images depict the conditions for high density of users within the bus

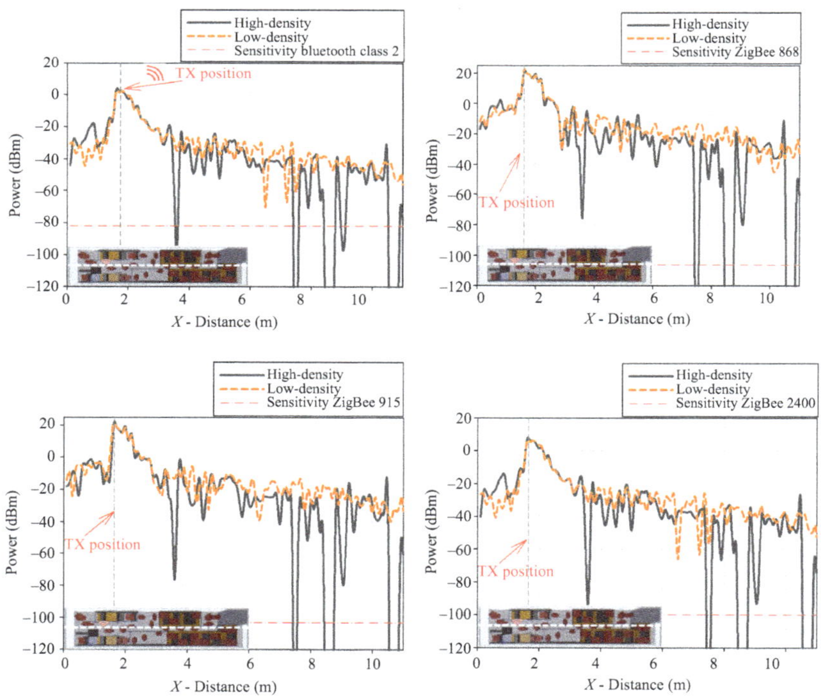

Figure 7.22 Linear distribution of received power for the rigid bus and comparison of passengers' densities, for Bluetooth Class 2 devices and ZigBee @ 868 MHz, @ 915 MHz and @2.4 GHz. The high-density and low-density user distribution follows the same distribution as described in Figure 4.8 in Chapter 4

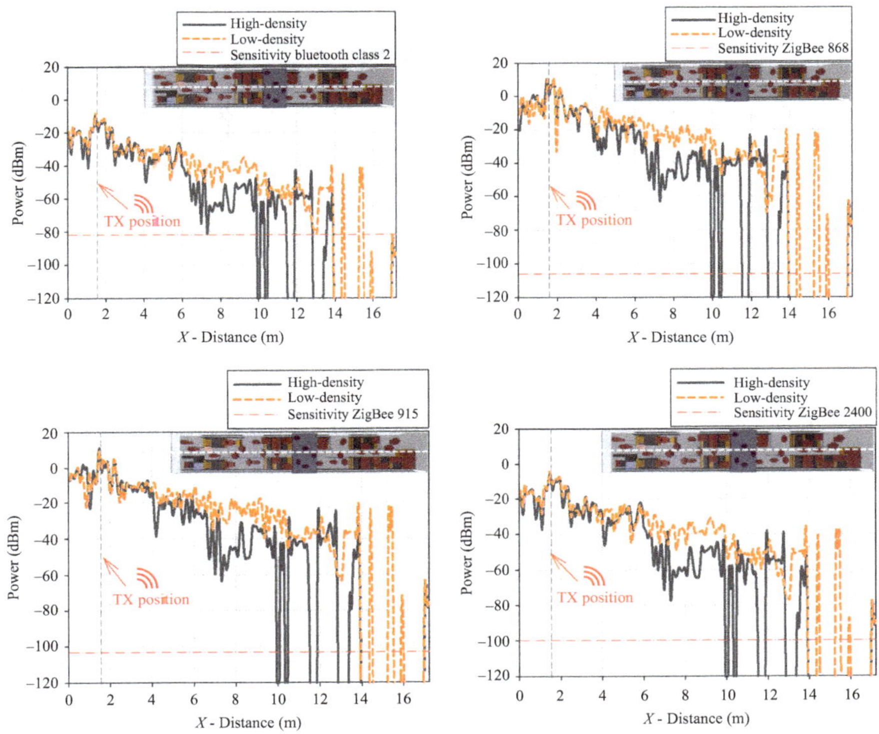

Figure 7.23 Linear distribution of received power for the articulated bus and comparison of passengers' densities, for Bluetooth Class 2 devices and ZigBee @ 868 MHz, @ 915 MHz and @2.4 GHz. The high-density and low-density user distribution follows the same distribution as described in Figure 4.8 in Chapter 4

presence of a larger number of scatterers within the scenario. In this case, the mid-rear part of the bus exhibits multiple regions in which coverage/capacity relations, considering a unique transmitter, are not fulfilled. A distributed wireless sensor network approach could then be adopted, which can inherently be implemented for the wireless systems that have been considered. Cell densification can be another option, in case that mesh networking isn't enabled, locating more than one transmit device within the scenario.

Another interesting communication environment to consider is given by train wagons, which can provide services for long-distance travel, as well as regional/urban-oriented services, such as suburban trains, subways or trams. In order to analyse these environments, a train wagon model with full detail has been modelled, with the schematic shown in Figure 7.24. In this case, the

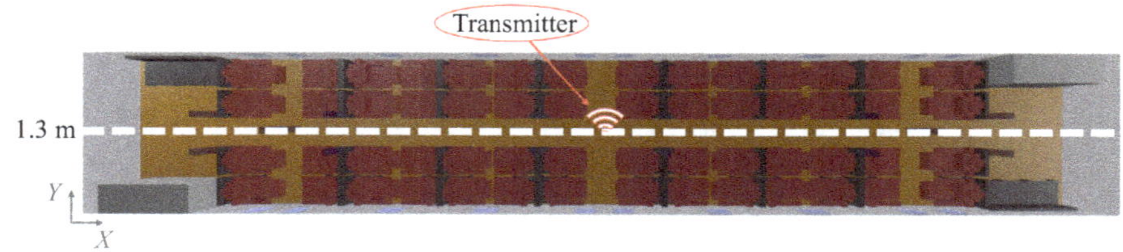

Figure 7.24 Radial lines considered for the train wagon car. The radial line considered is Y = 1.4 m. The image depicts the conditions for high density of users within the train wagon car

transmitter has been located in the centre of the train wagon at the wagon ceiling, considering the best-case approximation in terms of line-of-sight link conditions. As in the previous cases, ZigBee devices operating at frequencies of 868 MHz, 915 MHz and 2.4 GHz as well as Bluetooth Class 2 transmitters have been considered. The simulation results for these devices, considering once again low-density and high-density user occupation are depicted in Figure 7.25. It can be seen in this case that user density has a much lower impact than in case of the urban buses. This is given by the fact that users travel in general in seated positions, which in general provide line-of-sight conditions with the transmitter device if located within the train ceiling. This consideration would be very different in case of evaluating inter-user communication by means of body area networks, in which the presence of users, once again, would have a very relevant impact on path-loss estimation.

7.4 Considerations for inter-system operation

Cooperative communications can be enabled by different mechanisms, mainly based on traffic offload between systems, given by intra-system or inter-system handovers. In this way, as a function of coverage/capacity conditions, traffic can be adequately exchanged between systems, especially in critical operating conditions, given at cell edges or by highly variable fading. The corresponding coverage ranges are determined by receiver sensitivity thresholds, which vary as a function of adaptive multi-rate and coding parameters, which in a general way state that increasing transmission speed will reduce receiver sensitivity and, hence, coverage radius [1,2,30].

The resulting values of BER as a function of receiver locations for linear radials within the scenario under analysis depicted in Figure 7.2 have been obtained considering variations in the employed transmission rate. In case of 802.11p, transmission rates have been varied from 3 to 9 to 18 to 27 Mb/s, considering orthogonal frequency division multiplexing (QFDM) QAM modulation scheme. The results for the case of I2V and V2V communication 802.11p links for the urban

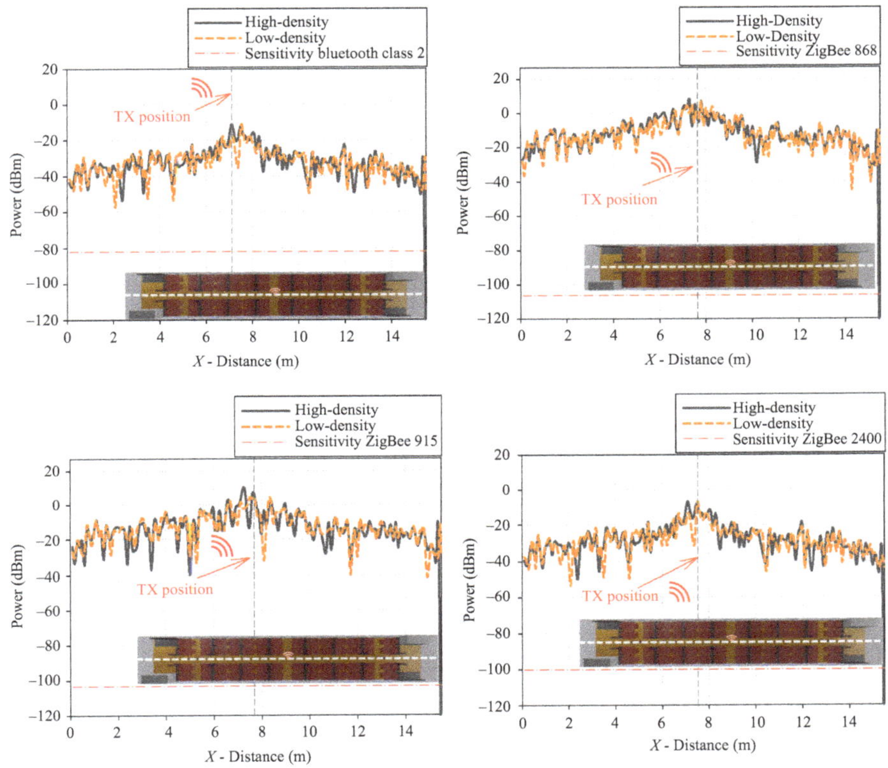

Figure 7.25 Linear distribution of received power for the train wagon and comparison of passengers' densities, for Bluetooth Class 2 devices and ZigBee @ 868 MHz, @ 915 MHz and @2.4 GHz. The high-density and low-density user distribution follows the same distribution as described in Figure 4.16 in Chapter 4

scenario are depicted in Figure 7.26. It can be observed that by increasing the requirements in transmission rate, BER values increase, with lower values in the coverage edges, owing to a decrease in received power levels. As expected I2V BER values are lower, owing to higher average received power levels, given by higher line-of-sight link conditions in I2V case as compared to V2V.

The same analysis has been performed for I2V and V2V communication links in case of suburban scenarios, with the results depicted in Figure 7.27. The same trend can be observed comparing I2V with V2V communication links, in which BER values are in general lower in case of I2V conditions. The values in general are lower in comparison with urban values, as expected by higher average received values, especially in case of I2V, as the density of scatterers is lower with increased link visibility and lower obstruction.

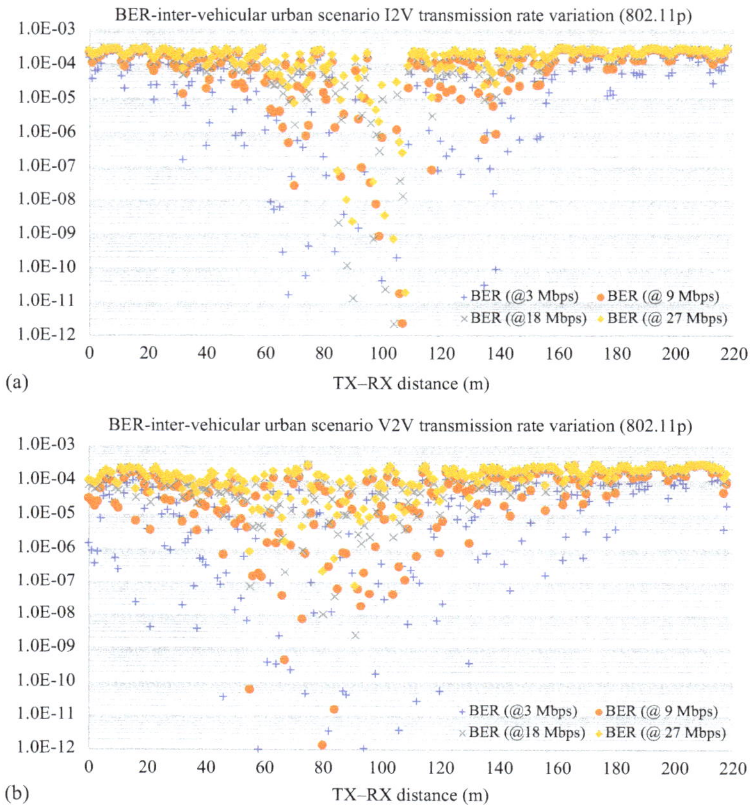

Figure 7.26 Estimation of BER in case of urban scenario for (a) I2V 802.11p and (b) V2V 802.11p communication links as function of transmission speeds (3, 9, 18, 27 Mb/s)

From the previous results, quality levels are strongly dependent on the given locations of the vehicles as well as the configuration of the surrounding environment. In order to maintain communication links, given specific quality metrics, such as required transmission rate or value of BER or BLER/FER (block error rate/frame error rate), intra-system as well as inter-system cooperative mechanisms can be employed. In this way, as a function of location and interference conditions, changes in bit rate/coding can enable communication continuity. As an example, BER thresholds and received power-level evolution for urban V2V are depicted in Figure 7.28. The distances in which values comply with required BER values increase as transmission rates decrease. Given the transition points, mechanisms

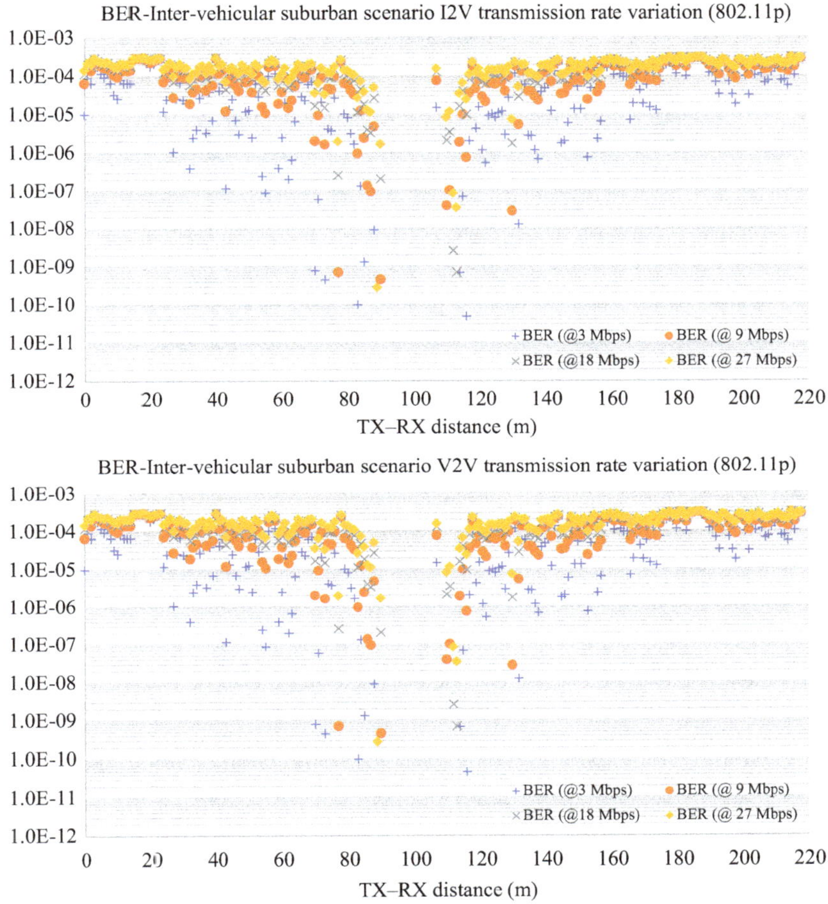

Figure 7.27 Estimation of BER in case of suburban scenario for (a) I2V 802.11p and (b) V2V 802.11p communication links as function of transmission speeds (3, 9, 18, 27 Mb/s)

such as adaptive coding schemes can be tuned, and coverage estimations can be employed in order to adequately dimension the scenario. In this case, for BER < 10^{-6}, transmission speeds of 3 Mb/s would comply almost to the corresponding cell edge, whereas in case of 27 Mb/s, this distance would be in the order of 30 m with respect to transmitter location. Therefore, if higher transmission speeds are required, higher node density or the use of other mechanisms, such as directional beams, should be employed.

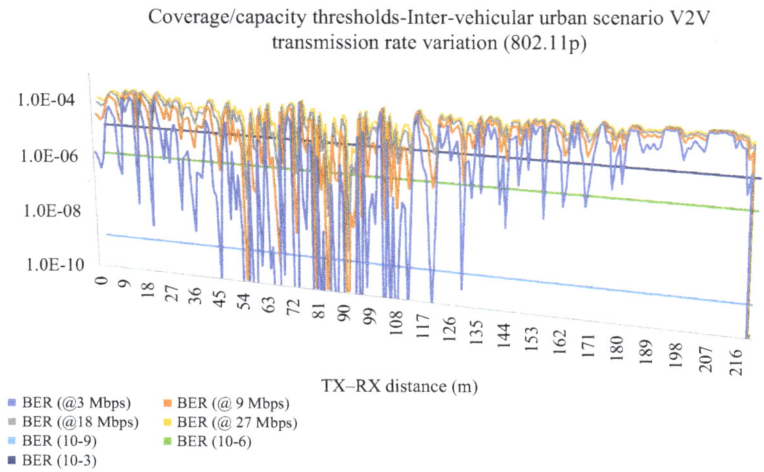

Figure 7.28 Evolution of received power levels for V2V 802.11p communication links in urban scenario conditions as function of transmission speeds (3, 9, 18, 27 Mb/s), compared with BER thresholds

References

[1] Tolstrup, M., *Indoor Radio Planning. A Practical Guide for 2G, 3G and 4G*, 3rd edn, Chichester, West Sussex, UK: John Wiley & Sons Ltd, 2015.

[2] Mishra, A.R., *Fundamentals of Network Planning and Optimisation 2G/3G/4G: Evolution to 5G*, 2nd edn, Hoboken, NJ, USA: John Wiley & Sons Ltd, 2018.

[3] Kukushkin, A., *Introduction to Mobile Network Engineering: GSM, 3G-WCDMA, LTE and the Road to 5G*, Hoboken, NJ, USA: Wiley Telecom, 2018.

[4] Nasri, N., Mnasri, S., and Val, T., '3D node deployment strategies prediction in wireless sensors network', *International Journal of Electronics*, 2020, 107 (5): 808–838.

[5] Mao, J., Jiang, X., and Zhang, X., 'Analysis of node deployment in wireless sensor networks in warehouse environment monitoring systems', *EURASIP Journal on Wireless Communications and Networking*, 2019, 288: 1–15.

[6] Frullone, M., Riva, G., Grazioso, P., and Falciasecca, G., 'Advanced planning criteria for cellular systems', *IEEE Personal Communications*, 1996, 10–15.

[7] Leese, R.A., 'A unified approach to the assignment of radio channels on a regular hexagonal grid', *IEEE Transactions on Vehicular Technology*, 1997, 45(4): 968–980.

[8] Veeravalli, V.V., and Sendonaris, A., 'The coverage–capacity tradeoff in cellular CDMA systems', *IEEE Transactions on Vehicular Technology*, 1999, 48(5): 1443–1450.

[9] Collmann, R.R., 'Evaluation of methods for determining the mobile traffic distribution in cellular radio networks', *IEEE Transactions on Vehicular Technology*, 2001, 50(6): 1629–1635.

[10] Haas, H., McLaughlin, S., and Povey, G.J.R., 'Capacity-coverage analysis of TDD and FDD mode in UMTS at 1920 MHz', *IEE Proc Commun.*, 2002, 149(1): 51–57.

[11] Lee, C.Y., and Kang, H.G., 'Cell planning with capacity expansion in mobile communications: a Tabu search approach', *IEEE Transactions on Vehicular Technology*, 2000, 49(5): 1678–1691.

[12] Binzer, T., and Landstorfer, F.M., 'Radio network planning with neural networks', *IEEE 52nd Vehicular Technology Conference*, Boston, September 2000.

[13] Cerri, G., and Russo, P., 'Application of an automatic tool for the planning of a cellular network in a real town', *IEEE Transactions on Antennas and Propagation*, 2006, 54(10).

[14] Bosio, S., Capone, A., and Cesana, M., 'Radio planning of wireless local area networks', *IEEE/ACM Transactions on Networking*, 2007, 15(6): 1414–1427.

[15] Valcarce, A., de la Roche, G., Nagy, L., Wagen, J.F., and Gorce, J.M., 'A new trend in propagation prediction. finite difference methods', *IEEE Vehicular Technology Magazine*, 2011, 73–81.

[16] Corre, Y., and Lostanlen, Y., 'Three-dimensional urban EM wave propagation model for radio network planning and optimization over large areas', *IEEE Transactions on Vehicular Technology*, 2009, 58(7): 3112–3123.

[17] Acuña, J.E., Cuiñas, I., Gómez, P., and Sanchez, M.G., 'Urban cellular network planning imbalances in wooded streets and parks', *IEEE Antennas and Propagation Magazine*, 2011, 53(5): 197–204.

[18] Wang, S., Zhao, W., and Wang, C., 'Budgeted cell planning for cellular networks with small cells', *IEEE Transactions on Vehicular Technology*, 2015, 64(10): 4797–48-06.

[19] Ghazzai, H., Yaacoub, E., Alouini, M.S., Dawy, Z., and Abu-Dayya, A., 'Optimized LTE cell planning with varying spatial and temporal user densities', *IEEE Transactions on Vehicular Technology*, 2016, 65(3): 1575–1589.

[20] Buenestadc, V., Toril, M., Luna-Ramirez, S., Ruiz-Aviles, J.M., and Mendo, A., 'Self-tuning of remote electrical tilts based on call traces for coverage and capacity optimization in LTE', *IEEE Transactions on Vehicular Technology*, 2017, 66(5): 4315–4326.

[21] Boukalov, A.O., and Häggman, S.G., 'System aspects of smart-antenna technology in cellular wireless communications – an overview', *IEEE Transactions on Microwave Theory and Techniques*, 2000, 48(6): 919–929.

[22] Saleh, A.B., Bulakci, Ö., Hämäläinen, J., Redana, S., and Raaf, B., 'Analysis of the impact of site planning on the performance of relay deployments', *IEEE Transactions on Vehicular Technology*, 2012, 61(7): 3139–3150.

[23] Wang, Y., and Zhu, X., 'A novel network planning algorithm of three-dimensional dense networks based on adaptive variable-length particle swarm optimization', *IEEE Access*, 2019, 7: 45940–45950.

[24] Athanasiadou, G.E., Fytampanis, P., Zarbouti, D.A., Tsoulos, G.V., Gkonis, P.K., and Kaklamani, D.I., 'Radio network planning towards 5G mmWave standalone small-cell architectures', *Electronics*, 2020, 9: 339.

[25] Kamel, M., Hamouda, W., and Youssef, A., 'Uplink coverage and capacity analysis of mMTC in ultra-dense networks', *IEEE Transactions on Vehicular Technology*, 2020, 69(1): 746–759.

[26] Al-Saedy, M., Al-Raweshidy, H., Al-Hmood, H., and Haider, F., 'Coverage and effective capacity in downlink MIMO multicell networks with power control: stochastic geometry modelling', *IEEE Access*, 2018, 6: 9173–9185.

[27] Tonini, F., Fiorani, M., Furdek, M., Raffaelli, C., Wosinska, L., and Monti, P., 'Radio and transport planning of centralized radio architectures in 5G indoor scenarios', *IEEE Journal on Selected Areas in Communications*, 2017, 35(8): 1837–1848.

[28] Zhang, S., Zhang, N., Fang, X., Yang, P., and Shen, X., 'Cost-effective vehicular network planning with cache-enabled green roadside units', *IEEE ICC 2017 Next Generation Networking and Internet Symposium*, Paris, May 2017.

[29] Xue, L., Yang, Y., and Dong, D., 'Roadside infrastructure planning scheme for the urban vehicular networks', *World Conference on Transport Research – WCTR 2016*, Shanghai, 10–15 July 2016.

[30] Bharati, S., and Zhuang, W., *Link-layer Cooperative Communication in Vehicular Networks*, Cham, Switzerland: Springer International Publishing AG, 2018.

Chapter 8
Applications and case studies*

In this chapter, different applications of vehicular communications will be shown. The different methodologies covered in previous chapters will be applied to specific scenarios with different applications, analysing the results of each of the different case studies, and presenting the advantages and disadvantages of each of them for different applications. The significance of radio wave propagation modelling will be described with different case studies of complex scenarios in vehicular environments.

First, location and tracking applications will be explained and analysed in different environments. Tunnels with different densities of vehicles and parking applications will also be analysed, as well as autonomous vehicles in different environments. Different applications in intelligent transportation systems (ITS) will be shown, as well as propagation properties of unmanned aerial vehicle (UAV) scenarios. The influence of different terrains and foliage will be analysed also for different applications of vehicular communications, with the aim of present safety and security applications.

Finally, the economic impact of vehicular communications technology and its regulations are analysed.

8.1 Location and tracking

In this section, we present the trends and evolution of location and tracking techniques in general and a brief discussion on the combined application in vehicular environments. First, localization will be discussed and afterwards the tracking techniques. At the end, some discussion on the combination of these for vehicular environments is presented.

The recent developments in high-end sensor environments, networks and systems that one finds in vehicular communications, together with the penetration of new services, user-sensing systems and wearables, can be used for new applications such as tracking, localization, navigation and location-based services in paradigms such as the ITS.

*This chapter has been co-authored with Raquel Fernández González, from the Department of Applied Economics, ERENEA-ECOBAS Research Group, University of Vigo.

When we see specifically the vehicular communications scenario where vehicle-to-vehicle (V2V) and vehicle-to-infrastructure (V2I) communication links need to be established, we will have to process and make decisions in real time for applications and services such as safety, accident or event information propagation, traffic balancing to decrease congestions, emergency services and entertaining applications, among many others. Vehicular ad-hoc networks (VANETs) will allow to develop these applications and others that will depend on location estimation and tracking. Even though global positioning systems (GPS) are available inside the vehicles, there are many situations in which such service is compromised, especially for accuracy and availability in tunnels, roads with high vegetation, environments with high density of buildings, etc. Thus, the accuracy and reliability of a location estimation and tracking system for vehicular networks is fundamental to opportunistically have relevant information for those applications in real time anywhere and at any time.

8.1.1 Localization

The localization problem refers to the estimation of the position of a device using some references. The references could be some devices in the same network measuring some parameters, or some of the elements of an infrastructure such as base stations, access points or roadside units (RSUs). At the same time, several of these devices measure the parameters. Depending on the parameters to be measured, one needs to use multilateration such as the classical triangulation.

Localization, especially for range-based methods, consists basically of three stages as follows:

1. **Measurement.** This stage is where sensors or different kinds of hardware are used to acquire measurements and determine the value of those parameters to be used for distance estimation. Some of the most used parameters are the following:
    1. the received power or received signal strength indicator (RSSI),
    2. time of arrival,
    3. time difference of arrival,
    4. angle of arrival or direction of arrival and
    5. Doppler shift.
2. **Distance estimation.** Once the parameters have been chosen and measurements have been taken, some techniques are used to estimate separation distance between the reference point and the device to be located. For example, one can use RSSI and determine the separation distance using the inverse function of a propagation model of the area where the devices are.
3. **Combination.** Once several references have calculated their separation distance, then a data fusion or processing centre needs to combine these distances. The combination is done usually optimizing or minimizing the distance error between the true position and the estimated position. One can use maximum

likelihood (ML), for example. The combination is usually a system of equations that need to be solved, and sometimes such system of equations is obtained by using geometry that describes the scenario and parameters used, such as having circles, ellipses and hyperbolas; see [1].

There are many different ways to carry out the stages just mentioned, and sometimes depend on the development of different standards or transmission techniques. For example, the method of dead reckoning is used mainly at sea where the number of references is small, hence no multilateration is performed and the use of ships allows to have more processing and computing power as well as more measuring power, which offers higher accuracy for the algorithms.

The development of localization techniques has evolved to consider not only scenarios on a plane, but also three-dimensional (3D) scenarios; see [2]. Another important evolution of the localization algorithms is with respect to mobility. Traditionally, the algorithms have been developed for static networks of devices, and recently, you can see works on mobile nodes, [3], where the error in the position estimation is based on the use of a random variable with beta distribution since the separation distance of the device to be located to the reference device when there is mobility has a beta distribution. Also, the mobility of drones or UAVs needs to be characterized in order to successfully track other network elements, where tracking of moving devices is more difficult due to the mobility of the trackers used such as those drones, [4–6].

8.1.2 Tracking

Tracking has been of interest for many years, where the focus was on multi-target environments [7–10], and also in cellular environments [11–14]. Many tracking techniques include the use of Bayes' rule and dynamic programming. Some develop joint probabilities with data association of likelihood of movement of targets with hypothesis testing.

In [4], tracking is performed for UAVs where specific paths are planned where the UAVs have a purpose of collecting or disseminating information. These techniques assist surveillance tasks, infrastructure monitoring and electronic countermeasures, among others. The UAVs are used to locate and track radiofrequency (RF) sources. In vehicular environments, the infrastructure for tracking and localization will mainly consists of sensors deployed in the areas where this service is needed. It is important to consider algorithms that are able to reduce energy consumption and use of the communication channel in order to have sensors with long lifetime. In [15], a distributed multi-target tracking method is proposed where a consensus filter is used that does a hypothesis testing. Sensors will share the information they obtained from their surroundings only if it is worth sharing. Sensors constantly check the local environment by comparing the last information they shared with the information shared by other sensors.

The use of GPS for localization is not convenient when you have mobility with sudden changes in direction or speed, and in order to deliver quality of service to the users, it is better to have tracking algorithms in combination with localization methods. In [14], technologies of localization such as range-free and range-estimation techniques are discussed, as well as application towards tracking in many different environments including those of 5G. In particular, for cellular networks, it is presented a method where the estimation of the mobility state is the main objective for navigation and tracking purposes. The architectures of the network that are relevant for localization purposes are also introduced, and solutions that have been developed by academia and those that are commercial solutions are compared.

In [10], a framework for tracking is proposed, where space-time sampling is used in collaborative algorithms for tracking single and multiple targets. The framework helps with the detection, the target localization and the further tracking by using a dynamic model for prediction of future locations. First the detection phase is performed, and then a classification takes place, especially when multiple targets need to be tracked. Classification algorithms such as support vector machine, ML and k-nearest neighbours are considered in the solutions proposed.

In [13], a method for tracking mobile devices in cellular networks is introduced. The authors use robust control fundamentals to address the problem by formulating a state model for mobility where the dynamic motion of the devices is represented as a linear system. RSSI is the parameter they use for distance estimation and they develop a model based on hidden semi-Markov models to estimate the command process which is the input signal to the linear system.

8.2 Tunnels

Reliable vehicle-to-everything (V2X) communications within tunnels are highly important for different applications, such as safety applications in vehicular communications, within which the following can be mentioned:

- lane change assistance;
- cooperative forward collision warning.
- slow vehicle warning;
- connected autonomous vehicles;
- disaster relief forces;
- public protection;
- management and maintenance of road and railway tunnels;
- trustworthy communications during car accidents and fires in tunnels, where the network infrastructure can be damaged and public safety users are forced to communicate directly with each other without using communication infrastructure;
- ensure personnel safety within the tunnel;
- sense earth movements or landslides;
- sense gas leaks;
- mine companies;
- optimize the mining process.

However, tunnel environments can be considered as high-risky environments for wireless communications, as electromagnetic wave propagation in these environments differs strongly from the open-air situation. The semi-enclosed environment gives rise to more severe multipath fading than terrestrial wireless links, since the wireless signals are confined to the internal space of the tunnel, and signal reflections and diffractions from the tunnel surfaces increase [16]. The multipath fading can even be more severe in underground tunnels with vehicular traffic, as the mobile vehicles gives rise to more signal reflections and diffractions, in addition to those of the tunnel walls [17]. These facts confirm that it is crucial to have a deep understanding of radio wave propagation in the tunnel environment for having reliable vehicular communication systems within it.

Given these facts, before a deployment of a wireless communication system within a tunnel environment, it is highly important to consider the following points which considerably affect radio wave propagation in these semi-enclosed environments: the tunnel geometry, the system operation frequency, the tunnel walls electromagnetic properties, the different types of vehicles and other obstructions within it, and the antenna parameters and location [16–18]. The impact of these parameters in the deployment of wireless communication systems within a tunnel environment are detailed in the following.

1. *Tunnel geometry*

   The tunnel cross section has an important impact on the signal attenuation in this type of environment. Typical tunnel cross sections can be classified in rectangular or circular shapes, as it is presented in Figure 8.1, although modern road tunnel cross sections can be slightly different from these two ideal types.

   In order to calculate radio signal attenuation within a tunnel, a general equation which considers different shapes of tunnel cross sections is presented in [19]:

$$\alpha = k\lambda^2 \left[\frac{\varepsilon_r}{a^3\sqrt{\varepsilon_r - 1}} + \frac{1}{b^3\sqrt{\varepsilon_r - 1}} \right] \quad (8.1)$$

   where a and b are the maximum tunnel width and height, respectively; ε_r is the relative permittivity of the tunnel walls and floors, as it is shown in Figure 8.1. The coefficient k indicates the tunnel shape, with values $k = 5.09$, 4.343, 5.13 and 4.45 for a circular, rectangular, arched and oval tunnels, respectively.

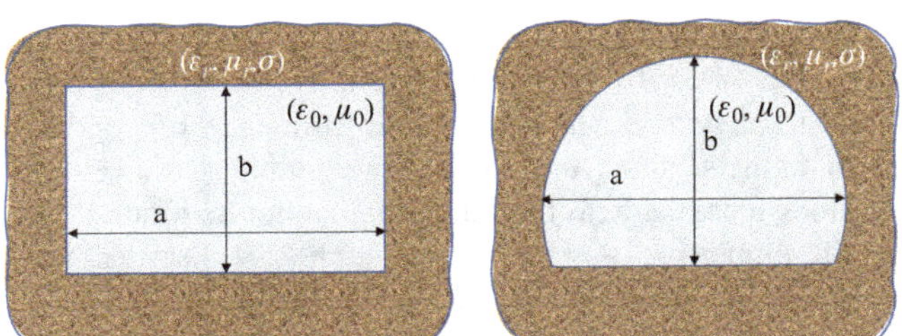

Figure 8.1 Rectangular and circular cross sections for road tunnels

The impact of cross section dimensions in a rectangular tunnel has been analysed in [20,21], showing that different heights of the tunnel impact mostly on the vertical polarization mode, while different widths have an effect on the horizontal polarization mode. Curves within tunnels can have a significant impact on radio wave propagation, since curves can provoke non-line-of-sight (NLoS) situations between the transmitter and receiver, thus producing higher attenuation or even link loss. Reference [22] presents a rise in the delay spread (DS) and a fall in the received power by the decrease of the curve radius within tunnels. Under NLoS situations, the received signal strength is due only to reflections and diffractions of the signal. Thus, in order to have reliable communications within a tunnel, if several NLoS situations are presented due to tunnel curves, it is recommended to install additional antennas to maintain direct visibility (line of sight, LoS) between adjacent antennas [22]. It has been demonstrated in [23] that the tunnel geometry, especially the cross sections' shapes, has greater impact on signal propagation than electromagnetic properties of materials within the tunnel. Additionally, the authors in [19] present that when the largest tunnel cross-sectional dimension is approximately 15 times larger than the wavelength, the attenuation rate no longer depends on the shape and cross-sectional area; thus, for this case, signal propagation is the same as free space.

2. *System operation frequency*

   The operation frequency of the system within the tunnel, conjointly with the tunnel geometry described above, has also a significant impact on radio wave propagation within this type of environments. In straight tunnels, signal-level attenuation decreases with increased frequency while in curved tunnels it increases [22]. The tunnel geometry has more influence when signal frequency increases, as the impact of curves, slopes and additional branches is emphasized. Chiba showed in [24] that a tunnel can be considered a transmission channel of high-pass type, showing that the radio signal attenuation in a straight tunnel decreases when the signal frequency and the transverse dimensions increase.

3. *Electromagnetic properties of tunnel walls*

   Electromagnetic properties of the tunnel walls are characterized by parameters such as the relative permeability μ_r, the relative permittivity ε_r and the conductivity σ [25]. The relative permeability of most rocks (except those of the ferromagnetic group with a high concentration of metals) is very close to that of the vacuum $\mu_0 = 4\pi \times 10^{-7}$ H/m, but the relative permittivity and conductivity can be highly changeable. In general, the influence of conductivity can be neglected in tunnels, since it is almost negligible [26]. However, the conductivity values increase with humidity and cannot be ignored when the water content in the tunnel walls is sufficiently high [27]. Besides, in general, conductivity decreases with frequency [25]. The permittivity values also increase with water content on the tunnel walls and decrease with frequency [25]. The signal attenuation due to penetration of lossy tunnel dielectric walls can be reduced by a metallic shield placed on the wall surface. The metallic air ducts or heating-ventilating-and-air-conditioning systems used to move air within the tunnel may enrich the wireless propagation. Despite the fact that conductivity and relative permittivity may have a high variability within the tunnel walls, some studies such as [27,28] show that for ultra high frequency (UHF) and above, electromagnetic properties of tunnel materials show weaker impact in radio wave propagation.

4. *Vehicles and other obstructions*

   When vehicles and people are moving inside the tunnel occupying the tunnel cross section, they can cause radio propagation environment time-varying and unpredictable. Obstacles and traffic within tunnels increase propagation loss and the RMS DS of the radio signal [16]. Larger dimension vehicles, such as buses or trucks, cause additional shadowing loss. For example, additional path loss can reach up to 50 dB for the same transmitter and receiver position with heavy traffic in a tunnel with about 150 trucks per hour at the frequency around 900 MHz [16]. In an experimental study, a difference of 78 ns was found from a vacant tunnel, occupied for the same transmitter and receiver position [29]. Measurements performed in tunnels show that the radio signal propagation path losses caused by vehicles depend mainly on the number of vehicles rather than on their relative position within the tunnel [30]. It is important to emphasize that the traffic flow and the different vehicles densities within the tunnel have a high impact on radio signal propagation. A theoretical mode-based analysis which predicts signal propagation and field distribution at any point in a road tunnel filled with cars is presented in [17], showing a closed form expression with accurate results.

5. *Antenna parameters and location*

   The antenna parameters such as the radiation pattern, polarization and placement of transmitter and receiver within confined environments such as tunnels have a great impact on wireless propagation. The tunnel size and the operation frequency mostly determined the attenuation rate, but the power distribution is governed by the position of the transmitter antenna [27]. Thus, the signal power loss can be reduced by using the appropriate antennas with the suitable locations. For example, higher power can be obtained by placing the transmitter antenna at the centre of the tunnel, rather than off-centre. Conversely, studies have shown that for multiple-input-multiple-output (MIMO) systems, positioning the antenna off-centre of the tunnel cross-section, would offer higher capacity [31]. The radiation pattern of the antennas also impacts signal attenuation within confined spaces, as omnidirectional antennas designed to operate in open air may not perform correctly in underground tunnels due to the large number of reflections from the walls, which has been regarded as insertion losses. So, directional antennas within tunnels decrease the insertion loss. For antennas with horizontal polarization, the tunnel width has a more significant role because the reflection coefficients on the horizontal ceiling/floor are larger than those on the vertical walls. Likewise, the tunnel height is more significant for vertically polarized antennas [27].

Table 8.1 Parameters affecting V2X radio propagation in tunnels

		Attenuation/loss	RMS Delay Spread (DS)	References
Tunnel geometry	Cross section	↑ Cross section → ↑ Loss	↑ Cross section → ↓ RMS DS	[17]
	Curves	↑ Curves → ↑ Loss	↑ Curves → ↑ RMS DS	[14,17]
Operation frequency		↑ Frequency → ↑ Loss	↑ Frequency → ↓ RMS DS	[17]
Wall material properties		No significant impact.	No significant impact.	[12,13]
Vehicles and obstructions		↑ Vehicles → ↑ Loss	↑ Vehicles → ↑ RMS DS	[14]
Antenna parameters	Location	Centre of the tunnel → ↓ Loss. Off-centre → ↑ Loss	Centre of the tunnel → ↑ RMS DS Off-centre → ↓ RMS DS	[12]
	Radiation pattern	Directional → ↓ Insertion loss (LoS) Omni-directional → ↑ Insertion loss (LoS)	Directional → ↓ RMS DS Omni-directional → ↑ RMS DS	[18,19]
	Polarization	Horizontal → ↓ Loss.	Horizontal → ↑ RMS DS	[12,14]

Table 8.1 presents the parameters described above affecting V2X radio propagation in tunnels and its impact with two main channel impairments, the power loss and the RMS DS. The analysis of these parameters can be a great aid for vehicular wireless system design within confined spaces such as tunnels.

8.3 Autonomous vehicles

The transport sector is going through a period of global change. This is because more and more economic actors participate in it and the factors that make it up are increasingly changing and unstable. The massification and increasing complexity in the transport network is a challenge at the socio-economic level to which the academy has dedicated efforts to find possible solutions. The most immediate and effective solution is to use new technologies in this sector. This implementation is necessary as long as it improves road safety, congestion, loss of productivity, pollution or loss of mobility, among other problems [35]. Given this objective, in recent decades, the use of various technological applications such as the telematic collection of tolls, the issuance of changing messages on the roads, on-board controls implemented in vehicles, and cameras and radars are extended to capture the conditions of the environment.

With the effective implementation of technology in vehicles, the next step is to create a network of cooperation [37]. For this, it is necessary to create protocols, with the same language, to carry out an appropriate exchange of information [38]. Mobile computing, wireless communication and remote sensing have made vehicular communications technology possible. This new concept includes not only the existence and expansion of intelligent and autonomous vehicles, but also the possibility for them to collect information about themselves and their environment and share it in real time with other vehicles [36,39]. There are three types of associated technologies: V2V, V2I and V2X [40]. Through these systems it is possible to detect in a larger physical–temporal space any inconvenience that may arise during driving, thus reducing the risk associated with mobility.

However, it must be borne in mind that such a profound change in a sector as the automotive industry will influence the markets related to this industry. As soon as a reliable and affordable product is developed, it will penetrate the market with force, constituting a substitute product for traditional vehicles. Moreover, its impact will have consequences on industries directly or indirectly related to it [41]. And not only on the supply side, but also on the demand side, since the existence of a new consumption paradigm will affect the behaviour of agents in various sectors. For example, a reduction in the accident rate will reduce the use of medical services, reduce car repairs and most likely affect the insurance sector. On the other hand, the transport of goods by road could reduce its price by not using drivers or collective transport – taxis or buses – would be replaced by fleets of autonomous vehicles. The positive environmental effects of a decrease in fuel consumption and CO_2 emissions should also be taken into account [41–43]. A comprehensive study of the economic impact of vehicular technology is presented in Section 8.10 of this chapter.

8.4 Intelligent transportation systems

One of the key elements in the progressive adoption of Smart City/Smart Region paradigms is given by ITS. Within the scope of ITS multiple applications have been envisaged and a great deal of research and development effort has been done, focusing on global aspects such as increasing safety, reducing environmental impact, improving multimodal operation and increasing regional cohesion [44]. Multiple initiatives are underway with the aim of designing, deploying and testing multiple ITS functionalities and solutions such as:

- ITS Joint Program Office (ITSJPO, US Department of Transportation) [45]: this research programme is focused in exploring ITS solutions, mainly related with vehicle connectivity and automation. There are multiple active research lines within (ITSJPO), spanning from safety enhancement, mobility management, connected vehicle implementation and testing, environmentally friendly vehicle driving and maintenance and smart city integration. Within the US, multiple regional initiatives focused on the adoption of ITS are active, spanning different applications related with traffic monitoring, weather condition alerts, route planning or road management and maintenance, among others [46–51]. Different research laboratories have also implemented test beds in order to trial different vehicular scenarios and applications [52–55]. In case of Canada, there is a specific nationwide ITS initiative [56], in order to promote

ITS-related applications and test beds, as well as specific initiatives, such as the case of the region of Alberta [57].

- ITS, mobility and transportation, European Commission [58,59]: multiple initiatives are being fostered within the European Union (EU), in multiple aspects of ITS such as emission and traffic congestion reduction, safe and sustainable person and freight transportation, multimodal transportation and the adoption of the Digital Single Market strategy. Different initiatives are being carried out in order to analyse the application of ITS in the areas of cooperative, connected and automated mobility and multimodal transportation (road, train, air and maritime navigation). National level initiatives are also promoting the use and extension of ITS [60–64].
- Australasian Regions: Different initiatives have been proposed in order to foster ITS-based applications in different countries within Asia/Oceania region. In case of Japan [65,66], multiple systems and applications have been implemented, such as in communications (Vehicular Information and Communication System, VICS, based on the combined use of RF-infrared (IR) beacons and frequency modulation (FM) radio broadcast information; Universal Traffic Management Systems, UTMS, with the aim of increasing communication capabilities and reducing environmental impact of traffic). Different actions have also been undertaken in other countries, such as India, New Zealand or Malaysia, among others [67–71].

Among the elements employed in order to enhance the operation of ITS is the use of big data processing and analytics, in areas such as operation efficiency or safety levels [72]. Data can be collected by multiple sources, such as smart cards from public transportation systems [73,74], global navigation satellite systems [75,76], sensor systems [77] or VANETs or other passive collection systems [78–81]. Multiple techniques can be employed, based on machine learning techniques (further classified as supervised, unsupervised and reinforcement learning algorithms [72]). As previously stated, multiple applications benefit from the use of big data and big data analytics in ITS, such as road traffic accident analysis (multivariate analysis of traffic accident patterns in urban scenarios [82] or non-parametric analysis in case of rural scenarios [83]), traffic flow prediction following different algorithms and models, as well as route planning [84–86] and application in rail transportation, for improving time table efficiency, reducing parking error and reducing energy consumption [87–89]. Moreover, big data framework can be combined with social data sources, such as different types of social networking platforms or web logs, with applications, for example, in cyber-physical social systems, related with electric vehicles and smart grid [90]

8.5 UAV systems

Utilization of commercial UAVs has lately seen remarkable progress and this growth is expected to continue for the incoming future. This is due to different factors such as the uncomplicated operability, the convenient multiple flight controls, the high manoeuvrability and the increasing payload weight that have aid for the use of many real-time civilian applications. Some of these applications are the following [91]:

- remote surveillance of areas with complicated access;
- vehicular traffic control;
- remote data acquisition;
- remote surveillance and monitoring;
- disaster relief;
- filming for different applications;
- transport of goods;
- communication relaying;
- recreation;
- package delivery;
- support broadband cellular communications, and wireless local area networks (WLANS) such as IEEE 802.11. With the aid of UAVs, cellular communications could be used in hotspot areas during peak demand events, and in cases of natural disasters or relief situations such as accidents, where the current communication infrastructure is damaged;
- it is anticipated that UAVs will be used as autonomous communicating nodes, for giving low latency and highly reliable communications, in the future 5G deployment. This will be used also for vehicular communications and autonomous vehicles. In this sense, several companies have started with tests to analyse the operability of UAVs for current LTE and future 5G cellular applications, such as Qualcomm [92]. Facebook and Google have also explored the potentiality of using UAVs for Internet connectivity in remote areas [93].

Depending of the application, different types of UAVs can be used, and distinct channel links will be created. The air-to-ground (AG) channel shows clearly different characteristics from those of other well studied terrestrial communication channels, such as the urban channel. Some of the most important issues that needs to be considered for wireless system design when considering UAVs are explained in the following.

8.5.1 Parameters that impact UAV AG channel

A crucial element in the reliability and safety of these UAVs is the AG communication link. The AG communication channel is defined as the link between the ground station and a human-made flying platform [94]. The most important parameters affecting radio wave propagation in the AG channel are the following [91]:

- Operation frequency: Typically, two frequency bands are used for commercial UAVs (2.4 and 5.8 GHz), although recently it is expected that other frequencies will be used, such as 3.4 GHz for other uses, such as video transfer with higher data rates. For lower interference, the 5.8 GHz frequency band is preferable. For these frequency bands, the tropospheric attenuation is almost negligible, which is not the case if higher frequencies are used (i.e. millimetre frequency bands), which can suffer larger free-space path loss. So, millimetre frequency bands are preferably for short-range AG links.

- To maintain a reliable link, a direct vision between the transmitter and receiver is preferably (LoS). Having LoS in an AG link has higher likelihood than terrestrial communications. This gives the advantage of reducing the transmit power requirements and increase the link reliability.
- Due to the UAVs different velocities, the AG channel has higher rates of change than terrestrial communications.
- The arbitrary mobility patterns of UAVs provoke a time-varying channel, and the non-stationarity will depend on the height of the UAV and the proximity to scatterers, such as buildings and direct surroundings.
- As the UAV is an aerial node, characteristics such as mechanical and electronic noise of the device must be considered.
- The movement of the UAV also provokes effects on the Doppler shift and Doppler spread that must be fully characterized.
- Finally, the antenna characteristics including size, orientation and array operation (for MIMO systems) must be also considered.

The AG channel has been historically far less comprehensively characterized than other types of channel, i.e. the cellular channel. Table 8.2 presents a historical AG channel modelling publication overview. The presented models are derived primarily from measurements and range from the simplest results from the earliest literature, without explicit description models for the channels, to more sophisticated channel models which become necessary to analyse performance requirements of modern digital communication systems. The two primary parameters of interest which have been analysed are DS and Doppler spread.

Historically, the signals used for AG communications had narrowband nature. Because of that, the earliest literature shows narrowband approaches with simple results. One of the earliest references is [95], in which the authors employed

Table 8.2 Historical AG channel modelling publication overview

Year	References	Channel modelling approach		Parameters analysed	
				DS	Doppler spread
1973	[95]	Narrowband	Geometrical	No	No
1992	[96]		Stochastical	No	Yes
1998	[97]	Wideband	Two-ray	Yes	No
1999	[98]		Statistical	Yes	Yes
2006	[99]		Ray tracing	No	No
2011	[100]		Statistical	Yes	No
2016	[101]	MIMO	Statistical	Yes	Yes
2019	[102]		GBSM*	Yes	Yes

*GBSM: Geometry-based stochastic channel model

geometry and mathematical approximations to model the flat-fading effect. Reference [96] shows the first complete stochastic description of the channel, showing an analytical description for the Doppler spectrum. In 1998, reference [97] represents the first paper on wideband modelling, using a two-ray approach based on measurements. Other works which model the AG wideband channel model are [98–100], based on different approaches, statistical and ray-tracing methods. The most recent works [101,102] show the measurement and characterization of high capacity communications for an AG MIMO channel.

For more information about AG propagation channel modelling of UAVs, reference [91] presents a comprehensive survey on available AG channel measurement campaigns, large- and small-scale fading channel models, their limitations and future research directions for UAV communication scenarios.

8.6 Terrain and vegetation

As seen so far, the radio propagation in vehicular channels can potentially be obstructed by different obstacles present in the application environment. Therefore, as indicated in Section 6.3, channel models for vehicular environments outcome from a combination of four main features specific of the environment and related to the potential use case [103] that includes the type of surrounding objects.

Vehicular radio channel obstructions due to static (buildings, other human-made structures, terrain, vegetation, roadside elements, etc.) and moving objects (basically vehicles) present in the environment destroy the homogeneity of the channel rendering it anisotropic [104,105]. The result is shadowing, dispersion, and time variability due to the mobility of the surrounding objects. Vehicular channel modelling tries to reflect all these impairments.

In Section 6.4, it was described that geometry-based deterministic model (GBDM) characterize V2V physical channel parameters in a completely deterministic manner [106–108] by replicating the actual physical radio propagation mechanisms within a specific environment, with singular stress on the roadside surroundings (buildings, parked cars, road signs, vegetation, etc.). For this reason, 2D or 3D ray-tracing models [105,108] that make up the majority group within the GBDMs for the vehicular channel modelling require a detailed description of the propagation environment to replicate the actual physical propagation mechanisms in order to accurately calculate the channel statistics [106–108].

Empirical measurements performed by [109] demonstrated that the propagation features of vehicular radio channel are decidedly reliant on the presence of LoS path, the case of V2V being more critical. As described in [110], the NLoS condition represents a serious problem for IEEE 802.11p systems. The NLoS condition results in signal attenuation leading to smaller effective coverage areas, and quality loss due to the disruption.

In urban or rural environments, the NLoS conditions are more frequent than expected [110,111]: buildings located at the crossroads, vegetation, steep crest and vehicles moving between emitter and receiver. In all these cases, the NLoS condition represents a major problem to the communication blocking it in many cases to the point of making it impossible or fail due to high packet loss rate [110,112]. Then the communication connectivity is lost once the LoS is affected.

In addition to the path loss that occurs in LoS situations, the additional vegetation-induced attenuation has to be considered in the evaluation of V2V communication performance, since vegetation is very frequent in vehicular environments [110], both urban and rural.

In [113], it is validated the spatial channel model proposed by 3GPP for NR-V2X systems [114] (Section 6.4.5) to operate in bands above 6 GHz. This model supports deployment scenarios for urban/highway propagation, and incorporates the effects of path loss, shadowing, LoS probability and static/dynamic blockage attenuation [113]. Realistic system-level simulation assumptions for typical scenarios were considered to embody the impact of several automotive-specific parameters on the overall network performance [113]. The work concluded the much stronger impact of static/environmental blockages like buildings or vegetation, compared to dynamic obstructions like pedestrians and cars, on the received signal strength [113].

This fact reinforces the need for specific modelling of vegetation-induced obstruction. The obstruction produced by foliage must be characterized beyond a static object that produces communication disruption and reduce the coverage area. Nonetheless most of research carried out to characterize the radio channel impairments produced by the presence of vegetation cannot be extrapolated to V2X environments due to different conditions under which were conducted [115].

There is not much research literature focused on modelling the vehicular radio obstruction due to vegetation in any of its forms: bushes (short, high, dense, thin), trees (isolated, clusters, forest), or crop fields. The main research work, Tchouankem *et al.* [115], found that vegetation obstructing the LoS produces a strong impact on V2V communication performance. In [116], the LoS obstruction due to vegetation is experimentally measured for V2I case. Significant differences and effects are produced on the communication, in terms of signal attenuation and data rate, for each type of vehicular link.

8.6.1 Effects on V2V radio link

For V2V communication, radio link involves low antenna height above the ground level leading to easier, more likely and more critical signal obstruction due to the presence of trees or bushes. In [115,117], it is analysed and modelled the effect of vegetation obstruction on V2V communication performance at 5.9 GHz. A measurement campaign is described, carried out at different urban and rural intersections, conducted in summer and winter, for different types of vegetation present in the intersections and under NLoS conditions between communicating vehicles [115].

The impact of vegetation on V2X communication performance is evaluated in terms of received power levels and packet delivery rates under varying conditions with respect to vehicle positioning, intersection layout and vegetation types [115].

In [115], it is concluded that different vegetation types and structures influence the communication link and determine the link range. Three types of intersections were analysed in [115]:

- *Type I*: Within a park area with mixed vegetation pattern dominated by trees and bushes of unequal height with different foliage densities, types of wood and leaf sizes. As the area is not fully covered with vegetation, the LoS between transmitter and receiver is only partially obstructed.
- *Type II*: In a rural area with densely growing trees and bushes, thick growing trees, dense underbrush and high grass up to 1 m tall.
- *Type III*: A cornfield completely obstructing LoS between vehicles on a rural road. For this case, connectivity was lost.

Additionally, results obtained established that weather conditions due to seasonal effects also affect quality and range of the vehicular communication [115]: for sparse and dry vegetation areas, attenuation can be approximately two times smaller. As a final point, a vegetation-induced path loss model for *Type I* intersection was derived [115]:

$$L_{\mathrm{veg}}[dB] = 2.6d^{0.42} \tag{8.2}$$

where d is the distance between transmitter and receiver in metres. This model is valid for distance d up to 200 m. The model in (8.2) was derived by applying a linear fitting to the measured additional path loss [115], only for summer received power values. The general validity of this model is limited due to the use of a specific antenna radiation pattern, antenna heights and transmitted frequency.

By comparison of the model in (8.2) with other existing models, it is concluded that ITU-R [118] and COST [119] models significantly overestimate the attenuation, whereas the modified exponential decay (MED) model [120] provides a reliably close fit [115]. However, this result is not consistent with the analysed scenario given that the MED model fits the situation where the LoS path is blocked by dense, dry and leafy trees. As many other models, MED model was determined as not valid for thin layers of vegetation or isolated clusters of thin foliage [121]. It is precisely this type of vegetation that is found in urban areas. Models should try to analyse the undergoing propagation mechanism. The model presented in [122] provides an empirical model that combines three components of signal [115]: edge diffraction, ground reflection and a direct through vegetation. Further research should be completed in the same line.

Generally, most of the works focused on evaluating the impact of vegetation on V2X communication performance in terms of received power levels and packet delivery rates. Different metrics should be provided for completing the radio channel modelling: multipath dispersion, angular power distribution, DS and Doppler shift. To this aim, the wideband channel characterization (Section 6.2.2.2) should be accomplished.

The intersections are not the only scenario where vehicular communication is affected by foliage induced obstruction of the radio link. The NLoS situation can be also produced by the roadside vegetation in tight curves [110,111]. Otherwise the lines of trees or dense brushes interposed between transmitter and receiver vehicles can likewise lead to packet loss or radio link outage [110,112]. Experimental characterization of line of trees was performed in [123,124] resulting in excess attenuation of up to 12 dB even that for frequency bands not corresponding to 802.11p.

The modelling of lines of trees should be completed by considering the statements provided in [121,125]: in the vicinity of the tree a region of anomalous signal, power decay occurs due to the predominance of diffuse scattering, and hence path loss predictions by ITU-R, MED and COST models overestimate the attenuation. Many research works [121,122,125,126] identify that diffraction is also a dominant mechanism when foliage obstruction occurs, and variations of knife-edge diffraction model have been derived [121,126].

Given that stationarity in vehicular channel is not guaranteed beyond a short time window, as described in Section 2.3.2, time variability of signal attenuation should be also determined. In [127–129], attenuations of up to 8 dB have been measured even that for other frequency bands different from V2X. Time variability becomes critical for millimetre frequencies due to the depolarization induced as experimentally determined in [130]. Therefore, vegetation cannot be modelled as mere surrounding static objects introducing a simple numerical value of signal attenuation, as in buildings or walls. In [131], the combination of multiple scattering theory and transmission coefficient is proposed as the realistic option to include in the ray-tracing simulators for vegetation. As pointed in [125,132], diffraction must be also considered.

Most of the models neglect the fact that the LoS condition is lost before the obstruction due to vegetation. Hence near ground assumption seems more suitable as a basis for modelling the foliage-induced channel obstruction [133]. As pointed out in [110], to compensate for signal loss due to NLoS situations techniques as diversity [134, Ch. 7] or MIMO [134, Ch. 10] could be considered. Both techniques take advantage of the multipath environment created around foliage [125], while compensate for signal fading and shadowing. Diversity can be implemented with antennas mounted on different car spots [110]: outside (under and over the, integrated with side mirrors) or inside (integrated with rear-view mirrors) the car.

8.6.2 V2I communications

The presence of vegetation is quite common both in urban and rural environments. Then, the impact of vegetation on V2I communications should be correctly determined to adequately define the RSU deployment conditions and the communications settings according to the quality requirements of IEEE 802.11p V2I communications [116].

In Section 6.4, it was explained that channel models obtained from conventional cellular systems cannot be directly applied to V2I systems. The reason for this is found in the differences between mobile cellular and vehicular channels summarized as follows:

1. V2I communicating antenna heights and their surroundings are similar. The opposite occurs in mobile cellular links. Then, the dominant propagation mechanisms lead to different multipath features.
2. In V2I scattering mainly occurs around link ends, whereas for mobile cellular channels scattering only affects to the mobile user.
3. The link distance is considerably smaller in V2I channels than for typical mobile cellular scenarios: few metres vs. ~1 km.
4. In cellular channels, only the user ends is mobile, and moving scatterers impact is low; as a result, the signal fluctuations are faster in the V2I channels than in the cellular case.
5. V2I channels exhibit larger signal attenuation, and specific propagation processes like diffraction are less efficient than in cellular radio.
6. V2I communications are highly dependent on the existence of the LoS path.

If the available research works are scarce in V2V, for the case V2I, they are absolutely minority. Some examples can be found in [123,124,127–129]; however, they can hardly be extrapolated to V2X environments due to different conditions under which were conducted [135], as frequency, antenna heights and mobility of radio link ends.

One of the main researches works regarding vegetation presence in V2I for IEEE 802.11p at 5.9 GHz is [116] that introduces a very extensive field-testing campaign conducted in the city of Bologna. The communications performance was measured in terms of the packet delivery ratio and the received power. Main conclusions derived from [116] are the following:

- The V2I communication performance is improved by using low RSU antenna heights: the optimum RSU antenna height in terms of maximum radio coverage heavily depends on the presence of trees and vegetation, and their characteristics (large and dense trees, their crown started at around 3–5 m from the ground) [116]. The conducted tests have shown that high antenna heights can increase the communications range when no vegetation is present. The possibility of reducing the impact of the tree's foliage through the use of low antenna heights is shown in [116].
- Nevertheless, reducing the antenna height significantly increased the link's variability due to the higher obstruction caused by surrounding vehicles: when high dense trees partially reduce the visibility conditions, the conducted experiments have shown that reducing the RSU antenna height can help overcome the propagation losses resulting from the tree's foliage at the expense of a higher V2I link's variability under high traffic density conditions [116].

8.6.3 UAV to ground

The NLoS condition of the AG radio link becomes critical if a UAV is involved. For this type of vehicular system, the elements shipped on them must be necessarily reduced, and so including MIMO or antenna diversity may not remain a priority for these aircraft in whose design battery life prevails over other criteria.

The worse radio propagation conditions for AG links are represented by slant paths with elevation angles under 10° and foliage obstruction. Some research works have modelled this scenario:

- In [126], path loss caused by a leafless, mid-Atlantic deciduous forest on slant paths was modelled by using a drone-based measurement system operating at gigahertz, large foliage depths up to 500 m. The proposed model is a variation of that in [121], consisting of a modified version of the knife-edge diffraction model.
- In [121], path loss caused by isolated trees with sparse foliage is modelled inside anechoic chamber conditions for the X frequency band (11–18 GHz) and elevation angles under 5°. The results demonstrated the presence of three regions as the mobile receiver moves towards the tree:

1. The region far away from the tree is considered as a free-space zone where only LoS condition occurs, and the Friis transmission equation applies. The free-space component is dominant.
2. As the receiver moves linearly towards the tree, it reaches a distance, the channel response changes to a response typical for a multipath or scattering situation. This middle zone is assumed to be a diffraction-dominant zone characterized by a modified version of the knife-edge model, which requires a corrected obstacle height value empirically defined from the measurements.
3. As the receiver moves towards the tree, the near-tree region is a diffuse scattering-dominant region due to the signal scattered by the tree trunk. The near zone is described by a CDF due to signal variations that depend on the random positions of the set of tree branches and leaves. The cumulative distribution function (CDF) indicates the probability of signal deep fading.

In [125], path loss caused by an isolated tree located within a forest, with foliage depth up to 8 m, was modelled at 2.4 GHz. The antenna heights were identical for transmitter and receiver. Results found similar characterization as in [121]. Measurements were also performed for dense foliage trees surrounded by forest trees, determining the importance of the diffraction.

In the works mentioned above, the transmitter stays at a fixed location, so that the extrapolation of the results to vehicular scenarios with both communication ends mobile may not be accurate.

8.6.4 Terrain

The presence of elements or variations in the terrain elevation influence the visibility conditions, and thereby the link connectivity [136]. Terrain features and variations need to be characterized in order to take full advantage of the V2I coverage and a correct deployment of RSUs. Many elements can produce a significant variation in the terrain elevation or in the terrain shape impacting the V2X connectivity. The main elements of these are the following:

- *Ramps*: Scenarios with ramps are the most analysed cases [136–138].
- *Bridges, over-pass roads*: Connectivity loss [136]; in [139], measurement-based channel model is verified by a geometry-based stochastic channel modelling in which height dimension is first introduced to characterize the diffraction phenomenon in V2V communication.
- *Viaducts*: With or without soundproof walls [136–138].
- *Sharp curves*: Strong NLoS conditions [136].
- *Multiple road levels*: Usually planar environment rather than multi-level traffic scenarios are considered for simulators and experimental-based channel modelling. However, as described in [135], near-vertical signal paths and obstructions due to floor slabs negatively impact the reception conditions affecting a significant proportion of delivered data packets [135]. Even antenna gains vary widely due to changing elevation angles [135].
- *Parking garages*: Typical above-ground parking garages common in urban or suburban locations that require an indoor propagation environment characterization for vehicular communication systems. Statistical tapped delay line models with Markov chains describing the birth/death of multipath components are provided in [140].
- Scenarios with significant terrain elevation variations.
- In general, any spot elevation above the sea level higher than the elevation at the RSU or vehicle location.

The availability of radio channel characterization for the above cases can help to determine if high requirements demanding applications can be deployed, as those dedicated to safety. For the ramp case, an example is [137,138] that focuses on V2V radio channel properties under ramp scenarios categorized according to different construction structures into viaduct ramp with soundproof walls in an urban area, and a general ramp without soundproof walls in a suburban region.

It is concluded in [136] that Weibull and Ricean distribution has a good match in NLoS and LoS region, respectively. In [137,138], it is indicated that soundproof walls produce abrupt fluctuation of the received signal level in the urban viaduct ramp scenario, and suburban ramp scenarios continuous changes of received signal level (RSL) and different adding depth values.

In [137,138], it is proposed that for a terrain that interferes the radio link, considering whether the LoS is available, the propagation link of the radio signal maybe separated into various propagation zones. Then, the statistical characteristics of DS are fitted using a generalized extreme value model with a good fit [137,138]. Furthermore, propagation path loss is modelled, demonstrating the difference of path loss values in the transition region owing to the impact of soundproof walls [137,138].

Additionally, the materials present in the terrain also influence the V2X communication due to the ground-reflected path that can become dominant propagation mechanisms. The analyses presented in [136] indicate very good reflective properties of the asphalt surface and demonstrate them by the small difference between the direct and the reflected multi path component (MPC) magnitudes in the 60 GHz band based on frequency-domain channel measurements. Materials as mortar or concrete also show good reflective properties at 5 GHz as characterized in [141].

8.7 Roundabouts

Another interesting case study related with vehicular handling is management of roundabouts. These road elements play a key role in regulating vehicular traffic in different situations, such as incoming/outcoming traffic from motorways and speedways or urban traffic organization. Traffic handling within roundabouts is a complex problem, owing to different factors, such as variable traffic flows, accurately estimating driver behaviour or other unexpected factors, such as interaction with pedestrians [142]. This has a direct impact in elements such as driver assistance systems and autonomous driving, in relation with operations such as lane management and displacement [143]. One of the most complex issues in case of driving conditions in case of human drivers is to foresee their behaviour, in order to obtain models applicable for driver assistance systems as well as for inference in autonomous driving conditions [144]. This can also be extended to the case in which we consider the interaction of pedestrians, especially in case of roundabouts, owing to higher probability of accidents occurring involving vulnerable road users [145,146].

As previously stated, traffic regulation within roundabouts is a complex problem, which has direct impact on overall traffic flow. Solutions have been presented, based on the use of different types of inductive magnetic loops to detect vehicles for handling multi-level roundabouts [147], or to characterize vehicle entry delay [148]. The case of autonomous vehicles has also been analysed, in order to provide optimal models to consider traffic handling within roundabouts [149], including the use of virtual vehicle models [150] and game theory [151] to enhance driving conditions. Examples have also been proposed in order to increase security in non-signalized intersections, by means of distributed communication schemes [152].

8.8 Safety and security applications

It is well known that 5G network infrastructure presents an important medium through which information will flow servicing different applications and use cases, among which we can see the Internet of things (IoT), vehicular networks, industry 4.0, digital government, e-health, digital entertainment, autonomous driving and many others. Within the IoT and the integration of cyber-physical systems (CPS), transportation and vehicular networks play an important role in future smart environments. The emergence of highly automated and autonomous driving vehicles, with rigorous requirements in terms of responsiveness, and security are needed to enable new services that improve efficiency and safety on the road.

Vehicular networks will depend on data sharing to carry out tasks such as proximity and authentication. Connectivity will be vital for vehicles, but it will also provide open doors and access to systems with potential cyberthreats that could compromise the operation of vehicles and everyone's safety.

In vehicular systems, safety tasks such as those of proximity, accident avoidance and alert systems (change of lane, blind spot, etc.) are full of sensors and systems such as radar that work at high frequencies usually in the 70 GHz band, and the solutions need good propagation conditions in order to assist the driver to make opportune decisions. In case of autonomous driving, then a vision system is also added to the vehicle and it is necessary that the infrastructure provides a communication service of the highest quality in parameters such as latency and data rate.

Other safety tasks such as efficient use of energy and minimization of travel time depend not only on the network infrastructure, but also on cloud services or edge-computing functions, where calculations of new and alternate routes, information of accidents and the management of vehicle speed will help in our safety and fuel economy.

Vehicular communications and networks are essential to make these ideas a reality, but wireless communication does not guarantee security. As part of the ITS, the prevention of accidents, energy efficiency and the deployment of infrastructure are essential for safety. Vehicle coordination is necessary for this to work and mitigate traffic problems where information sharing as well as joint decision-making is promoted. The design of cooperating networks of vehicles will demand a harmonized operation of control, communications and sensing with accuracy, inside and outside vehicles.

These systems will have a huge anticipated impact on society and the economy being disruptive by changing the way people do things. Thus, it is of interest to have systems that are robust to physical and remote attacks including terrorist attacks that use vehicles for their purposes. The research community needs to have a deeper understanding of cybersecurity threats, physical attacks and unintentional human errors that could impair driver performance and undermine road traffic safety.

In order to achieve the right level of safety and security, it is imperative that we design solutions that together with the software-based network infrastructure will use a combination of machine learning and signal-processing methods to develop smart networking techniques that will integrate communication and computing with a perspective of security and safety, so that the system can adapt and be strong against cyberattacks, such as falsification attacks. The techniques will form a network learning environment that will adapt through the distributed detection of intrusions and will be more robust by using coordination and cooperation in the network elements such as vehicles, sensors and systems as well as the transportation infrastructure.

The safety and security system that needs to be developed will define a base on which information about cyberattacks could be recorded in detail with the countermeasures that are effective to mitigate the negative effects they cause, so that learning will be continuous. Alternative methods could be implemented (e.g.

blockchain) to integrate vehicle metrics, driver biometrics and network status for transaction tracking in the network.

5G network infrastructure needs to be designed, planned and deployed in order to support IoT's huge amounts of data, because IoT is reversing the flow of information from the edges towards the core because of the large amounts of data produced at the end user's side, and the need to process such data before being shared. IoT will integrate CPS which are the next generation of engineered systems that require interoperability of computation and communication systems, together with the physical world to achieve stability, security, reliability, robustness and efficiency in the system [153]. The physical world includes buildings (homes, schools, offices, factories, etc.), utility networks (electricity, gas, water, etc.), transportation networks (roads, railways, airports, harbours, etc.), transportation vehicles (cars, rails, planes, etc.) healthcare systems and information technology networks, among others. Within the transportation paradigm, vehicular communications and networks are essential to make these ideas a reality.

A networked vehicle is connected through the use of V2I and V2V communication links to assist the driver or the autonomous driving system with opportune information needed for decision-making actions. Security, safety and privacy are involved in ensuring the correct operation, demanding data with minimum latency [154–157]. Thus, high-speed links are crucial for a successful connected vehicle to operate properly in real time through traffic. This communication includes on-board sensors, in-vehicle networked systems and external communications through standards such as IEEE802.11p, cognitive radio and 3GPP (4G LTE, 5G), among others. Communications among vehicles and infrastructure involved will be affected by the wireless channel impairments and variability, mobility, limited bandwidth/power, high congestion and interference environment, available technologies or standards, and required service-based performance metrics. Vehicle coordination is necessary for security and in order to work properly, detect and mitigate security problems; the design of cooperating networks of vehicles will demand a harmonized operation of control, communications and sensing systems with accuracy [158].

Some of the issues that have been identified and need to be addressed are the establishment of secure sessions through authentication or identity validation, detection of attacks/intruders, protection against attacks and trust to determine computing tasks nodes [159,160]. Solutions to these problems can be proposed using a smart platform created by the network where learning takes place and the network is adapted with functions that are based on software. Thus, the detection of malicious nodes (vehicles) in a network, especially when cooperation and coordination are needed tasks, is of highest importance for the secure operation of the system. These actions of detection need to be implemented within cooperative tasks, not only due to the security functions as they are, but also due to propagation environment where network coverage could be compromised due to shadowing and obstacles.

The 5G network infrastructure needs to have solutions where an intrusion detection system with emphasis on falsification attacks is developed. The vehicular network in particular will need to adapt and integrate intelligence by using machine learning algorithms and diffusion strategies [161]. The result will be a highly robust system that changes dynamically and where decision-making is carried out in a distributed way. The innovative ideas will be in the combination of machine learning techniques used in IoT [160] and signal processing techniques that capture dynamics of systems and communication network, and use it for learning [162–164]. The signal processing techniques will allow us to develop an intrusion detection system that works dynamically and in a distributed fashion to facilitate decision-making processes for intrusions.

8.9 Evolution within future 5G scenarios

When trying to talk about future directions that technologies take, one has a limited point of view because many different issues and problems from different sectors must be considered, and now even unpredictable events such as pandemics must be taken into account due to the economic impact that affects straightforward the technological development due to closure of some markets and the creation of new markets.

Thus, if we consider that our technological development is in a steady state, and we emphasize the point of view on the needs and wish lists of 5G, then we can concentrate the evolution in two main directions: the first one being the development of new use cases or markets, and the second one being the development of the 5G technology to be evolving towards becoming a platform where 6G can depend upon.

8.9.1 5G scenarios

What is clear so far is that 5G will become a communications platform that will support an ample set of new services, applications and use cases that will bring new leverage to different economies by creating and deploying these future scenarios.

5G will bring connectivity of things and people through different use cases including transportation, home, office, data centres, cities, entertainment, health, industry in general and government services, among others. 5G technology is a platform that empowers connectivity with a new radio technology that provides quality of service in terms of latency, high data rates and massive networking of people and things. 5G is the platform on which the IoT will become a reality for everyone.

Some of the future scenarios that will become a reality are already projects under development, for example, as those shown in 5G Barcelona, [165], that are grouped or classified as vertical industries of Energy, Entertainment, Health, Financial, Transportation, Industry, Rural and Security/defence. Also, the 5G-PPP, [166], from the European Union has its point of view and classifies these in a similar way defining the groups of Energy, Media & Entertainment, e-Health, Factories of the Future and Automotive. A brief description of these groups is the following:

- **Energy.** This sector has seen increased costs of generation, distribution and transmission which have been translated into an increased cost in the end user's side. The main concerns in this sector are with respect to the development of the smart grid, from the end user's access and the distribution network or backbone of the grid. This implies the organization at the network level of the energy distribution and service access where a large group of sensors will be coordinated to achieve connectivity to provide and measure energy consumption. Smart home/building solutions that provide connectivity are the platforms for the implementation of access to the smart grid. Homes will be able to monitor and control within certain limits the energy consumption of gas and electricity, as well as of resources such as water through smart meters by adapting the lighting system and using smart plugs and appliances. At some point, buildings belonging to industries and homes as well will be generating energy through the use of alternative energy sources. Thus, the solutions of the smart grid need to consider that energy will flow in both directions at some point and that this situation will imply the use of new communications, intelligence in the smart grid and the communications network, and new and different ways for business cases. 5G will play a fundamental role in the provision of an infrastructure for the smart grid. The new paradigm of the 5G network is to become a software-based infrastructure that will adapt to organize users in networks at different levels such as neighbourhood area network that collects the measuring data from the end users and connects such meters to a data centre for its distribution. Among the problems to be considered in this sector is the user density, the wide coverage needs, the reliability of the service provisioning as well as of the communications, and the security and integrity of the information to provide a service that is trusted by its users. The communications needs will not be stringent since the energy sector will require low data rates and latencies will not need to be low.

- **Industry or smart factories.** With respect to 5G communication requirements, the services to be provided to this sector need to consider connectivity and remote access, remote control that in some cases could be critical and coverage. As in the energy sector, the communications in general, the sensors and machines to be connected and the basic remote-control operations will be for a low density of devices with no need for strict low latencies and moderate data rates. In contrast, the industry has sectors where there are processes with critical situations that need to use real-time control and automation, for which the 5G network infrastructure must provide low latencies, moderate to high data rates and reliability. The main objective of this sector is to have an increased efficiency in production lines and value chains that could be enhanced by the use of smart devices and robotics that work in cooperative ways. The use of 5G will not only need to be seen from the perspective of providing access and transporting information to and from the smart factories and industry sites, but also as a way to connect data from the value chains, production lines, consumer trends, sales, etc., to the smart factory to adapt, grow and change according to demand to satisfy efficiency schemes and scalable scenarios. The 5G network will need to adapt in order to provide such services with quality and reliability with connectivity to and from the cloud and the potential use of new paradigms such as edge computing that will be of support to the decision-making processes involved in the smart factories of the future.

- **Entertainment.** This sector contains many different kinds of applications and services with also many different uses, and as a consequence, many different quality requirements that are needed from the 5G infrastructure. For example, streaming services (video, audio, movies, live TV, etc.) have different latency needs compared to online gaming, virtual reality, augmented reality. This sector is challenging because its diverse set of requirements that will be demanded to the 5G network infrastructure. For example, in terms of latency, it will be needed a very low latency for real-time applications for virtual reality, augmented reality, online gaming; moderate latencies for live events and cooperative uses; and not so strict latencies for content generated by users. With respect to the density of users that will be connected to network resources, we have high density of users for live events, streaming services and user generated media; and low to moderate densities for online gaming and immersive media such as virtual reality. Data rates will be from moderate to high for this industry. All these services in this sector will have a continuous growth that will produce an increasing demand that will be seen by the 5G infrastructure as an increasing density of devices, increasing data rate or bandwidth demand, support for an increasing number of simultaneous connections that will affect other quality of service indicators such as latency and reliable delivery, and the 5G network has to adapt to accommodate these continuous changes.
- **Health.** The economic growth of the market for sensors has triggered many different and wide sets of applications in the health sector. The social impact as well as the challenges of health solutions will be of upmost importance for the 5G network infrastructure development. Applications and services of connectivity to achieve assistive living, remote monitoring of health parameters, procedures followed by first responders in critical situations and assistance in emergencies such as the current pandemic are only a few of a vast set of possibilities that affect economies, technology development and infrastructure requirements. At the end, the use of these technologies has the purpose of increasing efficiency, reducing costs and reducing times, and all of these purposes impact the development of the 5G infrastructure coverage since we still have huge amounts of people and places without access to the internet. 5G will support paradigms such as remote and personal medicine, assistive living and the possibility of having remote intervention through surgeries for example that need precision, real-time and extra-low latency of data, video and audio to control and actuate machine and devices such as specialized robots. The massive amount of devices and sensors involved will generate huge amounts of data that need to be transmitted, stored, processed and visualized together with the integration of data from different sources and possibly from distant places and from different networks. So, from the sense of using robotics, latency must be extremely low, bandwidth and data rate must be extremely high, and

reliability must be the highest. In contrast, remote monitoring of health parameters does not need low latencies, nor high data rates, but it needs accuracy in positioning, mobility and coverage.

- **Transportation.** Whether we talk about autonomous driving, vehicular networks, V2X communications, road safety or provisioning of services inside different transportation means, the transportation sector needs from the 5G infrastructure are a reliable service with low latencies to extremely low latencies with high bandwidth and data rates to support mobility at very different speeds for scenarios with high densities in some instances, accurate positioning and ample coverage to achieve quality. Driving assistance services need to have such needs satisfied to provide with safety, with the possibility of autonomous driving and collaborative driving to improve and impact traffic in the cities, pollution, accidents that could be avoided, alternative routing for destinations, etc. There are important challenges; for example, the communications in the transportation sector must be available even if the network coverage is compromised, e.g. due to shadowing or obstructions or scenarios such as those inside tunnels. Communications must take place through concepts of vehicular networks with V2V links and eventually VANETs that must be established on the fly as the scenarios demand. This adaptability of the network is more critical in this transportation sector due to these situations and the speeds and mobility that needs to be managed. In these critical needs, we also have the emergency vehicles involved that need to have low latencies with high data rates at high speeds with large coverage areas. In all these scenarios that involve communications of vehicles of many kinds, signal propagation characterization is extremely relevant to network planning and deployment since it would help to determine at some point the number of transmitting antennas, their positions and the transmission characteristics to be used in order to achieve coverage and quality for the service delivery.

8.10 Economic impact of vehicular communication technology

8.10.1 Evolution of the automotive market

The magnitude of the evolution of technology in the automotive sector means that society is facing one of its greatest changes. In recent years, the major telecommunications companies – Google, Uber and Apple – and major automobile manufacturers have been interested in the materialization of intelligent and autonomous cars. Parallel to the autonomous vehicles will develop the market of vehicle communications, which is estimated that in 2024 will reach the figure of \$18.3 billion, presenting an annual compound growth rate of 19% between 2019 and 2024 [167].

The year 2016 was a turning point for innovations in automatic vehicles. Proof of this is the succession of announcements of new projects and pilot tests. In Colorado (USA), an autonomous truck delivered unmanned goods for the first time.

The passenger transport sector also launched various experiments. While in Helsinki (Finland), an urban minibus was launched; in Singapore and Pittsburgh (USA), the autonomous taxi made its first journeys. As for air transport, in Cambridge (UK), the first flight of an unmanned aircraft was successfully carried out. However, not all of them were positive experiences. Also, in the same year, 2016, the first fatal accidents were recorded in China and the United States, proof that there is still room for improvement in this type of transport [168].

Despite the problems that have emerged in the first years of the sector's development, its evolution will be exponential in the next 10 years. We will move from the current scenario where Level 0, 1 and 2 cars – no automation, driver assistance and partial automation vehicles – will be the majority to another where Level 3, 4 and 5 vehicles – conditional automation, high automation and full automation – will lead sales in the market. As can be seen in Figure 8.2, the use of non-drive vehicles will be a reality, with an estimated 10 million Level 5 vehicle units in circulation worldwide by 2030 [168]. In total, in that year, the market for partially or fully self-contained privately owned vehicles will reach $60.01 Billion.

The market introduction of automated vehicles will be progressive. By 2019, Level 3 and 4 vehicles will be marketed to the general public, but it will be Level 4 vehicles that will have the greatest acceptance. The reason is that its higher level of performance at a competitive price – compared to Level 3 – will make it an affordable and reliable product, consolidating its market niche even when the fully automatic vehicle begins to be marketed in 2025.

With regard to the geographic location of sales, it is estimated that America – North America in particular – will be the area where up to 25% of sales of Level 4 and 5 vehicles will be concentrated until 2035 (Table 8.3). The development of a comprehensive institutional framework that includes the attribution of

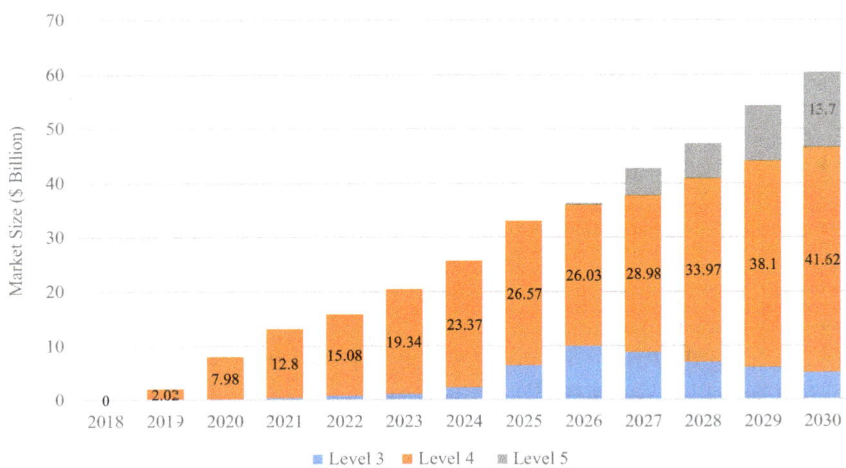

Figure 8.2 Total autonomous vehicle market size (ownership), worldwide, 2018–30 [169]

Table 8.3 Level 5 autonomous vehicle market penetration by per cent

	2020	2025	2030	2035
America	0.010%	1.1%	5.4%	19.8%
Europe, Middle East and Africa	0.000%	0.5%	3.9%	13.2%
Asia-Pacific	0.001%	0.3%	3.0%	8.0%

Source: Author elaboration based on [168]

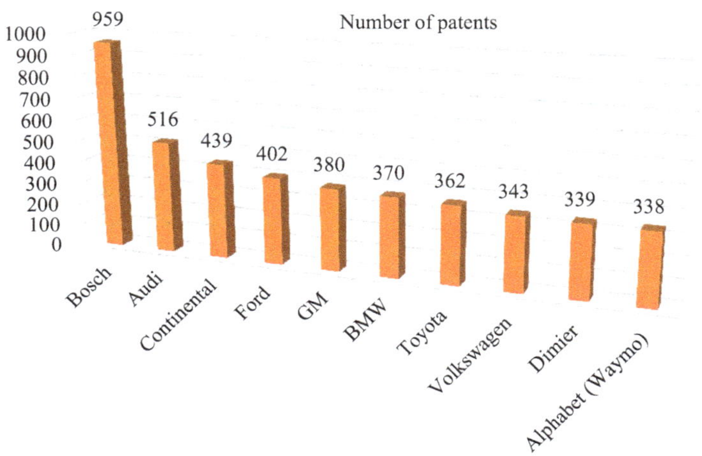

Figure 8.3 Autonomous driving patent race, worldwide, until 2017 [169]

responsibilities will be the key to the development of the market. At the same time, Europe, the Middle East and the Asia-Pacific region will show very similar demand behaviour. It is from 2035 onwards that China will lead the sales of automatic cars, with special emphasis on fully automatic vehicles. This turning point will be due to the country's consumption restrictions – a restrictive energy regime imposed by the executive branch – and demographic characteristics – high population density increasingly urbanized. Another of the most important markets would also be located on the Asian continent, India [169].

As stated above, the market for autonomous vehicles is a sector with a high innovation component. Therefore, there is a high competition between their companies for being the first to develop a profitable prototype. Thus, in the last decade the registration of patents has skyrocketed, with German companies being the most prominent, accumulating 52% of global patents (Figure 8.3). Among the areas that accumulate more registrations are connectivity technology, artificial intelligence and human machine interface. It is important to point out that in Figure 8.3, where only the ten companies with the highest number of patents are represented, there is only one company from the technology sector (Waymo). This fact should not detract from the work of these companies, since even if they file fewer patents, these are of great importance, such as the autonomous navigation system belonging to Apple, Inc. [169].

The creation of new patents will lead to a higher degree of innovation and the higher the degree of innovation, the cheaper the components and the development of economies of scale. From 2020, large-scale production of autonomous vehicles will be financially viable. The phenomenon of offshoring will also be replicated in this market, where China and India are considered to become the largest producing countries, exporting their production mostly to the rest of the Middle East-Asia region and Africa. However, the globalization of the vehicular communications market will affect not only the domestic vehicles, but also the shared – private or public – transportation [169].

8.10.2 Economic assessment of the consequences of AVs

The economic consequences of the expansion of autonomous vehicles are numerous and varied. The first and most obvious one concerns the vehicle technology components industry. The software will be larger and more complex, combining the task of entertaining and controlling traffic. However, there are many other aspects that will benefit. Driving AVs, which is more respectful with the law and the environment, will generate changes in demand in sectors such as vehicle repair, medical services, insurance or legal assistance.

8.10.2.1 Insurance industry

The insurance industry will be deeply affected. On the one hand, the first and most complex change will be to specify the liability arising from traffic accidents. A legal framework should be clearly defined to cover as many contingencies as possible, as this would reduce the costs of potential litigation. Unless the passenger takes irresponsible decisions, such as authorizing driving in adverse weather conditions, he can be held responsible. Otherwise, when driving depends only on a computer and not on a human being, the responsibility is transferred to the manufacturer of that software, so the company itself should have insurance against possible technical failures [170].

On the other hand, the use of autonomous vehicles will reduce the number of accidents. With the expansion of AVs to Levels 4 or 5, the decrease could be as much as 80%, as it is this percentage of accidents that is attributed to human failure. Due to increased road safety in the United States, the insurance industry could be reduced by up to 60%. Currently, this sector has a turnover of approximately $180 billion that could decrease to $ 72 billion, a change that would result in savings of $339 per capita [41].

8.10.2.2 Medical sector

The vehicular communications system will allow to indicate, in real time, the position and circumstances of each one of the vehicles that pass through the road. The 360° awareness that will be integrated in the vehicles will result in greater

Table 8.4 Decrease in the cost and number of accidents in Europe and USA according to the different market shares

		Low market share	**Middle market share**	**High market share**
Number of accidents avoided	EU	249,000	866,000	1,222,000
Number of accidents avoided	USA	211,000	1,880,000	4,220,000
Economic cost savings	EU	$16.56 billion	$58.9 billion	$84.9 billion
Economic cost savings	USA	$5.5 billion	$48.8 billion	$109.7 billion

Source: [170,171]

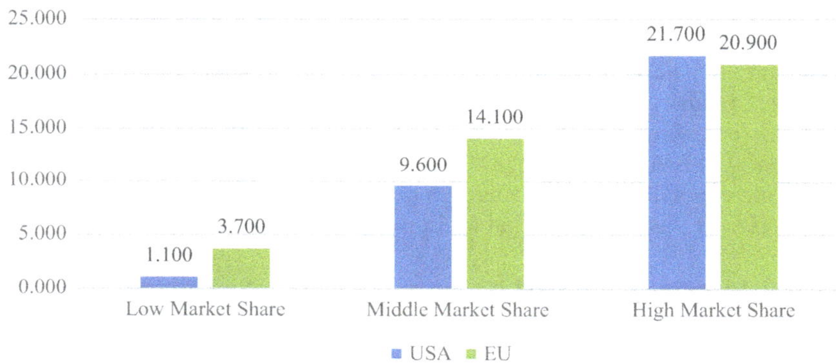

Figure 8.4 Lives saved (per year) in Europe and the USA according to the different market shares [170,171]

safety for passengers and pedestrians. Consequently, the number of accidents and resulting personal injuries will be contracted and medical services will decrease their demand. Currently, in the United States, nearly two million health services and more than 240,000 hospital admissions are due to traffic accidents. In a society made up of 90% autonomous vehicles, the total savings (Table 8.4) will be significant [170]. However, since the use of services can derive from many other circumstances, the estimated savings for the consumer would be much smaller ($36), as the cost of health insurance in the USA would not decrease too much [41].

A very similar scenario is also being considered in Europe, with a high percentage of accidents avoided. In this way, the European health system would save, by minor injuries, costs from €2.9 billion to €15.2 billion. As far as serious injuries are concerned, the savings would increase from €8.8 billion to €41.5 billion depending on the degree of market penetration of autonomous vehicles [171]. However, the most remarkable consequence is attributed to the number of lives saved (Figure 8.4). Between the USA and Europe, 42,600 deaths would be avoided every year [170,171].

Table 8.5 Environmental benefits of using AVs in Europe according to the different market shares

	Low market share		Middle market share		High market share	
	Quantity	**€ Saving (billion)**	**Quantity**	**€ Saving (billion)**	**Quantity**	**€ Saving (billion)**
CO_2 emission reduced (1,000 tonnes)	12,000	€0.7	54,000	€3.2	87,000	€5.3
PM reduced (tonnes)	44	€0.001	−1,397	−€0.039	−3,029	−€0.1
NO_X reduced (tonnes)	8,234	€0.1	29,878	€0.2	46,321	€0.4
VOC reduced (tonnes)	1,518	€0.002	4,925	€0.005	7,343	€0.008
Total environmental benefit	n/a	€0.8	n/a	€3.4	n/a	€5.6

Source: [170,171]

8.10.2.3 Oil and gas industry

Concern for environmental issues is growing. The high growth of emerging economies and the production model established in an increasingly globalized world economy have led to higher and higher emissions of CO_2 and other polluting gases. One of the sectors that generate the most pollution is land transport. The widespread use of automatic cars would lead to a reduction in these emissions by reducing the consumption of fuel, either because it is lighter or because it is more efficient [172]. Studies carried out for the European case show that the benefits are considerable (Table 8.5).

In the United States, this reduction would be between 8% and 13%, depending on the degree of penetration of stand-alone technology. With stand-alone cars, congestion would be reduced, leading to savings of 7,192 million litres of fuel in the United States alone. In addition, thanks to the drag coefficient, energy savings could be up to 20% [170]. Another factor with positive effects would be the reduction in the need for parking, a process in which there is overconsumption. All this would considerably increase the environmental benefit.

However, even if the expansion of AVs is more environmentally friendly and achieves greater fuel efficiency, in general terms the oil and gas industry will not see its profits eroded. This is because the fleet of land vehicles, whether public or private, will increase in order to meet mobility needs. In fact, estimates for the United States predict that the sector will increase its profits by 5%, which at the microeconomic level means that consumers will spend $44 more per year on fuel [41].

8.10.3 Legislative framework of autonomous technology

Technological changes are a challenge in society. A scenario where autonomous vehicles are a reality must be accompanied by a legislative change. A summary of the actions initiated in three different regions of the world – the United States of America, Europe and Asia – is featured next. In this way, it will be possible to compose a rigorous and holistic vision of what is the current legislative situation, concerning autonomous technology, in the world.

8.10.3.1 The United States of America

In the USA, the approval of laws that regulate the activity of autonomous vehicles has followed a constant trajectory. Since 2012, 29 states have passed laws. In the USA, the States have the power to allow the driving of autonomous vehicles. Therefore, there are very different scenarios between states [173].

The one that has developed a more permissive legislation has been California, where Silicon Valley is located. However, in September 2016, the US Department of Transportation (DOT) and the National Highway Traffic Safety Administration (NHTSA), in order to unify and give common guidelines for regulations related to vehicle communications, published the 'Federal Automated Vehicle Policy', which was renewed in September 2019. In this way, even with many existing differences, the states follow the same principle: autonomous driving is authorized while there is a person who can take control of the vehicle in case of risk. Then, in Table 8.6, the legislation in force in the United States is detailed, both at national and state levels [174,175].

8.10.3.2 Europe

Creating a single and fully agreed regulation in the EU is challenging. There are three main pitfalls here. The first is related to the 1969 Vienna Convention, of which a good part of the EU countries are signatories. This document shows that vehicles without drivers cannot drive more than 10 km/h, which must be modified. The second obstacle is the dispersion of leadership as there are three different European Commissioners dealing with traffic matters. The third problem is the lack of consensus on issues such as civil liability, data privacy or cybersecurity [173,176]. In Table 8.7, the regulatory framework for the testing of autonomous vehicles in Europe is detailed.

8.10.3.3 Asia

Asia is where some of the major producers in the automotive industry are based. In the future, it is expected that this continent will see the highest sales of autonomous vehicles worldwide. Therefore, the involvement of the governments of the region has been great, identifying the autonomous vehicle industry as a strategic sector [176,177]. In Table 8.8, the regulatory framework for the testing of autonomous vehicles in Asia is detailed.

Table 8.6 Regulatory framework for the testing of autonomous vehicles in the USA

Territorial administration level	Legislation / regulation
Federal action	In October 2018, the new guide for automated driving systems (ADS) by the NHTSA was approved. This guide replaces the one published in 2016 and is structured in two sections: voluntary guidance and technical assistance to the states. In addition, the approval of two complementary state laws was attempted. The first one would be the Self-Drive Act in the House, and the second one would be the AV Start Act in the Senate. Both were rejected in their approval process.
State action	
Alabama	SJR 81 (2016): Created the Legislative Committee responsible for creating the regulation applicable to autonomous vehicles. SB 125 (2018): Regulates autonomous truck driving. SB 47 (2019): Authorizes automatic commercial vehicles to operate without a driver. However, they must respect the traffic laws in force and have a liability insurance of $2,000,000, with the vehicle operator being the responsible agent.
Arkansas	HB 1754 (2017): Regulates the tests of autonomous vehicles that have the truck squad systems. HB 1561 (2019): Allows the implementation of pilot programmes for the circulation of fully or partially autonomous vehicles. HB 1822 (2019): Amendment to bill HB 1561, which specifies the obligation to comply with railroad crossing regulations.
California	SB 1298 (2012): Authorization for the circulation of autonomous vehicles on public roads, with the requirement that the California Highway Patrol Department guarantees road safety. AB 1592 (2016): Authorization for an autonomous vehicle pilot project. AB 669 (2017): Authorization for the truck squad system until January 2020. AB 1444 (2017): Authorization for the Livermore Amador Valley Transit Authority to launch a project on autonomous vehicles. SB 145 (2017): Elimination of the procedure by which the Department of Motor Vehicles notifies the legislature of the request to operate with autonomous vehicles. SB 1 (2017): Authorization to use the Road Maintenance and Rehabilitation Program funds to use advanced communication systems. AB 87 (2018): Authorization for the forces of law and order to immobilize autonomous cars without valid permission. AB 1184 (2018): Regulation authorizing the city of San Francisco to charge a fee for trips made with autonomous cars.
Colorado	SB 213 (2017): Allows automated driving. In case of not having all the permits, the authorities must give their permission and subsequently submit a report.

(Continues)

Table 8.6 (Continued)

Territorial administration level	Legislation / regulation
Connecticut	SB 260 (2017): Defines what is considered an autonomous vehicle and what are the requirements to perform tests with it (take out insurance of at least $5 million and the presence of a possible driver in the vehicle is necessary).
Florida	HB 1207 (2012): Defines what is an autonomous vehicle and autonomous technology. It also specifies that insurance, security and a person are needed inside the vehicle to perform traffic tests with autonomous vehicles. HB 599 (2012): Update to bill HB 1207. HB 7027 (2016): Authorizes the circulation of autonomous vehicles on public roads without a driver and without being on test trips as long as they comply with federal safety regulations. HB 7061 (2016): Collects the definition of autonomous technology and platoon technology. Authorizes a study on the movement in platoon and, if this were successful, a pilot project is allowed.
Georgia	HB 472 (2017): Collects the definition of coordinated platoon. SB 219 (2017): Collects the definition of automated driving and set the conditions for driving without a driver.
Illinois	HB 791 (2017): Defines automated driving and authorizes it in test mode.
Indiana	HB 1290 (2018): Collects the definition of coordinated platoon and establishes the need to present a plan for platoon testing.
Kentucky	SB 116 (2018): The movement in platoon is authorized with a driver with a valid commercial driver's licence under the authorization of the Department of Vehicle Regulation .
Louisiana	HB 1143 (2016): Collects the definition of autonomous technology. HB 308 (2018): Collects the definition of platoon, and platoon trips are authorized if the Department of Public Safety and Corrections, the office of state police, and the Department of Transportation and Development approve the operational plan.
Maine	HP 1204 (2018): Creation of the commission of autonomous vehicles to regulate and allow the circulation of autonomous vehicles.
Michigan	SB 169 (2013): Collects the definition of automated technology and authorize the testing of automatic vehicles under certain circumstances. SB 663 (2013): Limitation of the manufacturer's responsibility when there are failures due to modifications of the vehicle by third parties. SB 995 (2016): Allows platoon delivery vehicles not to respect the minimum distance of 500 feet apart. SB 996 (2016): Authorizes the movement of autonomous vehicles without any person inside under certain circumstances. SB 997 (2016): Collects the definition of automated driving and

(Continues)

Table 8.6 (*Continued*)

Territorial administration level	Legislation / regulation
	authorizes the creation of mobility research centres. SB 998 (2016): Exempts auto repair shops from liability. HB 5335 (2018): Creation of the Michigan Infrastructure Council belonging to the Department of the Treasury, which can finance autonomous communication technologies.
Mississippi	HB 1343 (2018): Collects the definition of platoon, and platoon trips are authorized if the Department of Transportation supports its circulation plan.
Nebraska	LB 989 (2018): Collects the definitions related to automated driving and allows the circulation of automatic vehicles without drivers under certain conditions.
Nevada	AB 511 (2011): Collects the definitions related to automated driving and allows the circulation of automatic vehicles with a special driving licence . SB 140 (2011): Allows the use of mobile phones in autonomous vehicles. SB 313 (2013): Specifies the conditions for the circulation of autonomous vehicles and stable that the manufacturer of the vehicle that has been converted by a third party into autonomous is not responsible for the injuries that may occur. AB 69 (2017): Allows the use of platoon technologies for testing.
New York	SB 2005 (2017): Allows autonomous vehicle tests and determines the conditions to perform them. AB 9508 (2018): Update of the bill SB 2005 (2017).
North Carolina	HB 469 (2017): Authorizes the circulation of autonomous vehicles in the state without a special driver's licence. HB 716 (2017): Allows platoon delivery vehicles to not respect the minimum distance of 500 feet apart.
North Dakota	HB 1065 (2015): Provides a study on autonomous vehicles. HB 1202 (2017): Authorizes a study of the Department of transport on vehicles with an automated system.
Oregon	HB 4059 (2018): Exempts the driver of an autonomous vehicle from responsibility if it has an automatic braking system. HB 4063 (2018): Foundation of a working group on autonomous technology under the direction of the State Department of Transportation.
Pennsylvania	SB 1267 (2016): Assign a $40 million item for research in autonomous technology. HB 1958 (2018): Defines the use of platoon technologies.
South Carolina	HB 3289 (2017): Allows platoon delivery vehicles to not respect the minimum distance of 500 feet apart.
Tennessee	SB 598 (2015): Prohibits local governments from authorizing the movement of autonomous vehicles. SB 2333 (2016): Authorizes the use, in the autonomous vehicle, of an integrated screen by the operator.

(Continues)

Table 8.6 (Continued)

Territorial administration level	Legislation / regulation
	SB 1561 (2016): Defines the concepts 'autonomous technology', 'driving mode' and 'dynamic driving task'. SB 676 (2017): Allows the use of platoon technologies under transport department and the security department permission. SB 151 (2017): Automated Vehicles Law. It allows autonomous vehicles to drive without a driver on the state roads.
Texas	HB 1791 (2017): Authorizes the use of the connected braking system. SB 2205 (2017): Defines the terms related to automated technology.
Utah	HB 373 (2015): Authorization for an autonomous technology testing programme by the Department of Transportation. HB 280 (2016): Authorization to conduct a study of good practices of autonomous vehicles. SB 56 (2018): Modification of HB 373 (2015) to include the definition of connected platoon system.
Virginia	HB 454 (2016): Authorizes the operation of a visual screen while driving an autonomous vehicle.
Vermont	HB 494 (2017): Permission to hold meetings with stakeholders in the autonomous technology sector.
Washington	HB 2970 (2018): Creation of a working group to develop recommendations on the use of autonomous technology.
Washington D.C.	DC B 19-0931 (2012): Defines the terms related to automated technology. DC B22-0901 (2018): The district's Department of Transportation will conduct a study to develop recommendations on the use of autonomous technology.
Wisconsin	SB 695 (2018): Authorizes the tests of the circulation of vehicles in platoon as long as they respect the minimum distance of 500 feet between them.

Table 8.7 Regulatory framework for the testing of autonomous vehicles in Europe

European country	Legislation/regulation
Austria	In June 2016, Austria amended its national road traffic law, the 'Kraftfahrgesetz', which dated from 1967. In order to update it, the Federal Ministry of Traffic, innovation and technology (bmvit) approved a plan to legislate on driving automated and therefore published an amendment, the '33. KFG-Novelle'. This amendment added two paragraphs to the 'Kraftfahrgesetz' whereby, in an automated test drive, the driver can delegate part of the driving decisions to the vehicle system as long as he can retake them at any time.

(Continues)

Table 8.7 (*Continued*)

European country	Legislation/regulation
	On the other hand, the Ministry of Transport also decided to create the AustriaTech agency to stimulate the advancement of autonomous technology.
Finland	TRAFI, the Finnish Transport Safety agency, allows automated driving of all levels, even without a driver in the vehicle. The relevant requirements, which must be granted by the agency itself, are to have a special registration for the tests and have an authorization to perform the tests.
France	France was one of the first European countries to incorporate the recommendations contained in the Vienna Convention. On 3 August 2016, article 38 of Law 2015-992 was amended. In this way, in France the tests of automatic vehicles on public roads are authorized provided that the tests are carried out by the ITS manufacturers and there are guarantees on the responsibility of that driving.
Germany	Germany has been one of the first European countries to develop an institutional framework to promote automated driving. In 2013, the German Federal Ministry of Transport and Digital Infrastructure founded the 'Round Table Automated Driving', in order to promote adequate regulation of automated technology. In addition, the same ministry created a test track in Bavier, the 'Digital Testfeld Autobahn'. Finally, in March 2017, the German transport law, the 'Straßenverkehrsgesetz', is modified to allow automatic driving on public roads, under the condition that international conventions signed by the country are respected (such as the 'Vienna Convention').
Spain	In 2015, Spain regulated the necessary conditions for the realization of autonomous vehicle tests through its state traffic agency 'Dirección general de tráfico'. It is the agency 'Dirección general de tráfico' itself that grants a 2-year permit, extendable, to perform the tests on certain routes. The prerequisites include possessing civil liability insurance and having the accreditation of the national accreditation body (ENAC).
Sweden	In 2015, the 'drive Sweden' programme was established, which promotes the automation and digitalization of transport.
The Netherlands	In 2015, the three institutions related to the regulation of traffic in the Netherlands (Rijkswaterstaat – National Road Authority, Dutch Vehicle Authority and the Ministry of Infrastructure and Environment) participated in the necessary law change to authorize automatic large-scale test driving. The Dutch Vehicle Authority is the agency responsible for analysing applications one by one and granting them permission.
The United kingdom	In July 2015, the UK Department of Transport published 'The Pathway to Driverless Cars: A Code of Practice for testing.' This manual includes the requirements for automatic driving tests, such as the presence of a driver in the vehicle, the possession of insurance or the type of licence needed. In the first quarter of 2017, the UK government presented the 'Modern Transport Bill' to promote research and investment in autonomous technology and regulate the characteristics of insurance applicable to this technology.

Table 8.8 Regulatory framework for the testing of autonomous vehicles in Asia

European country	Legislation/regulation
China	China has included the automated transport sector as part of its 'Made in China, 2025' strategy, which represents a clear commitment of the country's government for this sector. In April 2018, the Ministry of Industry and Information Technology, the Ministry of Public Security and the Ministry of Transport prepared the 'Intelligent and Connected Vehicle Test Management Practices' report, which contains the requirements for Chinese cities to authorize autonomous vehicle tests.
Japan	In April 2018, the Government Cabinet published the 'Charter for Improvement of Legal System and Environment for Autonomous Driving Systems' to specify the legislative framework to promote autonomous technology. In June 2018, reinforcing the country's commitment to this sector, the autonomous driving industry is included in the 'Future Investment Strategy 2018'. Finally, in December 2018, Japan's National Police Agency makes public the draft of the bill that updates the road traffic law, which includes tests of autonomous vehicles of Level 3 and the obligation to incorporate a 'black box' in the vehicle.
Singapore	The development of the autonomous car in this country is included in one of the six strategic projects of the country, concentrated in the 'Smart Urban Mobility'. It is the government of the country itself that promotes and finances research in this sector and, to facilitate its development, in February 2017 the parliament made amendments to the national traffic law. Already in January 2019, technical references 68 (TR 68) have been approved, which regulate four key aspects of autonomous technology: vehicle behaviour, vehicle functional safety, cybersecurity and data formats.
South Korea	The South Korean government is very interested in promoting autonomous technology. Specifically, it has approved different measures, including facilitating the issuance of permits for autonomous car tests or creating a test track called 'K-city'. Currently, the country's executive is working on an autonomous traffic system with a new road map and finances a test project for buses and autonomous trucks.

References

[1] Munoz, D., Bouchereau, F., Vargas, C., and Enriquez-Caldera, R., *Position Location Techniques and Applications*, Cambridge, Massachusetts, USA: Academic Press, Elsevier, 2009.

[2] Villalpando-Hernandez, R., Munoz-Rodriguez, D., Vargas-Rosales, C., and Rizo-Dominguez, L., '3-D position location in ad-hoc networks: a Manhattanized space', *IEEE Communications Letters*, 2017, 21(1): 124–127.

[3] Mass-Sanchez, J., Vargas-Rosales, C., Ruiz-Ibarra, E., Garcia-Berumen, A., and Espinoza-Ruiz, A., 'Localization based on probabilistic multilateration

approach for mobile wireless sensor networks', *IEEE Access*, (Early Access), 2020.

[4] Koohifar, F., Kumbhar, A., and Guvenc, I., 'Receding horizon multi-UAV cooperative tracking of moving RF source', *IEEE Communications Letters*, 2017, 21(6): 124–127.

[5] Li, L., Chang, T., and Cai, S., 'UAV positioning and power control for two-way wireless relaying', *IEEE Transactions on Wireless Communications*, 2020, 19(2): 1008–1024.

[6] Tahat, A., Kaddoum, G., Yousefi, S., Valaee, S., and Gagnon, F., 'A look at the recent wireless positioning techniques with a focus on algorithms for moving receivers', *IEEE Access*, 2016, 4: 6652–6680.

[7] Bar-Shalom, Y., 'Tracking methods in a multitarget environment', *IEEE Transactions on Automatic Control*, 1978, 23(4): 618–626.

[8] Reid, D., 'An algorithm for tracking multiple targets', *IEEE Transactions on Automatic Control*, 1979, 24(6): 843–854.

[9] Chang, C.-B., and Tabaczynski, J., 'Application of state estimation to target tracking', *IEEE Transactions on Automatic Control*, 1984, 29(2): 98–109.

[10] Li , D., Wong, K.D., Hu , Y.H., and Sayeed, A.M., 'Detection, classification, and tracking of targets', *IEEE Signal Processing Magazine*, 2002, 19(2): 17–29.

[11] Godsill, S.J., Vermaak, J., Ng, W., and Li, J.F., 'Models and algorithms for tracking of maneuvering objects using variable rate particle filters', *Proceedings of the IEEE*, 2007, 95(5): 925–952.

[12] Mihaylova, L., Angelova, D., Honary, S., Bull, D.R., Canagarajah, C.N., and Ristic, B., 'Mobility tracking in cellular networks using particle filtering', *IEEE Transactions on Wireless Communications*, 2007, 6(10): 3589–3599.

[13] Mark, B.L., and Zaidi, Z.R., 'Robust mobility tracking for cellular networks', *2002 IEEE International Conference on Communications. Conference Proceedings. ICC 2002 (Cat. No.02CH37333)*, New York, NY, USA, vol. 1, pp. 445–449, 2002.

[14] Laoudias, C., Moreira, A., Kim, S., Lee, S., Wirola, L., and Fischione, C., 'A survey of enabling technologies for network localization, tracking, and navigation', *IEEE Communications Surveys & Tutorials*, 2018, 20(4): 3607–3644.

[15] Gao, L., Battistelli, G., and Chisci, L., 'Event-triggered distributed multi-target tracking', *IEEE Transactions on Signal and Information Processing over Networks*, 2019, 5(3): 570–584.

[16] Hrovat, A., Kandus, G., and Javornik, T., 'A survey of radio propagation modeling for tunnels', *IEEE Communications Surveys & Tutorials*, 2014, 16 (2): 658–669.

[17] Sun, Z., and Akyildiz, I., 'A mode-based approach for channel modeling in underground tunnels under the impact of vehicular traffic flow', *IEEE Trans. Wireless Commun.*, 2011, 10(10): 3222–3231.

[18] Forooshani, A.E., Bashir, S., Michelson, D.G., and Noghanian, S., 'A survey of wireless communications and propagation modeling in underground mines', *IEEE Communications Surveys & Tutorials*, 2013, 15(4): 1524–1545.

[19] Changsen, Z., and Yan, M., 'Effects of cross section of mine tunnel on the propagation characteristics of UHF radio wave', *Antennas, Propagation EM Theory, 2006. ISAPE '06. 7th International Symposium On*, pp. 1–5, 2006.

[20] Wang, S., 'Radio wave attenuation character in the confined environments of rectangular mine tunnel', *Modern Applied Science,* 2010, 4(2): 65–70.

[21] Cheng, L., and Zhang, P., 'Influence of dimension change on radio wave propagation in rectangular tunnels', *Wireless Communications, Networking and Mobile Computing, 2009. WiCom '09. 5th International Conference On*, pp. 1–3, 2009.

[22] Yang, X., and Lu, Y., 'Research on propagation characteristics of millimeter wave in tunnels', *International J. Infrared and Millimeter Waves*, 2007, 28 (10): 901–909, http://dx.doi.org/10.1007/s10762-007-9268-y

[23] Didascalou, D., Maurer, J., and Wiesbeck, W., 'Subway tunnel guided electromagnetic wave propagation at mobile communications frequencies', *IEEE Trans. Antennas Propagat.*, 2001, 49(11): 1590–1596.

[24] Chiba, J., Inaba, T., Kuwamoto, Y., Banno, O., and Sato, R., 'Radio communication in tunnels', *IEEE Trans. Microw. Theory Tech.*, 1978, 26(6): 439–443.

[25] Delogne, P., *Leaky Feeders and Subsurface Radio Communications*, London, UK: Peter Peregrinus, 1982.

[26] Sun, J., Cheng, L., and Liu, X., 'Influence of electrical parameters on UHF radio propagation in tunnels', *Communications, 2004 and the 5th International Symposium on Multi-Dimensional Mobile Communications Proceedings. The 2004 Joint Conference of the 10th Asia-Pacific Conference On*, vol. 1, pp. 436–438, 2004.

[27] Sun, Z., and Akyildiz, I., 'Channel modeling and analysis for wireless networks in underground mines and road tunnels', *IEEE Trans. Commun.*, 2010, 58(6): 1758–1768.

[28] Dudley, D.G., Lienard, M., Mahmoud, S.F., and Degauque, P., 'Wireless propagation in tunnels', *IEEE Antennas Propag. Mag.*, 2007, 49: 11–26.

[29] Zhang, Y.P., and Hwang, Y., 'Characterization of UHF radio propagation channels in tunnel environments for microcellular and personal communications', *IEEE Trans. Veh. Technol.*, 1998, 47: 283–296.

[30] Yamaguchi, Y., Abe, T., and Sekiguchi, T., 'Radio wave propagation loss in the VHF to microwave region due to vehicles in tunnels', *IEEE Trans. Electromagn. Compat.*, 1989, 31(1): 87–91.

[31] Molina-Garcia-Pardo, J., Lienard, M., Degauque, P., Dudley, D., and Juan-Llacer, L., 'Interpretation of MIMO channel characteristics in rectangular tunnels from modal theory', *IEEE Trans. Veh. Technol.*, 2008, 57: 1974–1979.

[32] Boutin, M., Benzakour, A., Despins, C.L., and Affes, S., 'Radio wave characterization and modeling in underground mine tunnels', *IEEE Trans. Antennas Propag.*, 2008, 56: 540–549.

[33] Emslie, A.G., Lagace, R.L., and Strong, P.F., 'Theory of the propagation of UHF radio waves in coal mine tunnels', *IEEE Trans. Antennas Propag.*, 1975, 23: 192–205.

[34] Nkakanou, B., Delisle, G.Y., and Hakem, N., 'Experimental characterization of ultra-wideband channel parameter measurements in an underground mine', *J. Comput. Netw. Commun.*, 2011, 2011: 1–7.

[35] Ministerio de Fomento, *Los Sistemas Inteligentes de Transporte: Su Aplicación a Los Modos Terrestre, Marítimo y Aéreo: Writing and Editing for Ministerio de Fomento*. Madrid: Centro de Publicaciones Secretaría General Técnica Ministerio de Fomento, 2010, http://www.fomento.gob.es/AZ.BBMF.Web/documentacion/pdf/R16714.pdf

[36] Papadimitratos, P., de La Fortelle, A., Evenssen, K., Brignolo, R., and Cosenza, S., 'Vehicular communication systems: enabling technologies, applications, and future outlook on intelligent transportation', *IEEE Commun. Mag.*, 2009, 47: 84–95.

[37] Martin-Vega, F.J., Soret, B., Aguayo-Torres, M.C., Kovacs, I.Z., and Gomez, G., 'Geolocation-based access for vehicular communications: analysis and optimization via stochastic geometry', *IEEE Trans. Veh. Technol.*, 2018, 67: 3069–3084.

[38] Arena, F., and Pau, G., 'An overview of vehicular communications', *Future Internet*, 2019, 11: 1–12.

[39] Stawiarska, E., and Sobczak, P., 'The impact of intelligent transportation system implementations on the sustainable growth of passenger transport in EU regions', *Sustainability*, 2018, 10: 1–32.

[40] Gozalvez, J., Turksma, S., Lan, L., *et al.*, 'iTETRIS: the framework for large-scale research on the impact of cooperative wireless vehicular communications systems in traffic efficiency', *ICT-MobileSummit*, 2009, 2009: 1–10.

[41] Clements, L.M., and Kockelman, K.M., 'Economic effects of automated vehicles', *Transp. Res. Rec. J. Transp. Res. Board*, 2017; 2026: 106–114.

[42] Alsabaan, M., Alasmary, W., Albasir, A., *et al.,* 'Vehicular networks for a greener environment: a survey', *IEEE Commun. Surv. Tutorials*, 2013, 15 (3): 1372–1388.

[43] Ma, K., and Kijewska, K., 'Influence of intelligent transportation systems on reduction of the environmental negative impact of urban freight transport based on Szczecin example', *Procedia - Soc. Behav. Sci.*, 2014, 151: 215–229.

[44] United Nations Economic Commission for Europe, https://www.unece.org/trans/themes/transtheme-its/intelligent-transport-systems.html

[45] Intelligent Transportation Systems Joint Program Office, US Department of Transportation, https://www.its.dot.gov/factsheets/ITSJPO_overview.htm

[46] https://www.dot.ny.gov/divisions/operating/oom/transportation-systems/sys-tems-optimization-section/ny-moves

[47] Intelligent Transportation System Site, Los Angeles County, https://pw.lacounty.gov/TNL/ITS/

[48] Intelligent Transportation Systems (ITS) Operations, Washington State Department of Transportation, https://www.wsdot.wa.gov/travel/operations-services/its

[49] Intelligent Transportation Systems, North Central Texas Council of Governments, https://www.nctcog.org/trans/manage/its
[50] Seattle Department of Transportation, Seattle Government, https://www.seattle.gov/transportation/projects-and-programs/programs/technology-program
[51] Intelligent Transportation System, North Carolina Department of Transportation, https://www.ncdot.gov/initiatives-policies/Transportation/safety-mobility/its/Pages/default.aspx
[52] Intelligent Transportation Systems Lab, Massachusetts Institute of Technology, https://its.mit.edu/research-projects
[53] NTNU Intelligent Transportation Systems, Norwegian University of Science and Technology, https://www.ntnu.edu/iik/its
[54] Urban Mobility and Intelligent Transportation Systems Laboratory, New York University, http://engineering.nyu.edu/citysmart/
[55] Intelligent Urban Transportation Systems Lab, University of Washington, http://faculty.washington.edu/banx/iUTS.html
[56] Intelligent Transportation Systems (ITS) Canada, https://www.itscanada.ca/
[57] Emerging transportation technologies, Alberta Government, https://www.alberta.ca/emerging-transportation-technologies.aspx
[58] European Commission, Mobility and Transport, Intelligent Transportation Systems, https://ec.europa.eu/transport/themes/its_en
[59] Report of the Intelligent Transport Systems (ITS) High Level Policy Roundtable, 23rd Intelligent Transport Systems (ITS) World Congress, Melbourne Australia, 2016, https://strasbourg2017.itsineurope.com/wp-content/uploads/2017/03/HLPRT-Report-Final.pdf
[60] Intelligent Transport Systems, Platform Beter Benutten, http://www.beter-benutten.nl/en/its
[61] Intelligent Transport Systems, Estonia Government, https://e-estonia.com/solutions/location-based-services/intelligent-transpor-tation-system
[62] Intelligent Transport Systems, TRL, https://trl.co.uk/solutions/its
[63] Intelligent transportation system, Luxembourg Government, https://logistics.public.lu/en/why-luxembourg/logistics-infrastructure/intelli-gent-transportation-system.html
[64] Aurora Intelligent Road, Traffic Management Finland, https://tmfg.fi/en/itm/aurora-intelligent-road
[65] Intelligent transportation system, Government of Japan, http://www.its-jp.org/english/what_its_e/
[66] Society of Automotive Engineers of Japan, 'Intelligent transportation systems', *Journal of Society of Automotive Engineers of Japan*, 2019, 70: 1–9.
[67] Intelligent transportation system, Transport Department, The Government of the Hong Kong Special Administration Region, https://www.td.gov.hk/en/transport_in_hong_kong/its/index.html
[68] Intelligent transportation system, Australian Government, https://www.its-australia.com.au/
[69] Selangor Intelligent Transport System, Government of Malaysia, https://www.malaysia.gov.my/portal/content/30745

[70] Intelligent Transport Systems, Delhi Integrated Multi-Modal Transit System Ltd., https://www.dimts.in/Services_Transportation_Intelligent_Transport_System.aspx
[71] Intelligent Transport Systems, Ministry of Transport Te Manutū Waka, https://www.transport.govt.nz/multi-modal/technology/intelligenttransportsystems/
[72] Zhu, L., Yu, F.R., Wang, Y., *et al.*, 'Big data analytics in intelligent transportation systems: a survery', *IEEE Trans. Int. Transportation Systems*, 2019, 20(1): 383–398.
[73] Pelletier, M.-P., Trépanier, M., and Morency, C., 'Smart card data use in public transit: a literature review', *Transp. Res. C, Emerg. Technol.*, 2011, 19(4): 557–568.
[74] Liu, Y., Weng, X., Wan, J., Yue, X., and Song, H., 'Exploring data validity in transportation systems for smart cities', *IEEE Commun. Mag.*, 2017, 55 (5): 26–33.
[75] Asensio, C., López, J., Pagán, R., Pavón, I., and Ausejo, M., 'GPS-based speed collection method for road traffic noise mapping', *Transp. Res. D, Transp. Environ.*, 2009, 14(5): 360–366.
[76] Lopes, J., Bento, J., Huang, E., Antoniou, C., and Ben-Akiva, M., 'Traffic and mobility data collection for real-time applications', *Proc. IEEE Intell. Transp. Syst. (ITSC)*, September 2010, pp. 216–223
[77] Wang, X., Wang, C., Zhang, J., Zhou, M., and Jiang, C., 'Improved rule installation for real-time query service in software-defined Internet of vehicles', *IEEE Trans. Intell. Transp. Syst.*, 2017, 18(2): 225–235.
[78] Contreras-Castillo, J., Zeadally, S., and Ibañez, J.A.G., 'Solving vehicular ad hoc network challenges with big data solutions', *IET Netw.*, 2016, 5(4): 81–84.
[79] Zeyu, J., Shuiping, Y., Mingduan, Z., Yongqiang, C., and Yi, L., 'Model study for intelligent transportation system with big data', *Proc. Comput. Sci.*, 2017, 107: 418–426.
[80] Liu, S., Cao, H., Li, L., and Zhou, M.C., 'Predicting stay time of mobile users with contextual information', *IEEE Trans. Automat. Sci. Eng.*, 2013, 10(4): 1026–1036.
[81] Golob, T.F., and Recker, W.W., 'Relationships among urban freeway accidents, traffic flow, weather, and lighting conditions', *J. Transp. Eng.*, 2003, 129(4): 342–353.
[82] Karlaftis, M.G., and Golias, I., 'Effects of road geometry and traffic volumes on rural roadway accident rates', *Accident Anal. Prevention*, 2002, 34(3): 357–365.
[83] Jeon, S., and Hong, B., 'Monte Carlo simulation-based traffic speed forecasting using historical big data', *Future Generat. Comput. Syst.*, 2016, 65: 182–195.
[84] Liu, X.-L., Jia, P., Wu, S.-H., and Yu, B., 'Short-term traffic flow forecasting based on multi-dimensional parameters', *J. Transp. Syst. Eng. Inf. Technol.*, 2011, 11(4): 140–146.

[85] Ferris, B., Watkins, K., and Borning, A., 'OneBusAway: results from providing real-time arrival information for public transit', *Proc. SIGCHI Conf. Human Factors Comput. Syst.*, pp. 1807–1816,, 2010.
[86] Faizrahnemoon, M., Schlote, A., Maggi, L., Crisostomi, E., and Shorten, R., 'A big-data model for multi-modal public transportation with application to macroscopic control and optimisation', *Int. J. Control*, 2015, 88(11): 2354–2368.
[87] Jiang, Z., Hsu, C.-H., Zhang, D., and Zou, X., 'Evaluating rail transit timetable using big passengers' data', *J. Comput. Syst. Sci.*, 2016, 82(1): 144–155.
[88] Zhou, J., 'Applications of machine learning methods in problem of precise train stopping', *Comput. Eng. Appl.*, 2010, 46(25): 226–230.
[89] Zheng, X., Chen, W., Wang, P., *et al.*, 'Big data for social transportation', *IEEE Trans. Int. Transport. Systems*, 2016, 17(3): 620–630.
[90] Cheng, X., Hu, X., Yang, L., *et al.*, 'Electrified vehicles and the smart grid: the ITS perspective', *IEEE Trans. Intell. Transp. Syst.*, 2014, 15(4): 1388–1404.
[91] Khawaja, W., Guvenc, I., Matolak, D.W., Fiebig, U., and Schneckenburger, N., 'A survey of air-to-ground propagation channel modeling for unmanned aerial vehicles', *IEEE Communications Surveys & Tutorials*, 2019, 21(3): 2361–2391.
[92] Qualcomm, Leading the World to 5G: Evolving Cellular Technologies for Safer Drone Operation, accessed 8 March 2020, https://www.qualcomm.com/invention/technologies/lte/advanced-pro/cellular-drone-communication
[93] Patterson, T., Google, Facebook, SpaceX, OneWeb Plan to Beam Internet Everywhere, accessed 8 March 2020, http://www.cnn.com/2015/10/30/tech/pioneers-google-facebook-spacex-oneweb-satellite-drone-balloon-internet/
[94] Matolak, D.W., 'Air-ground channels & models: comprehensive review and considerations for unmanned aircraft systems', *2012 IEEE Aerospace Conference*, MT, USA, 3–10 March 2012.
[95] Painter, J.H., Gupta, S.C., and Wilson, L.R., 'Multipath modeling for aeronautical communications', *IEEE Trans. Comm.*, 1973, 21(5): 658–662.
[96] Elnoubi, S., 'A simplified stochastic model for the aeronautical mobile radio channel', *Proc. IEEE Vehicular Tech. Conf.*, Denver, CO, 10–13 May 1992.
[97] Dyer, G., Gilbert, T., Hendrickson, S., and Sayadian, E., 'Mobile propagation measurements using CW and sliding correlator techniques', *Proc. IEEE Ant. Propagation Symposium*, Atlanta, GA, 1998.
[98] Hoeher, P., and Haas, E., 'Aeronautical channel modeling at VHF band', *Proc. IEEE Vehicular Tech. Conf.*, Amsterdam, The Netherlands, 19–22 September 1999.
[99] Feng, Q., McGeehan, J., Tameh, E.K., and Nix, R.R., 'Path loss models for air-to-ground radio channels in urban environments', *Proc. IEEE Spring Vehicular Tech. Conf.*, Melbourne, Australia, 7–10 May 2006.
[100] Kunisch, J., de la Torre, I., Winkelmann, A., Eube, M., and Fuss, T., 'Wideband time-variant air-to-ground radio channel measurements at

5 GHz', *Proc. 5th European Conf. on Antennas & Propagation*, Rome, Italy, 11–15 April 2011.

[101] Willink, T.J., Squires, C.C., Colman, G.W.K., and Muccio, M.T., 'Measurement and characterization of low-altitude air-to-ground MIMO channels', *IEEE Transactions on Vehicular Technology*, 2016, 65(4).

[102] Chang, H., Bian, J., Wang, C., Bai, Z., Zhou, W., and Aggoune, E.-H.M., 'A 3D non-stationary wideband GBSM for low-altitude UAV-to-ground V2V MIMO channels', *IEEE Access*, 2019, 7: 70719–70732.

[103] Michelson, D.G., and Ghassemzadeh, S.S., 'Measurement and modeling of wireless channels', In: Tarokh V. (eds) *New Directions in Wireless Communications Research*, Springer, Boston, MA. pp. 1–27, 2009.

[104] Matolak, D.W., 'Modeling the vehicle-to-vehicle propagation channel: a review', *Radio Science*, 2014, 49(9): 721–736.

[105] Parsons, J.D., *The Mobile Radio Propagation Channel*, Hoboken, New Jersey, USA: John Wiley & Sons, 2000.

[106] Yin, X., and Cheng, Xiang, *Propagation Channel Characterization, Parameter Estimation and Modelling for Wireless Communications*, John Wiley & Sons: Singapore, 2016.

[107] Wang, C.-X., Cheng, X., and Laurenson, D.I, 'Vehicle-to-vehicle channel modeling and measurements: recent advances and future challenges', *IEEE Communications*, 2009, 47(11): 96–103.

[108] Viriyasitavat, W., Boban, M., Tsai, H.-M., and Vasilakos, A.V., 'Vehicular communications: survey and challenges of channel and propagation models', *IEEE Vehicular Technology Magazine*, 2015, 10(2): 55–66.

[109] Tan, I., Tang, W., Laberteaux, K., and Bahai, A., 'Measurement and analysis of wireless channel impairments in DSRC vehicular communications', *IEEE International Conference on Communications (ICC)*, pp. 4882–4888, May 2008.

[110] Cailean, A., Cagneau, B., Chassagne, L., Popa, V., and Dimian, M., 'A survey on the usage of DSRC and VLC in communication-based vehicle safety applications', *21th IEEE Symposium on Communications and Vehicular Technology, SCVT*, Delft, Netherlands, pp. 69–74, 2014.

[111] Bohm, A., Lidström , K., Jonsson, M., and Larsson, T., 'Evaluating CALM M5-based vehicle-to-vehicle communication in various road settings through field trials', *Local Computer Networks (LCN), 2010 IEEE 35th Conference On*, pp. 613–620, 10–14 October 2010.

[112] Karlsson, K., Bergenhem, C., and Hedin, E., 'Field measurements of IEEE 802.11p communication in NLOS environments for a platooning application', *Vehicular Technology Conference (VTC Fall), 2012 IEEE*, pp.1–5, 3–6 September 2012.

[113] Giordani, M., Shimizu, T., Zanella, A., Higuchi, T., Altintas, O., and Zorzi, M., 'Path loss models for V2V mmWave communication: performance evaluation and open challenges', 2019 IEEE 2nd Connected and Automated Vehicles Symposium, Honolulu, HI, USA, 22–23 September 2019.

[114] Baum, D.S., Hansen, J., and Salo, J., 'An interim channel model for beyond-3G systems: extending the 3GPP spatial channel model (SCM)', *Proc. IEEE VTC-Spring*, Stockholm, Sweden, vol. 5., pp. 3132–3136, May 2005.

[115] Tchouankem, H., Zinchenko, T., Schumacher, H., and Wolf, L., 'Effects of vegetation on vehicle-to-vehicle communication performance at intersections', *IEEE 78th Vehicular Technology Conference*, pp. 1–6, 2013.

[116] Gozalvez, J., Sepulcre, M., and Bauza, R., 'IEEE 802.11p vehicle to infrastructure communications in urban environments', *IEEE Communications Magazine*, 2012, 50(5): 176–183.

[117] Schumacher, H., Tchouankem, H., Nuckelt, J., *et al.*, 'Vehicle-to-vehicle IEEE 802.11p performance measurements at urban intersections', *IEEE International Conference on Communications (ICC 2012)*, 2012.

[118] International Telecommunication Union (ITU), 'Attenuation in vegetation', ITU-R Recomm, P. 833–33, 2001.

[119] European Cooperation in Science and Technology (COST) 235, 'Radiowave propagation effects on next generation fixed services terrestrial telecommunications systems, final report', Commission of the European Union, 1996.

[120] Weissberger, M.A., 'An initial critical summary of models for predicting the attenuation of radio waves by foliage', Electromagnetic Compatibility Analysis Center, 1981.

[121] Cid, E.L., Alejos, A.V., and Sanchez, M.G., 'Signaling through scattered vegetation: empirical loss modeling for low elevation angle satellite paths obstructed by isolated thin trees', *IEEE Vehicular Technology Magazine*, 2016, 11(3): 22–28.

[122] Rogers, N., Seville, A., Richter, J., *et al.*, 'A generic model of 1–60 GHz radio propagation through vegetation – final report', *QinetiQ, Tech. Rep.*, May 2002.

[123] Cuiñas, I., Alejos , A.V., and Sánchez, M.G., 'Vegetal barriers for minimising electromagnetic pollution at cellular phone bands', *Electronics Letters*, 2005, 41(6): 340–341.

[124] Gómez, P., Cuiñas, I., Alejos, A.V., Sánchez, M.G., and Gay-Fernández, J. A., 'Analysis of the performance of vegetation barriers to reduce electromagnetic pollution', *IET Microwaves, Antennas & Propagation*, 2011, 5 (6): 651–663.

[125] Picallo, I., Klaina, H., Lopez-Iturri, P., *et al.*, 'Radio channel model for D2D communications blocked by isolated trees in forest environments', *Sensors*, 2019, 19(2): 1–20.

[126] Hampton, J.R., Yao, S., Kroeger, T., Carducci, C., and Murphy, C., 'Drone-based forest propagation measurements for ground-to-air EMI applications', *IEEE Antennas and Wireless Propagation Letters*, 2019, 18(12): 2627–2631.

[127] Cuiñas, I., Alejos , A.V., Sánchez, M.G., Gómez, P., and Caldeirinha, R.F. S., 'Wind effect on the scattering from vegetation at cellular phone

frequencies', *2007 IEEE International Geoscience and Remote Sensing Symposium*, pp. 369–372, 2007.

[128] Gómez, P., Cuiñas, I., Alejos, A.V., Sánchez, M.G., and Caldeirinha, R.F.S., 'Shrub-blown time variability in attenuation and scattering at cellular frequencies', *IET Microwaves, Antennas & Propagation*, 2010, 4(4): 526–542.

[129] Hashim, M.H., and Stavrou, S., 'Measurements and modelling of wind influence on radiowave propagation through vegetation', *IEEE Transactions on Wireless Communications*, 2006, 5(5): 1055–1064.

[130] Alejos, A.V., Sanchez, M.G., Cuinas, I., and Gomez, P., 'Depolarization effect by wind incidence at 40GHz', *The Second European Conference on Antennas and Propagation*, Edinburgh, UK, 11–16 November 2007.

[131] Wang, L., Ai , B., Yang , J., Qiu, H., Wang, W., and Guan, K., 'Vehicular channel in urban environments at 23 GHz for flexible access common spectrum application', *International Journal of Antennas and Propagation*, 2019, 2019, 13 pages, Article ID 5425703.

[132] Azpilicueta, F.G.L., Vargas-Rosales, C., Lopez-Iturri, P., *et al.*, 'Spatial characterization of radio propagation channel in urban vehicle-to-infrastructure environments to support WSNs deployment', *Sensors*, 2017, 17(6): 1313–1335.

[133] Klaina, H., Alejos, A.V., Aghzout, O., and Falcone, F., 'Narrowband characterization of near-ground radio channel for wireless sensors networks at 5G-IoT bands', *Sensors*, 2018, 18(8): 18–28.

[134] Goldsmith, A., *Wireless Communications*, Cambridge, England, UK: Cambridge University Press, ISBN 0521837162, 2005.

[135] Brummer, A., German, R., and Djanatliev, A., 'Modeling V2X communications across multiple road levels', *IEEE 2nd Connected and Automated Vehicles Symposium*, Honolulu (HI, USA), 22–23 September 2019.

[136] Prokes, A., Mikulasek, T., Waldecker, M., Engiz, B.K., and Blumenstein, J., 'Multipath propagation analysis for static urban environment at 60 GHz', *2019 International Conference on Electrical and Computing Technologies and Applications*, Ras Al Khaimah, United Arab Emirates, 19–21 November 2019.

[137] Li, C., Chen, W., Yu, J., Yang, K., Li, F., and Shui, Y., 'V2V radio channel properties at urban intersection and ramp on urban viaduct at 5.9 GHz', *IET Communications*, 2018, 12(17)

[138] Li, C., Yang, K., Yu, J., *et al.*, 'V2V radio channel performance based on measurements in ramp scenarios at 5.9 GHz', IEEE Access, 2018.

[139] Liu, P., Ai, B., Matolak, D.W., Sun, R., and Li, Y., '5-GHz vehicle-to-vehicle channel characterization for example overpass channels', *IEEE Transactions on Vehicular Technology*, 2016, 65(8): 5862–5873.

[140] Sun, S., Matolak, D.W., and Liu, P., '5-GHz V2V channel characteristics for parking garages', *IEEE Transactions on Vehicular Technology*, 2017, 66(5): 3538–3547.

[141] Cuinas, I., Sanchez, M.G., and Alejos, A.V., 'Depolarization due to scattering on walls in the 5 GHz band', *IEEE Transactions on Antennas and Propagation*, 2009, 57(6): 1804–1812.

[142] Okumura, B., James, M.R., Kanzawa, Y., *et al.*, 'Challenges in perception and decision making for intelligent automotive vehicles: a case study', *IEEE Transactions on Intelligent Vehicles*, 2016, 1(1): 20–32.

[143] Nilsson, J., Brännström, M., Fredriksson, J., and Coelingh, E., 'Longitudinal and lateral control for automated yielding maneuvers', *IEEE Transactions on Intelligent Transportation Systems*, 2016, 17(5): 1404–1414.

[144] Zyner, A., Worrall, S., and Nebot, E.M., 'ACFR five roundabouts dataset: naturalistic driving at unsignalized intersections', *IEEE Intelligent Transportation Systems Magazine*, 2019, 8–18.

[145] Zyner, A., Worrall, S., and Nebot, E.M., 'A recurrent neural network solution for predicting driver intention at unsignalized intersections', *IEEE Robotics and Automation Letters*, 2018, 3(3): 1759–1764.

[146] Völz, B., Mielenz, H., Gilitschenski, I., Siegwart, R., and Nieto, J., 'Inferring pedestrian motions at urban crosswalks', *IEEE Transactions on Intelligent Transportation Systems*, 2019, 20(2): 544–555.

[147] Xu, H., Zhang, K., Zheng, Q., and Yao, R., 'Multi-level pedestrian signalisation at large four-leg roundabouts', *IET Intelligent Transport Systems*, 2018, 12(8): 838–850.

[148] Duan, Y., Qu, X., Easa, S., and Yan, Y., 'Optimising total entry delay at roundabouts with unbalanced flow: a dynamic strategy for smart metering', *IET Intelligent Transportation Systems*, 2019, 13(3): 485–494.

[149] Rodrigues, M., McGordon, A., Gest, G., and Marco, J., 'Autonomous navigation in interaction-based environments – a case of non-signalized roundabouts', *IEEE Transactions on Intelligent Vehicles*, 2018, 3(4): 425–438.

[150] Debada, E., and Gillet, D., 'Virtual vehicle-based cooperative maneuver planning for connected automated vehicles at single-lane roundabouts', *IEEE Intelligent Transportation Systems Magazine*, 2018, 35–46.

[151] Ding, N., Meng, X., Xia, W., Wu, D., Xu, L., and Chen, B., 'Multivehicle coordinated lane change strategy in the roundabout under internet of vehicles based on game theory and cognitive computing', *IEEE Transactions on Industrial Informatics*, 2020, 16(8): 5435–5443.

[152] Córdoba, A., Astrain, J.J., Villadangos, J., *et al.*, 'SesToCross: semantic expert system to manage single-lane road crossing', *IEEE Transactions on Intelligent Transportation Systems*, 2017, 18(5): 1221–1233.

[153] Burg, A., Chattopadhyay, A., and Lam, K-Y., 'Wireless communication and security issues for cyber-physical systems and the Internet-of-things', *Proceedings of the IEEE*, 2018, 106(1).

[154] Strandberg, K., Olovsson, T., and Jonsson, E., 'Securing the connected car', *IEEE Vehicular Technology Magazine*, 2018, 13(1).

[155] Aghaei, A.S., Donmez, B., Liu, C.C., *et al.*, 'Smart driver monitoring: when signal processing meets human factors', *IEEE Signal Processing Magazine*, 2016, 33(6).

[156] Choi, Y., Han, S.I., Kong, S.H., *et al.*, 'Driver status monitoring systems for smart vehicles using physiological sensors', *IEEE Signal Processing Magazine*, 2016, 33(6).
[157] Rane, S., Wang, Y., Draper, S.C., *et al.*, 'Secure biometrics', *IEEE Signal Processing Magazine*, 2013, 30(5).
[158] Hult, R., Campos, G.R., Steinmetz, E., *et al.*, 'Coordination of cooperative autonomous vehicles', *IEEE Signal Processing Magazine*, 2016, 33(6).
[159] Kalkan, K., and Zeadally, S., 'Securing Internet of things with software defined networking', *IEEE Communications Magazine*, 2018, 56(9).
[160] Xiao, L., Wan, X., Lu, X., *et al.*, 'IoT security techniques based on machine learning', *IEEE Signal Processing Magazine*, 2018, 35(5).
[161] Sayed, A.H., Tu, S.Y., Chen, J., *et al.*, 'Diffusion strategies for adaptation and learning over networks', *IEEE Signal Processing Magazine*, 2013, 30 (3).
[162] Giannakis, G.B., Shen, Y., and Karanikolas, G.V., 'Topology identification and learning over graphs: accounting for nonlinearities and dynamics', *Proceedings of the IEEE*, 2018, 106(5).
[163] Ortega, A., Frossard, P., Kovacevic, J., Moura, J.M.F., and Vabdergheynst, P., 'Graph signal processing: overview, challenges, and applications', *Proceedings of the IEEE*, 2018, 106(5).
[164] Meng, W., Tischhauser, E.W., Wang, Q., *et al.*, 'When intrusion detection meets blockchain technology: a review', *IEEE Access*, 2018, 6.
[165] 5G Barcelona, https://5gbarcelona.org/
[166] 5G PPP https://5g-ppp.eu/
[167] Vehicular Communication Market Report: Trends, Forecast and Competitive Analysis, Writing and Editing for Lucintel. Dallas: Lucintel, 2019, https://www.lucintel.com/vehicular-communication-market.aspx
[168] Taming the Autonomous Vehicle. A Primer for Cities, Writing and Editing for Bloomberg Philanthropies, The Aspen Institute: New York City, 2017, https://www.bbhub.io/dotorg/sites/2/2017/05/TamingtheAutonomousVehicleSpreadsPDF.pdf
[169] Global Autonomous Driving Market Outlook, Writing and Editing for Frost & Sullivan, New York City, 2018, https://info.microsoft.com/rs/157-GQE-382/images/K24A-2018%20Frost%20%26%20Sullivan%20-%20Global%20Autonomous%20Driving%20Outlook.pdf
[170] Intelligent Transportation Systems Benefits, Costs, and Lessons Learned, Editing for US Department of Transportation, U.S. Department of Transportation: Washington DC, 2017, https://www.itsknowledgeresources.its.dot.gov/its/bcllupdate/pdf/BCLL_2017_Combined_JPO-FINALv6.pdf
[171] Support Study for Impact Assessment of Cooperative Intelligent Transport Systems, Writing and Editing for European Commission, European Commission: Brussels, 2018, https://ec.europa.eu/transport/sites/transport/files/studies/2018-support-study-for-impact-assessment-of-c-its.pdf

[172] Suthaputchakun, C., Sun, Z., and Dianati, M., 'Applications of vehicular communications for reducing fuel consumption and CO_2 emission: the state of the art and research challenges', *IEEE Commun. Mag.*, 2012, 50: 108–115.
[173] ITF: Automated and Autonomous Driving. Regulation under Uncertainty, Writing and Editing for International Transport Forum Policy Papers [homepage on the Internet], OECD : Paris, 2015, https://www.oecd-ilibrary.org/transport/automated-and-autonomous-driving_5jlwvzdfk640-en
[174] Autonomous Vehicles. Self-driving Vehicles Enacted Legislation, Writing and Editing for National Conference of State Legislatures [homepage on the Internet], NCSL: Washington, 2019, http://www.ncsl.org/research/transportation/autonomous-vehicles-self-driving-vehicles-enacted-legislation.aspx
[175] Cole, C., and McCullough, K., 'Regulatory issues related to autonomous vehicles'', *Journal of Insurance Regulation*, 2016, 35(7): 1–11.
[176] Arcade Project: Report on the State of the Art of Connected and Automated Driving in Europe (Final), Writing and Editing for Arcade Project [homepage on the Internet], Aachen, SCOUT Deliverable 3.2, 2017, https://connectedautomateddriving.eu/wp-content/uploads/2018/10/D3_2-Report-on-the-state-of-the-art-final.pdf
[177] The Race is on for Autonomous Vehicles, Global Policy Update, Writing and Editing for FTI Consulting [homepage on the Internet], FTI Consulting: Brussels, 2019, https://getready4.eu/wp/wp-content/uploads/2019/02/The-Race-is-on-for-Autonomous-Vehicles.pdf

Index

www.ingramcontent.com/pod-product-compliance
Lightning Source LLC
LaVergne TN
LVHW011837200926
848484LV00002B/366

* 9 7 8 1 7 8 5 6 1 8 2 3 9 *